"*Flattening the Medieval Earth* is an important study of two historical transformations: of terrestrial geography in Columbus's wake and of its later weaponisation in anti-religious rhetoric. Pablo de Felipe writes an engaging biography of a pervasive error – the belief that medieval Christendom had been flat-Earthbound."

John Hedley Brooke, *Oxford University, UK*

"*Flattening the Medieval Earth* is brilliant interdisciplinary research on how, in the 16th–17th centuries, several factors (including defending Copernicanism) led to the wrong idea of a medieval flat Earth, contributing to the prejudice of a permanent conflict between science and Christianity."

Miguel Á. Granada, *University of Barcelona, Spain*

"*Flattening the Medieval Earth* is a kaleidoscopic revelation of the richness of medieval discussions about the Earth's shape and population. With masterful command of the sources and keen analytical eye, Pablo de Felipe clarifies the origin of the Flat Error and much more."

Henrique Leitão, *University of Lisbon, Portugal*

"*Flattening the Medieval Earth* is a ground-breaking study of the medieval debate around the shape of the Earth and the Antipodes. Pablo de Felipe convincingly demonstrates that it arose from within Christianity and not from a dichotomy between 'science and faith'."

Karla Pollmann, *University of Tübingen, Germany*

Flattening the Medieval Earth

Flattening the Medieval Earth explores the origin of the 'flat error', i.e. the false accusation that ancient and medieval Christians believed in a flat Earth, and what this implies in terms of a conflict between science and Christianity. Engaging with scientific and religious debates, this book includes a reception study of two key figures of the 4th century AD, Lactantius and Augustine. This study demonstrates that the mistaken 'dark' image of medieval scholars as flat-earthers started very early, c. 1600, as an internal Christian debate in the context of new geographical and astronomical views. The author draws on extensive research including many primary and secondary sources from different countries and languages not previously put into conversation. Combining history of science, church history, science and religion, history of navigation, and historiography, this book gives the most updated explanation of the origin of the flat error, finding paradoxes and unexpected answers along the way to understand the past and to reflect on some current approaches in science and religion.

Pablo de Felipe is a researcher and lecturer, with a focus on the relations between science and Christianity, at the Protestant Faculty of Theology SEUT in Madrid (Spain). In the same city he also lectures at the Xavier Zubiri Foundation and is part of the advisory board of the Science, Technology and Religion Chair at the Comillas Pontifical University. He holds a PhD in Molecular Biology from the Universidad Autónoma de Madrid and a PhD in Religion and Theology from the University of Bristol (UK). He works as Head of Service (Virology) at the Spanish Agency of Medicines and Medical Devices (AEMPS) in Madrid.

Routledge Science and Religion Series

Series editors:
Michael S. Burdett, University of Nottingham, UK
Mark Harris, University of Edinburgh, UK

Science and religion have often been thought to be at loggerheads but much contemporary work in this flourishing interdisciplinary field suggests this is far from the case. The Science and Religion Series presents exciting new work to advance interdisciplinary study, research and debate across key themes in science and religion. Contemporary issues in philosophy and theology are debated, as are prevailing cultural assumptions. The series enables leading international authors from a range of different disciplinary perspectives to apply the insights of the various sciences, theology, philosophy and history in order to look at the relations between the different disciplines and the connections that can be made between them. These accessible, stimulating new contributions to key topics across science and religion will appeal particularly to individual academics and researchers, graduates, postgraduates and upper-undergraduate students.

New Frontiers in Islam and Evolution
Scriptures, Scholars, and Societies
edited by Shoaib Ahmed Malik and David Solomon Jalajel

Design Discourse in Abrahamic Traditions
History, Metaphysics and, Science
Edited by E. V. R. Kojonen & Shoaib A. Malik

Maximus the Confessor and Evolutionary Biology
The Phylogenetic Logoi
Andrew P. Jackson

Flattening the Medieval Earth
Seeking the Early Modern Origins of the Idea of an Historical Conflict between Science and Christianity
Pablo de Felipe

For more information and a full list of titles in the series, please visit: https://www.routledge.com/religion/series/ASCIREL

Flattening the Medieval Earth

Seeking the Early Modern Origins of the Idea of an Historical Conflict between Science and Christianity

Pablo de Felipe

Routledge
Taylor & Francis Group
LONDON AND NEW YORK

First published 2026
by Routledge
4 Park Square, Milton Park, Abingdon, Oxon OX14 4RN

and by Routledge
605 Third Avenue, New York, NY 10158

Routledge is an imprint of the Taylor & Francis Group, an informa business

British Library Cataloguing-in-Publication Data
A catalogue record for this book is available from the British Library

ISBN: 978-1-032-89306-8 (hbk)
ISBN: 978-1-032-89996-1 (pbk)
ISBN: 978-1-003-54568-2 (ebk)

DOI: 10.4324/9781003545682

Typeset in Times New Roman
by codeMantra

Access the Support Material: www.routledge.com/9781032893068

I dedicate this research to my wife Icíar Fernández and my daughter Sara de Felipe, life companions in the long voyage of this investigation.

Contents

Figures

Preface

From my undergraduate science years, I have been keenly interested in the history of science and its historical relationship with philosophy and religion. In the 1990s, I read extensively about the Copernican debates of the mid-16th to 17th centuries and the religious controversy they provoked. It was in 1998 that I first read about an ancient debate regarding the shape of the Earth in 6th-century Christian Alexandria. Therefore, it should not be surprising that when in the winter of 2001–2002 I read Jeffrey Burton Russell's *Inventing the Flat Earth* (1991), I did it with great interest.

That book shifted my interest from the ancient and medieval debates on the shape of the Earth towards their modern (mis)reception, which has led to the erroneous view of ancient and medieval Christians as flat-earthers who rejected ancient science. This erroneous depiction of history has been termed "flat error" by Russell, a term that I have retained throughout this book. Furthermore, this specific topic has influenced the broader topic of science and religion relations, which are often perceived to be at war.

In 2007, during a postdoctoral research period at the University of St Andrews, I had the privilege of meeting Prof. Karla Pollmann, who encouraged my studies and inspired me to investigate the reception of Augustine in greater depth and suggested to formalise my research interests in the history of science and Christianity. Years later this resulted in my undertaking a PhD in the humanities under her direction, which was awarded on 25 January 2022 at the University of Bristol.

The present work presents an abridged, revised and updated version of that research which seeks to introduce my investigations to a wider public. This book could be considered a specific investigation in the history of ideas or cultural history in the interphase of the history of science and church history. The title, *Flattening the Medieval Earth*, is derived from Russell's seminal work, where the term is employed once. In critiquing the historians who in the past accepted the flat error, Russell wrote, "Many other historians, however, acquiesced in flattening the medieval earth" (Russell, 1991, 46). He also referred to those who "flattened the medieval globe" (Russell, 1991, 27). However, my subtitle, *Seeking the Early Modern Origins of the Idea of an Historical Conflict Between Science and Christianity*, makes it clear that I disagree with Russell's core claim that this misconception originated in the 19th century. Instead, the aim of this book is to demonstrate that the flat error can be traced back to Early Modernity, more precisely to c. 1600. Furthermore, it is argued that the flat error has played an important part in the origin of the idea of an historical confrontation between science and Christianity (the so-called 'conflict thesis'). It should be noted that the two terms 'science' and 'Christianity' are used in an anachronistic way; however, their contextual meaning will become clearer in the following pages.

Finally, it is important to note that this book is not concerned with the recent revival of the flat Earth views. The content of this book will help us to understand the historical background of how the flat Earth views disappeared in Antiquity, the kind of debates that the sphericity of the Earth sparked up to the Renaissance and how they were forgotten and distorted in Early Modernity. The manner in which the sphericity of the Earth became newly disputed since the 19th century, and particularly in the early 21st century, is beyond the scope of this book and a topic of interest in itself.

Acknowledgements

I am very grateful for the encouragement and advice given by Prof. Karla Pollmann, who supervised my PhD thesis throughout the universities of Kent, Reading and Bristol. I also thank my second supervisor, Dr Jon Balserak, for his support at Bristol.

I have to thank Rev. Dr Robert D. Keay, whom I met at St Andrews, for more than two decades of continuous conversations surrounding the topic of this book. He has been always happy to talk about that, also reading portions of the following pages.

There is another friend from my years in St Andrews, Prof. Harry Hine, to whom I must express my deepest gratitude for his continuous help with translating crucial Latin texts for this research and kindly reading portions of this book. I also thank Dr Christa M. Gray, who was my second supervisor at the University of Reading, for kindly translating long sections of a Latin book of great importance for my investigation.

I owe a debt of gratitude to Dr Denis R. Alexander, Prof. Robert S. White, and Prof. Keith R. Fox for finding some financial support for the fees of my PhD. These funds came from a Templeton World Charity Foundation grant to The Faraday Institute for Science and Religion (Cambridge, UK). I also thank Dr Alexander for reading portions of this text.

In the last years of this research, I had the valuable help of Prof. John Hedley Brooke. His suggestions for new ideas, related bibliography and continued support were of great encouragement.

This research has been possible thanks to the help of an extensive array of family members, friends and colleagues. In addition to the names mentioned above, I would like to acknowledge others who helped me in several ways: Dr Mark Abbott, Prof. Onésimo Almeida, Dr Ruth Bancewicz, Fiona Barnard, Prof. Antoine Bret, Teresa Canale, Dr Dick L. Couprie, Pedro de Felipe, Dr Alberto Elena (†), Carlos Escaño, Josefa Fernández, Dr Filip Lim, José Luis Gómez, Prof. Miguel Ángel Granada, Prof. Céline Guillot, Jeelmer Heeren, Prof. Malcolm A. Jeeves, Dr Garry Luke, Dr Miguel Marañón, Dr Nathan Moser, María Jesús Prieto, Dr Sergio Rosell, Dr Ana Ruiz, Prof. David Rutherford, Juan Sánchez, Cristina Schoebel, Sir Richard Sorabji, Dr Mireia Vidal, David Vincent, Dr Klaus A. Vogel, Prof. Marianne Woollven and Dr Pedro Zamora.

I would also like to acknowledge several libraries from where I obtained documentation: Österreichische Nationalbibliothek (Austria), International Institute of Social History (Netherlands), Biblioteca Ambrosiana (Italy), New York Public Library (USA), University Libraries of St Andrews, Kent, Reading and Bristol (UK), and the Universidad Autónoma de Madrid and Universidad Pontificia Comillas, as well as the Protestant Faculty of Theology SEUT (Spain).

I found many of the books used here in large online databases and repositories: Early English Books Online, Gallica-Bibliothèque Nationale de France, Munich DigitiZation Center, Biblioteca Nacional de Portugal, Biblioteca Nacional de España, Biblioteca Virtual Miguel de Cervantes, Europeana, Google Books, Internet Archive, Project Gutenberg, Making of America, etc. I am very thankful for these valuable online resources, which allowed me the consultation of a large number of primary sources.

Finally, I would like to thank the publishers for the permissions to reproduce Figure 2.1 (Springer Nature BV) and Figures 2.2, 2.4, 2.5, 2.7 and 3.1 (Brill).

Permissions have also been granted to reproduce certain texts in the Appendix that is available as a downloadable file at: www.routledge.com/9781032893068 (details are given in that document).

Introduction

Columbus and the flat Earth

Flattening the medieval Earth

Arguably, the 1492 voyage of Columbus across the Atlantic to America is considered universally one of the most momentous events in human history. Telling this story typically includes as a feature that the great sailor had to confront Christian ecclesiastics who denied the voyage was possible because, it is claimed, they were flat-earthers. They allegedly believed the Earth[1] not to be a globe but rather to be flat and therefore feared the ships would fall from the edge of such a flat Earth. One recent example is the 'educational' material for children which states: "At the time, everyone believed that the Earth was square and flat, except Columbus, who believed it was round, and with this trip he intended to prove it."[2]

In fact, this is the dominant view in the media, among the well-educated public and even in university-level geography textbooks, such as this example:

> The Middle Ages were a dark period for the development of science in Europe. At best, scholars made accurate copies of the works of the ancients, rejecting anything that did not conform to the dogmas of the Christian Church. Such an intellectual environment stifled any development of scientific investigation. The earth became a flat disc with Jerusalem at its centre.
>
> (Holt-Jensen, 1981[2018], 24)

Amazingly, as we will discuss in Chapter 1, historians of the Middle Ages and the Renaissance know that this alleged confrontation is fictitious: Columbus did not have to face flat-earthers, as there were hardly any educated people at that time who believed so. We can then ask, where is this false story that everyone has heard coming from?

This question has haunted several historians for over a century. However, a complete picture of what happened from Columbus's times until our days is more complex than it seems. This book will reconstruct centuries of debates about the shape and habitability of the Earth to understand the intellectual knowledge accumulated by the time of Columbus in Western Europe. We will also chronicle how these ancient and medieval views were received, interpreted and judged in order to understand how, over the next century and a half, the distortion of

DOI: 10.4324/9781003545682-1

those age-old debates ended up in the dramatic false story that we have all heard about Columbus.

The 19th-century "flat error"

From the late 15th century to the early 19th century, a complex chain of events led to a denigratory portrait of ancient and medieval Christians, regarded as flat-earthers up to Columbus's times. This view has been termed "flat error" by the US historian Jeffrey Burton Russell (1934–2023), who defined it as: "The almost universal supposition that educated medieval people believed the earth to be flat" (Russell, 1991, xiii).

The flat error usually regards Columbus's voyages to America as the reason for the 'recovery' in Western Europe of the knowledge of the Earth's sphericity, which had been common in the Graeco-Roman world and was allegedly suppressed by Christianity. In this fictional account, Columbus defended the sphericity of the Earth and the feasibility of his ocean crossing to Asia against late 15th-century Spanish clerics, portrayed as a bunch of bigoted flat-earthers at the so-called 'council' of Salamanca, which in fact was a meeting of experts, not a religious tribunal.

The flat error spread in Western Europe and America during the late 19th century, particularly in the new science and Christianity conflict literature (such as Draper, 1874, and White, 1896) and academic texts of history, geography, etc., well into the 20th century. As we will describe in Chapter 1, in parallel, 20th-century historians started to abandon this error and attempted in many cases to reach the wider public and the educational system to remove such misconceptions, with little success even today.

Since the publication of Russell's book *Inventing the Flat Earth: Columbus and Modern Historians* in 1991, the thesis that the flat error is a recent slander has gained acceptance. According to Russell, it originated in the early 19th century from a fictionalised biography of Columbus by the US writer Washington Irving (1828) and a scholarly paper on the "cosmographical" views of the early Christian writers by the French historian Jean-Antoine Letronne (1834; on the term cosmography, see below). The flat error then became widespread towards the end of that century. Russell's localisation of the origins of the flat error in the early 19th century has become the standard explanation, accepted by many scholars from different disciplines over the last three decades (see Chapter 1, endnote 3).

Older roots

In spite of the apparent simplicity of the history of the flat error, which originated in the early 19th century with Irving and Letronne and spread in the late 19th century with Draper and White, things turn out to be more complex.

True, Columbus was not assaulted with quotes from 4th-century Christian flat-earther Lactantius, but as Russell recognised, Lactantius was mentioned by Copernicus in his letter dedicating to the Pope his landmark book, *De revolutionibus* (1543). Furthermore, a different but related topic, that of the existence of antipodes, was pointed out to Columbus in the meetings with the Castilian experts before his Atlantic crossing in 1492 and debated extensively in the centuries before that voyage. Both topics were discussed during the 16th and 17th centuries in the light of the new geographical and astronomical knowledge. Could the root of the flat error lie somewhere around that time by a misleading connection of those elements?

An initial problem is how to start such a search. Russell went backwards from present to past to arrive at Irving and Letronne. With the initial suspicion of an origin in the Early Modernity, but not wanting to fall short, here it was decided to invert the strategy, going from the past to the present.

As we will discuss in Chapter 1, Russell and other authors have been since the late 20th century pointing to earlier references to the flat error back to the 17th century or even the 16th century. While in some cases these identifications appear to be wrong when the primary sources are carefully discussed, in other cases they have a true value that we will incorporate in this research. In addition, much new ground has been broken here to gain a comprehensive view of the context of the 16th century – for the specific topics involved in this investigation – from Columbus's time to the first appearance of cases of flat error c. 1600. This research is a voyage where history meets geography, astronomy, theology, navigation, mapping, rhetoric and propaganda.

The long shadow of the flat Earth

The implications of this research are not merely of historical interest regarding a happening in the past, but they have direct relevance for contemporary attitudes and beliefs. Today the flat error is still alive, taught from schools to universities worldwide, playing an essential role in shaping people's wider historical views, particularly their highly negative view of the Western European science during the Middle Ages. The flat error is often found in the narratives about the alleged obscurantism of the medieval science, as one of its prime examples. Sadly, as we will see in Chapter 1, the efforts of historians for more than a century to counter this erroneous perception of the past – by accumulating medieval statements of the sphericity of the Earth – have not diminished the still widespread misconception that the Earth was believed to be flat during Late Antiquity and the Middle Ages. This battle to disseminate real information about what was the science of that period still continues from classrooms to the internet and the social networks.[3]

In parallel there is another front to tackle the flat error by reflecting on the question with which we opened this introduction and that was already posed by medievalist historian Charles W. Jones nearly a century ago: "Whence, then, arose our modern idea that in the Middle-Ages students believed the earth was flat?" (Jones, 1934, 304). Answering that is a good and additional way to show that the flat error is a modern invention and not a medieval fact, fake and not history. The aim of this book is to tackle this second issue and to investigate the older roots mentioned above.

Further than that, the flat error is also a key piece in the negative perception of the role of Christianity in the history of science and influences the general public perception of the relationship between science and Christianity in what has been named the 'conflict thesis'. Through its analysis of the origins and early history of the flat error, this book therefore has a contemporary relevance in depriving that thesis of one of its common pillars of support.

General outline of this book, from Antiquity to modern science

Much of this book is about context. If we want to understand how Early Modernity perceived scientific debates during the Middle Ages, engaging with early Christianity and Graeco-Roman views of the Earth, it is clear that we need to know with detail the ideas of those periods. Although this could render this book of a huge size, here we will restrict the investigation to the very specific necessary topics regarding the shape and population of the Earth, as well as the authors that are relevant for our discussion. In this way, when we reach c. 1600 in both Chapters 4 (on geography) and 5 (on astronomy), we can understand the issues that were at stake. To help the readers to focus in the aim of this research, Chapters 2–5 include final sections of 'conclusions' to summarise the key findings and their place in the overall argumentation.

In Chapter 1, we will discuss how, since the late 19th century, several scholars have sometimes suspected and at other times clearly exposed the falsity of the flat error. In addition, we will look into those that have wondered about the origin of such a profound historical distortion and analyse the responses they provided in trying to identify the 'culprits' of the flat error. In the light of such data, the plan of research for this work will be explained with a critical discussion of the 'culprit-search' methodology.

Although the main focus in this investigation is the period from the 15th to the 17th centuries, as indicated above, a proper context is needed, so our historical navigation begins with an introductory Chapter 2. There we sketch the first doubts in Greek culture of the ancient assumption of the flat nature of the Earth. Once the Earth was accepted to be spherical, the next debate arose about the existence of antipodean continents and inhabitants on the other side of the sphere, a debate that was never wholly closed in Roman culture. This lingering debate in the ancient world will help us to understand how it could reappear among late ancient Christian authors and persist well into the medieval and Renaissance debates. For the medieval Western Europe, the two key Christian authors concerning the debates about the sphericity of the Earth and the existence of the antipodes were Lactantius and Augustine. We will study their core texts relevant for our purposes and investigate their contemporary context pertaining to geography and Christian apologetics. Finally, we will follow their very different medieval reception. While Lactantius was nearly forgotten, being of little influence during the Middle Ages, in sharp contrast, Augustine became a key reference in the medieval debate over the antipodes and never lost popularity.

Chapter 3 chronicles how in the Early Renaissance Lactantius was 'rediscovered' with admiration for his literary quality, but soon his denials of the sphericity of the Earth and the antipodes attracted criticisms. We will focus on the early

15th-century criticisms of Lactantius that appear in authors as different as Cardinal Fillastre, Antonio da Rho and Galeotto Marzio. However, none of them criticised Augustine. In fact, he was 'protected' by scholars and ecclesiastics, despite having denied the antipodeans which were progressively discovered as the Portuguese navigated the African coast toward the south. We will analyse the strategies and reasons they had for his justification. In parallel, a late scholastic model of an eccentric Earth became the focus of the geographical debates, including the discussions surrounding Columbus at the famed meetings with experts in Portugal and Castile. Their objections to the planned voyage will be analysed in detail.

Chapter 4 will trace the impact of the revolutionary geographical discoveries of the late 15th and early 16th centuries that led to strong criticisms of the ancient geographers, some of them just recovered during the Renaissance. Their descriptions of the Earth did not fit with what sailors encountered on their voyages, in particular regarding the uncrossability of the Equator and the inexistence of the antipodes. Very soon, ancient Christian authors, especially Lactantius and Augustine (but soon others), came under fire for the same reasons. We can see here one of the first instances when Renaissance people, who used to venerate the wisdom of the ancients (both pagan and Christian), looked back to the classical past with contempt. We can observe a crucial turn in the evaluation of this previously idealised past. It was the first time that western Europeans had hard evidence – in an important field – that refuted what they had read in sources from the past. Thus, it was difficult to escape from the feeling that there was an unavoidable *conflict*, although the continuous defence of Augustine will show the strategies devised to avoid the damage caused by a difficult choice between the past and the present, *authority* and *experience*, not previously encountered. This complex scenario set the stage for the appearance of the 'flat error'.

In Chapter 5, we will see how it was, after more than half a century of criticisms of ancient geographers and early Christian writers, that Copernicus entered the stage and picked just one example as a target for his criticism. This was the most obvious of the disgraced early Christian authors: Lactantius, denier of both the sphericity of the Earth and the antipodes. The previous discussion in Chapters 2–4 will enable us to understand how a brief text from an old 4th-century Christian apologetic book ended up being mentioned – in a strongly negative way – in a 16th-century astronomical treatise composed by a canon of the Catholic Church. Intriguingly, Lactantius was not criticised in the context of a defence of the sphericity of the Earth – that appeared in book 1 of *De revolutionibus* with references to ancient Greek flat-earthers – but in the dedicatory letter to the Pope. The context was not a discussion of astronomy or geography, but a carefully crafted rhetorical defence of the independence of astronomy from theology. This was the novelty of the usage of Lactantius by Copernicus that set a precedent of mentioning Lactantius in the emerging discussions on the relationship between science and Christianity. Subsequent generations after Copernicus did not ignore this strategy to protect heliocentrism by pre-empting theological attacks. We will track the further usage of the references to Lactantius and the continuously increasing list of flat-earthers in early 17th-century Copernican books. In addition, we will investigate the rediscovery of what became an earlier precedent of science and Christianity conflict: the condemnation of the 8th-century cleric Virgilius of Salzburg for defending the antipodes, as well as the rising periodisation of history, leading to the modern invention of the concept of 'Dark Ages'. In this context, we will see the first clear instances of the flat error.

Finally, the General conclusions will summarise the preceding analysis, based on primary sources (some already well known while others nearly forgotten), and draw further conclusions. This research results in a new complex proposal for the origins of the flat error, taking all those data into account. In particular, it will be shown how during the late 16th and early 17th centuries, actual historical late ancient and medieval cosmographical debates, in particular relating to the existence of the antipodeans, were conflated with non-historical, non-existent debates about the flat Earth. These debates started in the geographical literature but were also discussed in theological treatises and early descriptions of newly found lands. The confusion grew when these topics were rhetorically used among Copernicans as a defensive tool from theological interference, with the situation becoming more polarised after the 1616 Catholic condemnation of Copernicanism. That led to the first instances of the flat error, which did not follow the confessional divide of the times, as it can be found in both Protestant and Catholic authors. However, this was not the unique line of arrival to the flat error, as other non-Copernican sources also contained elements of the flat error.

It is proposed that the idea that there have been an historical series of conflicts between science and Christianity – the so-called 'conflict thesis' – is not an invention of the late 19th century. Instead, this view can be regarded as an unintended consequence of an excessively self-critical attitude and confessional polemics that started with certain early modern Christian authors who became the pioneers of the flat error c. 1600. Over the centuries, these polemics accumulated and ended up backfiring when they were seized by truly anti-Christian propagandists to strengthen their previous arsenal of argumentation. This has created a situation today where many favour the model of a perennial conflict between an unstoppable, advancing science and an obscurantist, retreating Christianity.

A downloadable Appendix, with important primary sources referred to throughout the text, is available at www.routledge.com/9781032893068.

Clarification of the terminology used in this book

An important clarification is needed regarding the term 'flat error'. Russell used this term throughout his 1991 book to refer to the *erroneous* belief that flat-earthers were still active – as a result of Christian opposition to science – until Columbus in the late 15th century. Although other authors used different terminologies, throughout this book, we will follow Russell's terminology. The term 'flat-earther' will be used to denote someone who believes the Earth is flat, and the term 'flat error' to refer to the false claim that a person believes the Earth is flat. This is notwithstanding the fact that claiming the Earth is flat is also an error.[4]

As we associate here the flat error with the debates on the relation between science and Christianity, it is essential to highlight the critical transformations that the disciplines under the current label of 'science' underwent in the period covered by this book. The origin of the modern notion of 'science' (encompassing physics, chemistry, biology, etc.) is usually dated back to the 17th-century 'Scientific Revolution', which demoted the Aristotelian 'natural philosophy' and introduced a new 'scientific method' which replaced the philosophical approach. However, many authors consider science as having achieved its current status only about 200 years ago, when some claim a second Scientific Revolution took place.[5] An important term for this book is 'cosmography', a word popular between the 14th and 17th centuries, which then went out of fashion. This discipline was dedicated to the interconnections between the heavens and the Earth, encompassing topics of modern disciplines from geography/geology to astronomy (Biro, 2009). We will use this term when discussing this period; for earlier and later periods, 'cosmology' will be used. Strictly speaking, this is an anachronistic term, originating only in the 18th century, but often used in scholarship in relation to the view of the universe among ancient civilisations and ancient authors, such as Aristotle.

If the study of nature has changed over time, Christianity has also undergone important transformations in its 2,000 years of existence. Different theologians and church leaders displayed varied attitudes towards the study of nature, from amicable to confrontational (see p. 39). However, the dominant approach saw theology as the 'queen' of all 'knowledge', with all other disciplines as 'handmaidens', due to Augustine's influence (Lindberg and Numbers, 1986; Lindberg, 1987; 2003; 2010; 2013). Nevertheless, the theological boundaries of the authority of the Bible and of doctrinal positions limited this conciliatory approach. The shape of the Earth was outside of these strictures, except among some particular branches of Christianity, such as the theological school of Antioch. However, the existence of antipodeans, after Augustine, became an undesirable idea for most orthodox Christians.

Talking about 'science and Christianity' or 'science and religion (or faith)' as if these terms meant the same 500 years ago as in the present has an anachronistic component. On the other hand, this is probably the only succinct and convenient way of simplifying the terminology.[6]

Notes

1 Throughout this book, I have put Earth in capitals to refer to the planet, leaving 'earth' for the element in the ancient classification of elements (however, when quoting other authors, I will respect their capitalisation).

2 See https://happylearning.tv/en/discovery-of-america/, also available in https://www.youtube.com/watch?v=MHIfzBNnhqI (all the websites in this book have been consulted for a final time in autumn 2024).

3 See my contribution in Felipe (2017).

4 To demonstrate this usage with an example: to affirm that Lactantius was a flat-earther is not a flat error, as he clearly denied the sphericity of the Earth in his 4th-century writing (see Chapter 2). However, to claim that all, most, or many western medieval theologians believed the Earth was flat is a flat error, as this was indeed not the case (see Chapters 2

and 3). Christine Garwood has pointed to these two errors: "The error is double-sided: that the earth itself is flat, and that educated people who lived after the fifth century BC believed that it was so" (Garwood, 2007[2008], 350).

5 It is important to note that mathematics (which included astronomy and astrology) had an ancient separate existence, as well as medicine (in this book we frequently find the terms mathematics or astrology referring to what today is astronomy). Alchemy was an ancient practice, which was transformed during the 17th and 18th centuries into chemistry, in parallel with the abandonment of astrology by professional astronomers. Between the 16th and the 19th centuries, it was also common to group under 'natural history' the study of animals, plants and minerals. For details on these terms, see Shapin (1996, 5–6), Yeo (1999) and Ogilvie (2015, 263).

6 These terms are used in many books and papers and even in the name of institutions and organisations. On the evolution of these terms, see Peter Harrison (2015; 2024); see also his earlier discussion with leading historians on the terms 'science' and 'natural philosophy' (Harrison et al., 2006).

1 Searching for the origins and causes of an error

A century of searching for the origin of the flat error: Good hints and blind alleys

Although US medievalist Jeffrey Burton Russell's *Inventing the Flat Earth: Columbus and Modern Historians* is the most popular book on the topic of the flat error, and the one that has shaped the topic since its publication in 1991, the question of the origin of the flat error had been in the air for more than a century. In this section we will overview how this question has haunted generations of historians.

The identification of the flat error since the late 19th century

Washington Irving's depiction of the supposed "council" rejecting Columbus at Salamanca (1828) was criticised already a few decades later, as can be seen in the booklets by the librarian of the University of Salamanca, Domingo Doncel y Ordaz (1858), and by a Prof. of Law at the same, Modesto Falcón y Ozcoidi (1881). Also critical was an anonymous article published in *Catholic World* (1869). The paper by Letronne (1834) attracted even more immediate criticism in a book by Catholic Bishop Louis-Auguste Delalle (1835).

More critical voices against the flat error were connected with the 400th anniversary of Columbus's arrival in America. One was Edward John Payne, a British historian specialising in colonial America, who wrote about "the singular error that the doctrine of the earth's sphericity was generally discredited during the Middle Ages, and only revived after the exploit of Columbus" (Payne, 1892, 42, note 5). John Fiske, a US philosopher and historian, although he considered that the "views of Cosmas [...] represent beliefs current among the general public until after the time of Columbus" (Fiske, 1892, vol. 1, 267), noted that the denial of the antipodes "did not necessarily involve disbelief in the globular figure of the earth" and that "among the more enlightened clergy the views of the ancient astronomers were never quite forgotten" (Fiske, 1892, vol. 1, 371). And in relation to the depictions of Columbus as been opposed by the clergy, Fiske noted the "cordial support from the clergy" and denounced "the account in Draper's *Conflict between Science and Religion*" (Fiske, 1892, vol. 1, 413) – a key late 19th-century book promoting the idea of a science and religion conflict.

DOI: 10.4324/9781003545682-2

In the early 20th century, others criticised the assumption that the sphericity of the Earth was ignored before Columbus. Some were Catholics eager to defend ancient and medieval Christians from accusations of flat-earthism, such as the Prussian-US Jesuit Francis S. Betten in a 1923 paper.[1]

Renewed criticisms of late 19th-century proponents of the flat error can be found in the well-known work by US historian of medieval science Lynn Thorndike, severely criticising the British historian Sir Charles Raymond Beazley for "imply[ing] that a majority of the fathers, like Cosmas Indicopleustes, attacked the belief in the sphericity of the earth." Thorndike also noted: "I wonder if he is not following Letronne, *Des Opinions Cosmographiques des Pères*, without having examined the citations," to which he opposed the example of Basil's *Hexaemeron* (Thorndike, 1923, vol. 1, 480, note 1; he referred to Beazley, 1897, vol. 1, 274). Thorndike also noted that contemporaneous with the Christian flat-earther Lactantius was the pagan flat-earther Plutarch: "Thus we see that Christian fathers like Lactantius were not the first to ridicule the notion of the Antipodes; apparently as well educated and omnivorous a pagan reader as Plutarch could do the same" (Thorndike, 1923, vol. 1, 220; on Plutarch, see p. 32).

In James Y. Simpson, Prof. of Natural Science in Edinburgh, we find again criticisms of White and Draper – which works on the history of science and Christianity he aims to improve – for their selection of early Christian authors that are easy to criticise, pointing out that "of early Christian writers, Lactantius, [...], is again and again selected as a subject for attack" (Simpson, 1925, 96). However, it was Charles W. Jones, a US medievalist, well known for his research on Bede, who went a step further and after ridiculing the common school teaching regarding medieval times as believing in a flat Earth asked: "Whence, then, arose our modern idea that in the Middle-Ages students believed the earth was flat? I cannot answer, [...]" (Jones, 1934, 304). This is the key question that inspires us in this book.

Other authors interested in the flat error were researchers of Columbus's history, such as British historian Lionel Cecil, who affirmed that in the 15th century, the sphericity of the Earth was commonly accepted in Western Europe and: "There is no foundation for the assertion, which was once credited, that in Spain a contrary view was maintained by orthodox theologians and supported by religious prejudice" (Columbus, 1930, vol. 1, xxii; also 1933, vol. 2, lviii). The US Navy captain and historian Samuel Eliot Morison, in his well-known biography of Columbus, affirmed that "every educated man in his day" accepted the sphericity of the Earth and that "every European university" taught so, a fact that was also known to "seamen" (Morison, 1942, 33). He attacked Irving's fictitious debate at Salamanca as "pure moonshine," "misleading and mischievous nonsense" (Morison, 1942, 89). Morison continued dispelling the flat error in later works without offering further potential originators of it (Morison, 1971, 6; 1974, 38).

The British geographer and historian of science Eva Germaine Rimington Taylor, after denouncing the flat error in several papers (1931, 78; 1935, 5; 1937, 54–55, 57), wondered about the origins of this modern myth: "It is difficult to understand why the story has gained such ground that prior to Columbus's voyage it was generally believed that the world was flat. Yet it is widely current even to-day [...]" (Taylor, 1943, 9). She pointed at school education as an explanation, as the flat error "has not yet been completely expunged from history texts or from the classroom" (Taylor, 1937, 54). Twenty years later she repeated this observation (Taylor, 1957, 133), at a time when the flat error was arriving in German-speaking central Europe (Bernhard, 2013a; 2013b; 2017; 2019).

Jeffrey Burton Russell and the search for the roots of the flat error (1991)

This lead was followed in US textbooks from the 19th century onwards by Jeffrey Burton Russell in his 1991 *Inventing the Flat Earth*. He showed how the flat error in them was influenced by Irving. Russell not only denied that the flat Earth was characteristic of all early Christian and medieval thought, as many before him had, but he went further by trying to give a detailed answer to the same question Jones had asked in 1934 and could not answer.

According to Russell, the flat error "had been planted by Irving and Letronne in the nineteenth century" (Russell, 1991, 43; and 49, 61). He referred to Washington Irving's popular book *A History of the Life and Voyages of Christopher Columbus* (1828), and Jean-Antoine Letronne's scholarly paper *Des Opinions Cosmographiques des Pères de L'Eglise* (1834).

A third culprit was found in William Whewell, Master of Trinity College (Cambridge), priest of the Church of England and author of a landmark history of science book, *History of the Inductive Sciences* (1837).[2]

As far as I am aware, Russell is the first to attempt a systematic search for the source of the flat error. Russell claims that there is no flat error before Irving and Letronne in the 19th century. However, he did find the "seed" of the idea in the 16th century:

> By the time Copernicus had revolutionized the way people viewed the planets [...] the seed of the Flat Error had been planted, but it did not grow to choke the truth until much later. When did it triumph and why? Who was responsible? These are the main questions of this book. [...].
>
> (Russell, 1991, 5)

Later in the book, Russell explained that this seed:

> [...] was planted by Nicholas Copernicus (1473-1543) [...]. In the preface to his *De Revolutionibus*, Copernicus used Lactantius to illustrate how the ignorance of opponents of the round earth was comparable to that of those insisting on geocentricity in his own time.
>
> (Russell, 1991, 64)

Russell considered Copernicus's preface to his *De revolutionibus* as being "the basis for later writers using Lactantius as a villain illustrating a medieval belief in a flat earth," but he argued that there was no trace of the flat error in Copernicus himself (Russell, 1991, 64).

As we will see in Chapters 3–5, attacks against Lactantius and other early Christian writers for denying the sphericity and the antipodeans were routine in many books at Copernicus's time (as his casual reference suggests). Therefore, while Copernicus's book could have influenced astronomical books of Copernican persuasion, the full context is more complex than it seems.

How did Russell connect the flat error "seed," as early as Copernicus in the 16th century, with the fully fledged flat error found in Irving and Letronne in the 19th century? What happened in the intervening centuries? This connection was highly tenuous in Russell's 1991 book. He credited the Renaissance humanists with "inventing" the Middle Ages and the Protestants with "darkening the Middle Ages" (Russell, 1991, 65). He then stated: "The Humanists did not mention flatness, but as the idea grew that the people in the intervening ages were stupid, so did the assumption that nothing was too silly for them to believe" (Russell, 1991, 66). For Russell, this set the scene and the background for the flat error to arise. Nevertheless, there was still no mention of any actual concrete source for the flat error. In fact, a few pages earlier, he already stated that "Protestant opponents of Roman Catholicism avoided the [Flat] Error" (Russell, 1991, 62).

Russell also cleared later thinkers of the flat error: "Nor is there more than a hint of the Error in the sixteenth and early seventeenth centuries among the revolutionary cosmologists such as Copernicus, Galileo and Campanella" (Russell, 1991, 62). He did not find flat error in Kepler, Montaigne, Rabelais, Bruno and Bacon (Russell, 1991, 63), but he accused the Anglican clergyman and natural philosopher John Wilkins (1614–1672):

> [...]. John Wilkins was a rare exception to the rule, writing in 1708 explicitly about the flat earth. He [...] compared himself and his detractors to Columbus and the alleged flat-earthers. Wilkins offered an eccentrically long list of alleged flat-earthers, including Herodotus and Lucretius along with Chrysostom, Augustine, Procopius, and of course Lactantius.
>
> (Russell, 1991, 62; the 1708 book is a posthumous re-edition of a 1638 book; see p. 201)

Russell affirmed, with some examples, that the flat error was not found among the philosophers of the Enlightenment (Russell, 1991, 61). His 1991 book concluded, simplifying, that the founding fathers of the flat error were Irving and Letronne, and although he intended to identify earlier "seeds," he discredited their existence, with the possible exception of Wilkins. Russell still supported these views in a 1993 paper and in a brief article in 1997: "No one before the 1830s believed that medieval people thought that the earth was flat. The idea was established, almost contemporaneously, by a Frenchman and an American [...]" (Russell, 1997).

Russell's views quickly attracted considerable attention from scholars in different fields, and it is still today the standard view on the origins of the flat error that is referenced – almost inevitably – when this topic is discussed.[3]

Russell's latest identification of the roots of the flat error (2001)

Despite the well-crafted reasoning above, Russell's views seem to have changed in two short articles published in 2001 and 2002. In 2002 he wondered about the origin of the flat error in these terms:

> [...]. Copernicus was certainly not to blame. When he condemned the geocentric opponents of the heliocentric system as being as wrongheaded as those who once asserted that Earth was flat, he could cite only the one flat-earther that he knew of: Lactantius, who was, as Copernicus said, "hardly an astronomer." Neither Copernicus nor Galileo had to contend with a single flat-earther anywhere.
>
> Oblique hints of the Flat Error exist as early as the late 1500s and early 1600s in works by Thomas Hobbes, Thomas Campanella, Francis Bacon, Herbert Rosweide, and Bernard de Montfaucon. The earliest flat-out Flat Error I have found so far is in a book published in 1708 by John Wilkins [...].
>
> (Russell, 2002; an almost identical text in 2001, 19)

This text was innovative on several accounts: (i) Russell completely cleared Copernicus, (ii) several authors whom he had cleared of the flat error were here responsible for "oblique hints of the Flat Error" (adding Rosweide and Montfaucon, who were not mentioned before), (iii) Wilkins is here culprit of "[t]he earliest flat-out Flat Error" in 1708, (iv) Montfaucon and other 18th century authors establish a connection with the early 19th-century Irving and Letronne (in the lines following the above quotation).[4]

So now Russell attributed the seed for the flat error to Wilkins in 1708, although pointing to oblique hints before him, surprisingly not realising that this 1708 book was a later, posthumous edition of a book published in 1638. I will not deal with this book in detail, as I intend to show that the flat error was already launched by that time.[5]

Reinhard Krüger's search for the flat error roots (c. 2000)

A careful reader can discover small changes in the position of Russell before and after the turn of the millennium. They are evident in his reference in the articles from 2001 and 2002 to the "oblique hints of the Flat Error" in three authors he had mentioned before, but most importantly, he added two new authors: the Dutch Jesuit Rosweide (1569–1629) and the French Benedictine Montfaucon (1655–1741), both interested in ancient Christian writings.

Russell did not explain why he included them a decade after his original book; he simply referred to the German historian Reinhard Krüger. Much of Krüger's research in the late 1990s and early 2000s was concerned with a detailed study of the primary sources from ancient and medieval Christian authors, clearly showing that the flat-earthers were a tiny minority, nearly unknown in Western Christianity (Krüger, 2000a; 2000b; 2012). This is in line with studies on the same topic by a growing number of scholars in the late 20th and early 21st centuries dealing with the medieval views on the shape of the Earth.[6]

Krüger offered an essential contribution to the study of the sources of the flat error, identifying Rosweide and Montfaucon as stepping stones.[7] A few years before, the German philologist Rudolf Simek also pointed to the importance of Montfaucon's unearthing of Cosmas for the history of the flat error (Simek, 1992[1996], 37–38; 1997, 139–147). He expected to find the culprits of the flat error in the 18th century (Simek, 1997, 147).

Although we could see Rosweide's conflation of ideas as an early oblique hint of a flat error, it is hardly possible to see his isolated sentence, in the footnote of an obscure book with old legendary religious texts, as influential in the origin of the flat error. In addition, I have not seen this book mentioned by any other author discussing the shape of the Earth or the antipodes in the following centuries (on Rosweide, see p. 134). In contrast, Montfaucon was a truly influential author, but more in the *diffusion* of the flat error than in its origin (see pp. 204–205).

In an interview, Krüger mentioned the physician John Dryander as an early source of the flat error – in the 16th century – for considering Augustine as a flat-earther (Wolff, 2005). However, Dryander's text contains just a correct criticism of Lactantius and Augustine for their denial of the antipodes (see pp. 126–127); therefore, I discard this as an example of the flat error, as the shape of the Earth is not even discussed.

A new identification of the root of the flat error by James Hannam (2009)

A new origin for the flat error is offered by British historian James Hannam:

> The myth that a flat earth was part of Christian doctrine in the Middle Ages appears to have originated with Sir Francis Bacon (1562-1626), who wrongly claimed that geographers had been put on trial for impiety after asserting the contrary. [...].
>
> (Hannam, 2009, 35)

In 2010 and 2023, he expanded the accusation from Bacon to Protestants in general without providing more data (see pp. 210–211).[8] Hannam refers to a passage where Bacon mixes the sphericity of the Earth and the antipodes in *Thoughts and Conclusions* (*Cogitata et Visa*, 1607; see Appendix, no. 4, available at www.routledge.com/9781032893068). Curiously, Russell in 1991 had a similar quotation from the related *Filum Labyrinthi*, 1606 (see Appendix, no. 3) but considered that Bacon "showed little interest in flatness. Bacon's only reference is vague" (Russell, 1991, 63). Russell mentioned Bacon in 2001 and 2002 as a case of "oblique hints of the Flat Error."

Although identifying Bacon as the single originator of the flat error is an overstatement, his simple, brief and misleading remarks were probably part of the build-up to the flat error, in a wider context, as we will see in pp. 134–135 and discuss in Chapter 5 and the General conclusions.

Revival of the interest on the flat error in 2021

In 2021, when the first version of this book was under examination as a PhD thesis, three books appeared containing references to the flat error, one of them entirely devoted to this issue.

Published in early 2021, US theologian Derrick Peterson's *Flat Earths and Fake Footnotes* aims to explain, as indicated in its subtitle, "The strange tale of how the conflict of science and Christianity was written into history." The title of Peterson's book is taken from that of his chapter 7, which is the first of a section collecting false myths on science and Christianity used to buttress the model of conflict between them. Peterson points out how historical examples have been used to advance the idea of a science and Christianity conflict: "just like the myth that Christians believed in a flat earth, any number of memorable images, set pieces, and historical anecdotes are levied to exemplify warfare and Christian irrationalism" (Peterson, 2021, 7). When it comes to explain the origin of the flat error in chapter 7, the "real perpetrator" is Irving "at a more popular level," and Letronne "as an academic commonplace" (Peterson, 2021, 188–191).[9] As expected, Russell is his key source (Peterson, 2021, 179, 188–191, 195–198). Peterson connects the origin of the flat error with the origin of the idea of a conflict between science and Christianity. But he does so almost exclusively in the context of the 19th-century English-speaking culture, apart from the usual reference to Letronne. In particular, Peterson criticises the Victorian revival of flat-earthers (Peterson, 2021, 187–188, 194–195). As Russell did before,[10] he also denounced the linking of the 19th-century flat-earthers with the anti-Darwinists: "the image of a fanatical Christian touting the flat earth against scientists stuck. [...] after the publication of Darwin's *Origin of Species* the flat earth migrated into the polemical toolbox of evolutionists" (Peterson, 2021, 198; also 187–188, 191).

Near the end of 2021, British physicist David Hutchings and US historian of science and religion James C. Ungureanu published *Of Popes and Unicorns*, which in many ways covers a similar ground to Peterson's book. It also incorporates into a continuous narrative how the main examples used to advance the model of conflict between science and Christianity were integrated in the 19th century,[11] including the flat error. To explain its origins, the authors just quote Irving (Hutchings and Ungureanu, 2022, 68–69), and mention Russell (Hutchings and Ungureanu, 2022, 62–64, 67, 70). In addition, they also link the Victorian flat-earthers to their contemporary view of a science and Christianity conflict (Hutchings and Ungureanu, 2022, 53).

Completely different is the book by the French historians Violaine Giacomotto-Charra and Sylvie Nony, *La Terre Plate. Généalogie d'une idée fausse*, late in 2021, devoted entirely to the problem of the flat error. They open their Introduction by referring to Russell but want to go further, as in my case (Giacomotto-Charra and Nony, 2021, 5, 11). In fact, similarly to my approach in this book, they start discussing the Greek ideas about the sphericity of the Earth and advance through the Middle Ages to Modernity, later discussing the 19th-century invention of the flat error. They are able to find earlier roots than those of Irving and Letronne. However, it seems that their search is skewed toward French authors. They go back as far as the mid-17th century, "The legend, however, is older and timidly appears in the 17th and foremost in the 18th century, especially with Voltaire" (Giacomotto-Charra and Nony, 2021, 141)[12] and mention Cyrano (Giacomotto-Charra and Nony, 2021, 151), Montfaucon (Giacomotto-Charra and Nony, 2021, 147–148, 161) and Voltaire (Giacomotto-Charra and Nony, 2021, 149–150, 187). They also refer to earlier debates on the antipodes, the geographical discoveries and the Copernican controversies and suggest their connection with the flat error (see quotations in General conclusions, endnote 10), although they do not investigate these connections in detail, as it will be done here in Chapters 3–5.

The approach taken in this book to investigate the origin and causes of the flat error

Searching for the roots of the flat error through a reception study

As Lactantius and Augustine were the usual suspects of having launched the Christian medieval retardation in science to a flat earth belief, I decided to start with them, not without trying to understand before how the Greeks discovered the sphericity of the Earth and devised the concept of the antipodes. From that a detailed study of the key texts, Lactantius's *Divine Institutes* 3.24 and Augustine's *City of God* 16.9, was needed as well as their spatio-temporal, theological and scientific context (see Chapter 2). These factors place both Christian texts in an already established debate regarding the shape and population of the Earth in Roman culture, dismissing claims that they need to be attributed to specifically Christian ignorance or fanaticism. As we will see in Chapters 4 and 5, some 16th-century authors offered a more balanced exposition of early Christian authors than many of our contemporaries. Instead, these texts were much more connected with their cultural context than generally acknowledged, in particular in the case of Augustine.

The original context and purpose of these passages should be distinguished from their later, often decontextualised usages. I have been particularly attentive to creative usages of the ideas of Lactantius and Augustine, as "reception is not merely a matter of inheriting the past but a further reprocessing and restructuring of that inheritance," which in extreme cases means that "Augustine's thought was used, abused, or transformed by his readers" to produce "actualization(s) of a text by one or more reader(s)" (Pollmann, 2013, 4, 7, 8).

Therefore, I decided to undertake a reception study – from Late Antiquity up to the time of Columbus – of those texts by Lactantius and Augustine to understand how they were read, appropriated and influenced theology and science up to the Renaissance (Chapter 2).[13] Throughout this book we will encounter not only erudite treatises but also maps, drawings, letters, logbooks, poems and pieces of popular culture. Although my primary aim in this research is to discover the roots of the flat error, it is also apparent that the medieval reception of the texts of Lactantius and Augustine did not trigger a Western European medieval school of flat-earthers. In fact, the topic of the flat Earth was no longer a matter of debate, and Lactantius was almost forgotten until his recovery in the 15th century, while the real debate up to the c. 1500 was centred on Augustine's argumentation against the antipodes (Chapters 2 and 3).

The rise of a mentality favourable to the flat error

As I followed the historical development of the flat error, I found a multi-causal complex process. Simek (1992[1996]), already pointed to the confluence of three ideas responsible for the flat error: the flat Earth statements by Lactantius and Cosmas, a misunderstanding of the medieval debate on the antipodes and a confusion due to the round/disc shape of many medieval world maps, in particular the T-O maps (Simek, 1992[1996], 37–38; for these maps, see Figure 2.2 below).

Later Simek introduced a fourth idea, the importance of changes in the meaning of keywords, like *orbis*, in the common expression *orbis terrae* (Simek, 1997, 145–147; 2008, 798–799).[14] *Orbis* was ambiguous in the Middle Ages, since it could refer to a spherical shape (from where the current English word 'orb' acquires the meaning of 'globe'), but then acquired a restricted meaning during the Renaissance, inspired by classical Latin, to be understood as disc or slice. Betten had already warned about the original restricted meaning of this term:

> [...]. We must not be misled, however, by the phrase *orbis terrarum*, which occurs so frequently in classic literature. The word *orbis* commonly means a circle or a disc. Its use in connection with *terra* dates from the time when the Romans still thought of the earth as a plain. In the term *orbis terrarum* there was no reference to sphericity of the earth. Even a part of the earth's surface might be called *orbis*, as, *Orbis Romanus*, the Roman world. [...].
>
> (Betten, 1923, 75–76)

However, other authors also consider the usage of *orbis* in classical Latin ambiguous.[15]

This book deals in detail with Simek's first two topics to characterise the early steps of the development of the flat error. However, the focus will be on the reception of Lactantius and Augustine, as Cosmas was rediscovered in the early 18th century, a period outside the scope of this book, which focuses on earlier origins of the flat error c. 1600. Some attention is also given here to medieval mapping to understand the debate on the antipodes. However, the misunderstandings involved in Simek's last two topics became more important in debates since the 19th century.

At this point, it is important to go back to the original definition of flat error as "[t]he almost universal supposition that educated medieval people believed the earth to be flat" (Russell, 1991, xiii.). This neat definition of flat error is minimalist. As Russell himself explains, other false ideas can be found aggregated around it: the ancient Christians as the negative influence from which the medieval people draw the alleged flat Earth view, the alleged role of the Christian Church as the dark force supposedly asserting flat-earthism and Columbus as the dispeller of the alleged medieval view of the Earth as flat, leading to an alleged conflict with the Church. To the best of my knowledge, all these issues did not suddenly appear in a single text. This diversity of ideas points to a multiplicity of strands coming together to generate a spectrum of flat errors, already noted by Russell:

> [...]. First there is the flat-out Flat Error that *never* before Columbus did anyone know that the world was round. [...]. Another crude form of the Flat Error is the lurid embellishment that sailors feared that they would plunge off the edge of the flat earth if they voyaged too far out into the ocean. [...]. Another version of the Error is that the ancient Greeks may have known that the world was round, but the knowledge was lost (or suppressed) in medieval darkness. [...]. Many inconsistent varieties of this version exist: The knowledge was lost in the first century A.D., or the second, or the fifth, or the sixth, or the seventh; and on the other end it was lost until the fifteenth century, or the twelfth, or the eighth. [...]. Still another version is that almost everyone always believed the earth was flat, but in the darkness had shone a few, scattered lamps, held by Aristotle and Ptolemy and Bacon and Toscanelli. [...].
>
> (Russell, 1991, 27–29)

In addition to the reception of Lactantius and Augustine, to follow all those strands, we will delve into a broader study of the period from the 15th century to the early 17th century to better understand the changes in the perception of the ancient and medieval past (Chapters 3–5). Multiple authors and their interconnections are examined to investigate how this new mentality or climate of opinion, critical of the past, allowed the flat error to appear, take root and grow. In the words of historian of science Reijer Hooykaas, here we followed a "phenomenological approach [...] to revive the past, to enter into the minds of our predecessors" (Hooykaas, 1987, 454).

In Chapter 3, we will see how this change started in the 15th century with a new critical evaluation of past authors, which for the first time included criticisms of Lactantius's denial of the spherical Earth and the antipodes. In contrast, Augustine's denial of the antipodes was routinely excused, even after having been shown wrong, as scholars and ecclesiastics desired to retain his authority.

We will follow in Chapter 4 the impact of the new geographical explorations undertaken by the Portuguese in the 15th century and the geographical debates that followed. They introduced a new element of criticism in the mindset of late 15th- and early 16th-century scholars, because this was no longer just a bookish criticism of the past from pitting ancient authorities against each other. The new geographical knowledge facilitated an historical perspective of the past, which was better able to appreciate its shortcomings, giving rise to a critical historicist view – sometimes even bordering on patronising – of the progression of knowledge. This was an ideal milieu for the flat error.

The Scientific Revolution and the origins of the flat error

Although most historians of science have given prominence to the revolutionary changes in astronomy during the 'Scientific Revolution' of the 17th century, a significant change in worldview occurred in the century preceding Galileo and Newton. Kuhn affirmed that these voyages led to learning:

> [...] how wrong ancient descriptions of the earth could be. In particular, they learned how wrong Ptolemy could be, for Ptolemy had been the greatest geographer as well as the greatest astronomer and astrologer of antiquity. The astronomer's awareness – an awareness which we shall shortly discover in Copernicus himself – that Renaissance man could at last correct Ptolemy's geography prepared him for changes in his own closely related field.
>
> (Kuhn, 1957[1970], 124)

Sadly, this intellectual shift has been often overlooked in the secondary literature. However, there was a first geographical phase of the Scientific Revolution in the 15th and 16th centuries that we study here in Chapter 4. Hooykaas used the term "geographical revolution" (Hooykaas, 1987, 456, 472, 473), and historian Susanna de Beer placed it in the title of a 2013 book chapter. The geological implications of the geographical revolution have been recently noted (Leitão and Alvarez, 2011), something that should not be surprising, as both disciplines were bound together in the term 'cosmography' used at the time (see p. 7). In fact, other authors used "cosmographical revolution" (Flórez et al., 1988, 30; Vogel, 1995, 30).

The changes in the perception of the Earth were already perceived as revolutionary at the time. Humanist Volterra, in 1506, already noted that the navigators of his time "have opened up our globe in length and in breadth" (see p. 108). It has been pointed out that "to most people living in the Renaissance, the real scientific revolution was not the revelation of a new arrangement of the heavens, but the discovery of new worlds on earth" (Navarro and Eamon, 2007, 34; also, Hooykaas in Rheticus, 1984, 127–128). In fact, Francis Bacon himself recognised the importance of the geographical discoveries of the previous two centuries in his *Novum Organum* (1620).[16]

Furthermore, some authors have suggested that this geographical/geological scientific revolution cannot be separated from the Scientific Revolution in astronomy/physics of the 16th and 17th centuries, both as a crucial precedent to weaken the authority of the ancients, and as providing a new 'terraqueous' view of the Earth that helped the Copernican astronomy. In Chapter 5 we will explore the first point, how the old stories about Lactantius's and Augustine's cosmographical mistakes entered the Copernican polemical works c. 1600 as precedents to criticise the ancient geocentrism.

The consolidation of the confessional division of Europe after the Protestant Reformation and the Council of Trent generated another source of tension regarding the authority of past key Christian authors, including Augustine. This came together with a new impetus to the periodisation of history, started by humanists among late 16th-century Protestants against a medieval period increasingly seen as 'Catholic'.

My working hypothesis has been that this was a potential context in which to investigate the rise of the flat error. Therefore, in addition to the reception of the two key texts by Lactantius and Augustine, I have attempted to correlate such major cultural changes during the 16th century with the origins and early development of the flat error. In fact, the flat error could be seen as an overcritical product of "a Renaissance agenda of retrospective criticism," applying here the words of the historian of Christianity Willemien Otten (2013, 24), against the medieval period, which was carried over into Early Modernity in an even more radical form. Arguably, this is an antecedent of the famous 17th-century *querelle des anciens et modernes*. Historian of science Pietro Daniel Omodeo already placed Copernicus "as a protagonist of the *Querelle des Anciens et des Modernes*" (Omodeo, 2014, 2). Here I propose to find even earlier antecedents of this *Querelle* in the geographical/cosmographical field c. 1500, in line with historian Klaus A. Vogel (see p. 123).

The developments in geography, theology and finally astronomy progressively eroded the authority of the ancients and medievals, slowly but inevitably changing the perception of the late ancient and medieval past, which became increasingly seen as primitive and dark. The rise of this European feeling of superiority in the late 16th century will be followed in Chapters 4 and 5. This change in mindset explains why the flat error and the general negative view of the Middle Ages survive until today. Placing the origin of the flat error in this context is a radical change from Russell's contextualisation of its origin in the post-Enlightenment early 19th century and within the science and Christianity debates associated with Darwinism in the second half of the same century. All that will be discussed in the General conclusions.

The problem of seeking 'culprits' for the flat error

As we have seen above, several authors from the 16th to 17th centuries have been accused as originators of the flat error, complicating the explanation of the origin of such an idea. To clarify the situation, this book will study the Late Renaissance and Early Modernity primary sources accused of creating the flat error and other texts so far not discussed in the flat error bibliography (see Chapters 4 and 5). I will not disclose here the conclusion on individual authors; however, as we will see, the origin of the flat error is a much more complex matter than the passing, or even deliberate, 'error' of a single person. The study of previous scholarship on this topic – and the search for the roots of the flat error – leads to the following core question: what does it mean to search for early 'hints', 'origins', 'sources', 'roots', 'seeds' or even 'culprits' of the flat error?

The idea of an heroic history of science marked by 'precursors', 'forerunners', 'predecessors', etc. has been long discredited as Whiggish, using the expression derived from Herbert Butterfield's *The Whig Interpretation of History*, 1931. It is nowadays seen as an anachronistic interpretation of the past from a contemporary position that views history only insofar as it becomes a pathway to present views, a supposed linear process from darkness to the current light of true knowledge that divides the scientists of the past into progressive 'precursors' of truth and 'retrogrades' who opposed advances.[17]

In contrast, current history of science reconstructs the *complexity* of the historical developments in their own contexts. In this light, we can see the kind of research project undertaken by Russell (and some authors discussed above) as equivalent to the old search for enlightened 'precursors' of great scientific ideas but focusing instead on finding 'culprits' of erroneous views (such as the flat error). However, the expectation of finding one or a few originators of the flat error from which this error was transmitted throughout the centuries does not match up with what we encounter in the historical records.

This investigation aims to show how the mentality changed in Early Modernity to enable the flat error to become a recurrent idea in different formats and disparate contexts. This has led to paradoxical results which resist an easy simplification, as discussed in the General conclusions. The complexity of the origins of the flat error that this investigation unveils, undermines polemical 'conspiracy' theories which appear in popular accounts of the flat error today: for instance, the flat error has been portrayed as invented by anti-Christian authors to erode Christianity in the context of late 18th-century criticisms directed against Christianity or created in connection with 19th-century Darwinian debates or more general science and Christianity debates.[18] Although the flat error was indeed used in such contexts, we show here that its origins largely predate these later debates, paradoxically, in a *Christian* setting.

Notes

1 Cormack (1994, 373), indicates that a similar position was also taken earlier by another Catholic author, Schneid, in 1877, whose work I have not had the chance to study.

2 While there are grounds to criticise Irving and Letronne as promoters of the flat error, there is no strong case against Whewell. He recognised that the anti-antipodean argumentation in Augustine was "without denying the globular form of the earth" and even criticised the polemics created around Virgilius of Salzburg after the denunciation of his teaching about the antipodes to the Pope: "It does not however appear that this led to any severity; and the story of the deposition of Virgil[ius] from his bishopric, which is circulated by Kepler and more modern writers, is undoubtedly altogether false." However, he uncritically accepted Monfacucon's wrong depiction of Abulensis as a flat-earther (Whewell, 1837, vol. 1, 256). On the influence of the story of Virgilius for the rise of the flat error, see pp. 170, 172; on Abulensis, also named, Tostatus, pp. 201, 204–205; and on Montfaucon, pp. 13–14, 204–205.

3 As examples we can mention historians such as Gingerich (1992; 1993), Cormack (1994; 2009), Edson (1997, 167, note 7), Grant (2001, 341–345), Wolf (2004), Woods (2005), Aufgebauer (2006), Chapman (2013, 60–61), Ungureanu (2013); Hutchings and Ungureanu (2022) and Stevens (2020, 137), and scientists such as Gould (1994; 1999) and Singham (2007). Russell has also been followed by the theologians DeMar (1993[2005], 221–234; 2007, 13, note 37, 70, 90, 95, note 4), Moritz (2012) and Peterson (2018; 2021), the specialist in medieval and Renaissance literature Bishop (2008), the sociologist Stark (2016, 1, 2, 6) and the philologist Eco (1998, 4–7; 2004; 2005).

4 Despite these developments, in 'popular' versions of the story, Russell maintained a simplified version with Irving as the sole culprit (Russell, 2012, 142–144).

5 In addition, the accusation against Wilkins as defending the flat error derives from a misunderstanding (see p. 201 and Table 4.1).

6 Brincken (1976), Woodward (1985), Simek (1988), Lindgren (1990), Russell (1991), Brincken (1992), Simek (1992[1996]), Randles (1994[2000], a re-worked version of the earlier Randles, 1980, Spanish edition, 1990c), Cormack (1994), Vogel (1995), Hamel (1996), Edson (1997), Simek (1997; 2002), Hamel (2005), Bishop (2008), Simek (2008), Cormack (2009; 2015).

7 Krüger pointed to Montfaucon as a culprit of the flat error in Krüger (1998). A detailed analysis of the role of Montfaucon in the birth of the flat error can be found in Krüger (2001). A later reference to Montfaucon and Rosweide's role in the origin of the flat error appears in Krüger (2007).

8 Recently, historian of science Michael Newton Keas has followed Hannam in attributing to Bacon the origin of the flat error (Keas, 2019, 54–56), which he called "flat myth" (Keas, 2019, 42). Keas also mentioned the work of Russell on this topic (Keas, 2019, 53).

9 Peterson also mentioned Whewell (Peterson, 2021, 195) following Russell.

10 On Russell, see endnote 18. On this connection more generally, see p. 210.

11 On their decision of starting their analysis in 1780, see p. 209.

12 My translation from: "La légende, cependant, est plus ancienne et apparaît timidement au XVIIe et surtout au XVIIIe siècle, en particulier avec Voltaire."

13 On the reception theory and its terminology, see Holub (1984, xi–xiv). On these terms, applied to the work of Augustine, see Visser (2011, 5–8), Pollmann (2013, 8) and Saak (2013, 67).

14 Wolf (2004), has also indicated the confluence of several ideas in the rise of the flat error, as seen in Reinhardt (2007), with a more specific view of this process in the US.

15 See examples of ambiguity and contradictory usage among Roman authors in Randles (1994[2000], 12–16), and Stevens (2020, 129–131), who adds medieval examples. See also Lecoq (1996, 22, 25 and 35 [notes 46 and 48].

16 See references in Hooykaas (1987, 470–471) and Almeida (2009, 78–79); for parallels with other Spanish material, such as Acosta, see Gascoigne (2013, 223–226); Castilla Urbano (2021, 40) and for the strikingly similarity of *Novum Organum*'s famed cover to that of Céspedes (1606), see Pimentel (1999, both frontispieces on p. 54; 2000, frontispieces on p. 24); Cañizares-Esguerra (2004, 89–90; 2006, 15, 17), Brendecke (2009[2016], 14–16), Castilla Urbano (2021, 40–41). Such connections are strikingly absent in Grafton (2001[2009]).

17 Hooykaas (1987, 454–455), Shapin (1996, 6–8), Osler (2000), Moro and Pelayo (2010).

18 Russell seems to come to the first conclusion, for example, when he criticises Letronne by attacking his mentors, Mentelle and Gail. Russell is mainly concerned with Mentelle, described in rather negative terms as sceptical Voltairean and Enlightenment progressivist (Russell, 1991, 58). There is also an unsubstantiated attempt to connect Irving and Letronne (Russell, 1991, 57) and references to the flat error in the context of Darwinist controversies (Russell, 1991, 35).

2 From the Greek discovery of the spherical Earth to the medieval debate of the antipodes

From flat Earth to spherical Earth in ancient Greek and Hellenistic culture

The Earth acquires three dimensionality

It seems that most, if not all, known ancient cultures have had the common-sense view that we live on a flat surface, including ancient Mediterranean cultures. Usually, the image was that of a circle with a landmass or landmasses comprising the oikoumene (the inhabited land), a term that goes back to the time of Herodotus in the 5th century BC (Romm, 1992, 37). The oikoumene was surrounded by an ocean ring, often enclosed by a wall of mountains (on which rested the rim of a vaulted hemispherical skydome). Such a circular flat Earth can be found in Mesopotamia, Egypt, Greece and Israel.

Tradition attributed to Anaximander (c. 610–c. 546 BC) the first 'modern' map,[1] initiating a sequence of Ionian maps which divided the flat Earth disc into two 'symmetric' regions of equal size: Europe and 'Asia' (which included Africa), on opposite sides of an 'equatorial' line from Gibraltar to the Taurus mountains in Turkey, close to parallel 36° north (Figure 2.1).[2] The symmetry was also ethnographic, with peoples filling the map.[3]

Herodotus (c. 484–c. 425 BC) already described and criticised circular maps surrounded by the ocean and depicting Asia and Europe as having the same size in *History* 4.36.2 (see also 2.23 and 4.8), while Aristophanes made fun of maps in his play *Nubes* 200ff (432 BC, surviving in a revised version from 419 BC to 416 BC).

As new knowledge of the eastern regions was introduced in the Mediterranean, this equatorial line was extended further into Asia up to the Himalayas. The new enlarged Asia prompted a new division of the circular map into three regions: Europe, Asia and Africa. Such tripartite division was already known to Herodotus (*Histories* 2.16). A written depiction of this tripartite schema, separated by the Mediterranean, the Don and the Nile, can be found in the mid-1st century Pomponius Mela's *Description of the World* 1.8. Augustine, *City of God* 16.17, describes Asia as occupying the whole eastern semicircle of the circular oikoumene, while the western side is split between Europe and Africa (see p. 28). There we can find the remote Roman source of the tripartite medieval T-O maps (Figure 2.2; Arnaud, 2014, 42–43).

DOI: 10.4324/9781003545682-3

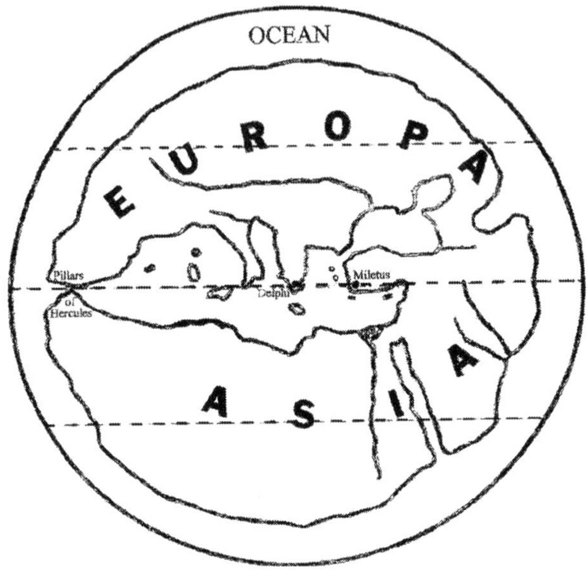

Figure 2.1 Reconstruction of Anaximander's Earth map according to Couprie (2011, 83, Figure 6.2, reproduced with permission of the Licensor through PLSclear).

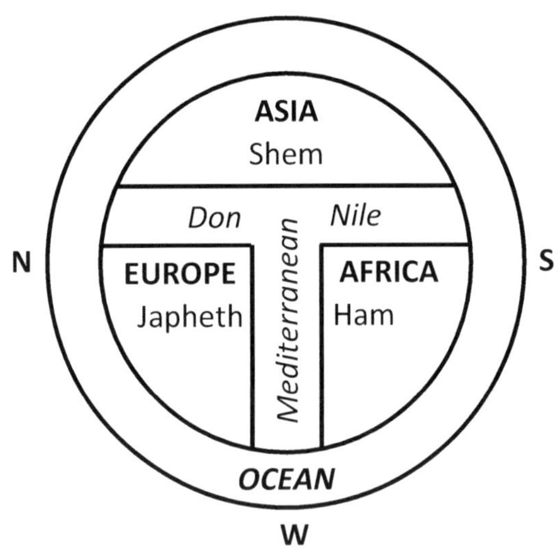

Figure 2.2 Typical medieval T-O map with the three continents of the oikoumene surrounded by the ocean. Note the names of the three sons of Noah associated with the three continents (from Felipe, 2012, 283, Figure 1B, used with permission).[4]

As mentioned above, there were those who criticised the arrangement of these circular maps, while others proposed a rectangular frame for the oikoumene (following the shape of the Mediterranean Sea). This parallelogram frame went back at least as early as Democritus (c. 460–c. 370 BC), and it was still influential in the 6th century with Christian author Cosmas Indicopleustes, who took it from

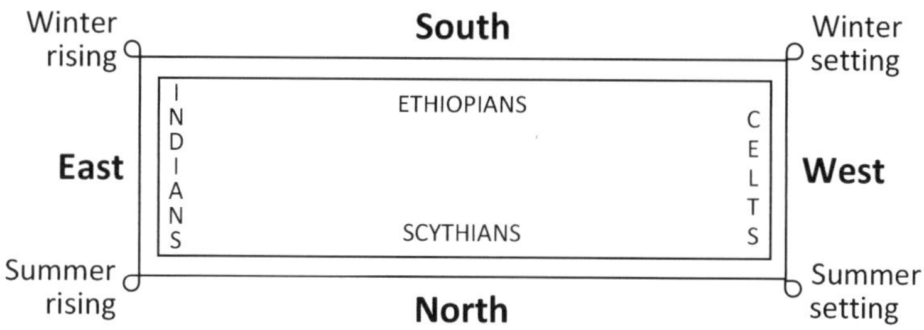

Figure 2.3 Ephorus's map according to Cosmas Indicopleustes's *Christian Topography*. The circles attached to the corners represent "the sunrises and sunsets of the summer and winter solstices" (Couprie, 2018, 23). Here south is at the top.[5]

historian Ephorus, c. 405–330 BC (see Figure 2.3, on this map see Kominko, 2013, 35–41).

The rectangle was centred in the Ionian equator with two parallel tropics on the north and the south and limited on the west by the Atlantic Ocean and by India on the east. As this was a map of just the known world, the oikoumene, it could be maintained as a representation of only a part of the Earth's surface in Hellenistic and Roman cultures, after the planet was found to be a sphere, as in the 1st century BC astronomer Geminus's *Introduction to Phenomena* 16.5. Indeed, in *Sketch of Geography* 1.1-2, the geographer Agathemerus (2nd–3rd centuries) described the progress from circular to rectangular mapping, not taking into account the change from a flat to a spherical Earth because he was only discussing the maps of the oikoumene, not the shape of the whole planet!

Some Greek pre-Socratic philosophers (6th–5th centuries BC) seem to have been the first to challenge the simple flat circular model of the Earth covered by a hemispheric sky containing the heavenly bodies nailed to it. It seems that the criticisms of this 'archaic' cosmology started not by focusing on the Earth but on the sky. Anaximander (c. 610–c. 546 BC) is credited with having destroyed the idea of a vaulted sky. He realised that heavenly bodies were placed at different distances from the Earth and went underneath a floating unsupported Earth, which he still believed to be like the flat section of a column (Couprie, 2011, 99–151; 2018, 47–98).

Afterwards, the sky was quite quickly agreed to be spherical. However, the shape of the Earth was a more difficult matter. Sadly, we do not have direct testimonies, but mostly fragmentary references from later doxographical tradition to the intense debates of that period. As this detailed discussion does not concern us here, we will focus our attention on the first references to a spherical Earth.[6]

The spherical Earth

Most Late Hellenistic and Roman authors traced back the first suggestion of a spherical Earth to Pythagoras (c. 569–c. 475 BC) or a member of his close circle (in particular Parmenides, c. 515–c. 450 BC). There were some who even dated it back earlier, to Hesiod or Thales. Sadly, as Otto Neugebauer noted, "we have no account of how the sphericity of the Earth was discovered" (Neugebauer, 1975, 576).

Ptolemy, in the 2nd century AD, introduced the explanation of the sphericity of the heavens and the Earth with the following cautious words: "It is plausible to suppose that the ancients got their first notions on these topics from the following kind of observations […]" (Ptolemy, *Almagest* 1.3, 1984, 38).

From the documentation we have, the oldest clear reference to the sphericity of the earth appears only in Plato (429–347 BC), probably inspired by Pythagorean circles, as already indicated. According to historian Dirk L. Couprie, Plato could represent a transition between the flat and spherical Earth. Plato first stated the sphericity of the planet with words that he put into the mouth of Socrates in his *Phaedo*:

> I've been convinced, first, that, if the earth is round [περιφερὴς] and in the middle of the heaven, it has no need of air or of any other such necessity to stop it falling, but the uniformity of the heaven on every side and the equilibrium of the earth itself are enough to hold it in place. The reason is that, if a thing in equilibrium is put in the middle of something uniform, it will not be able to lean more or less in any direction, but in its uniform state it will stay in place without leaning.
>
> (Plato, *Phaedo* 108e–109a, 2011, 106)

To put that into context, see *Phaedo* 97b–99d, where the interest of Socrates in the shape of the Earth is described, as well as his disappointment with the explanations he found in Anaxagoras. In 108c, Socrates attributes to "someone's authority" the new view of the 'spherical' Earth that he will describe thereafter (from 108d onwards). In *Phaedo,* "the arguments adduced to support the hypothesis are astronomical and geometrical rather than geographical" (Heidel, 1937[1976], 83). In fact, they are more philosophical-mathematical, which matches well what we can expect from Pythagorean argumentation (on this text and its potential Pythagorean influences, see Couprie, 2025).

Plato explained later that those living around the Mediterranean dwell in one of many concavities in the surface of the planet (109b). He described our planet as a dodecahedron (a ball [σφαῖρα] made up of 12 pieces of leather, 110b-e, 111c).[7] Couprie rejects a common and simple interpretation of the dodecahedron as an approximation to the sphere, on the assumption that the curvature of the pentagonal faces was convex (Couprie, 2005, 196–198; 2011, 206). He rather interprets this as a conception of the Earth where the 'oecumenical' concavities were 'excavated' on the surface of the flat pentagonal faces of the dodecahedron, in accordance with Plato's description in 109b and in line with the traditional concave oikoumene. For Couprie, here Plato "combines, or reconciles, the ancient conception of the flat earth and the new one of a spherical earth" (Couprie, 2011, 208).[8]

A simpler view of a truly spherical Earth can be found in Plato's disciple Aristotle (384–322 BC). He discussed the shape of the Earth at length in the second book of *On the Heavens*.[9] He first stated it: "The shape of the heaven is of necessity spherical; for that is the shape most appropriate to its substance and also by nature primary" (Aristotle, *On the Heavens* 286b10, 1984[1995], 1039). When he later tackled the shape of the Earth, he started reviewing the ideas of his predecessors (Aristotle, *On the Heavens* 293b34–294b31), in some cases having pains to counter elaborated arguments from the pre-Socratic flat-earthers, such as Anaxagoras. He then put forward several arguments to defend the sphericity of the Earth. Nevertheless, the key piece there was not empirical evidence but a metaphysical argument based on his 'theory of falling': his view of the 'gravity' that pulls the 'heavy' element earth 'down' towards its 'natural place' at the centre of a spherical universe where the planet Earth is situated, therefore forming a sphere, central and immobile (Aristotle, *On the Heavens* 294b32–297b22). We can see there are similarities with Plato's text quoted above (p. 26). Only as a complement, Aristotle offered empirical astronomical arguments, which in fact are not conclusive: the circular shadow of the Earth during lunar eclipses (Aristotle, *On the Heavens* 297b23–297b31),[10] and the observation that there are different stars in the sky as one moves north-south, which pointed to a north-south curvature (Aristotle, *On the Heavens* 297b32–298a8).[11]

As Couprie has emphasised, a convex Earth would be a model which would be very difficult to distinguish from a complete sphere, even with the range of *empirical* arguments put forward by Aristotle and expanded by his successors, as the only region of the Earth's surface known to ancient Greeks was in fact a small portion of the sphere that could be surmised to be a small convexity.[12] Eventually the Aristotelian view of a spherical sky with a central concentric spherical Earth became well established. On top of it, there were concentric spheres of the other three lighter elements: water, air and fire,[13] and then layer upon layer of heavenly spheres, among which the planets were embedded, all made of a fifth 'heavenly' element or 'quintessence'. Finally, the whole universe was wrapped with an external sphere that had the fixed stars attached onto it. The Stoics also supported the sphericity of the central Earth (Sambursky, 1959, 108–115). However, the rival school of the Epicureans was, apparently, dismissive of the sphericity of the Earth (see p. 31 on Lucretius).

Our place on the spherical surface of the Earth

Once the sphericity of the Earth was accepted, the conclusions from such a configuration were quickly established. Parmenides was credited with having divided the Earth into five perpendicular zones along a north-south axis: two uninhabitable frigid ones limited by the polar circles, two habitable temperate ones from there to the tropics and one uninhabitable Torrid Zone between the tropics centred in the Equator.[14] As with the sphericity of the Earth, it is difficult to attribute that to Parmenides with certainty, as Plato, in presenting the first extant clear statement of the sphericity of the Earth, shows no knowledge of this division of the Earth's surface in zones. Couprie (2011, 64), also dismisses claims of the zones ascribed to Thales and Pythagoras. Another attribution is made to Plato's contemporary Eudoxus,[15] and – in any case – they can be found already in Aristotle, *Meteorology* 362a32-b9 (see Heidel, 1937[1976], 85–91). The division into zones was quickly refined and gave birth in time to the zonal maps, which became common in the Middle Ages (Figure 2.4A).

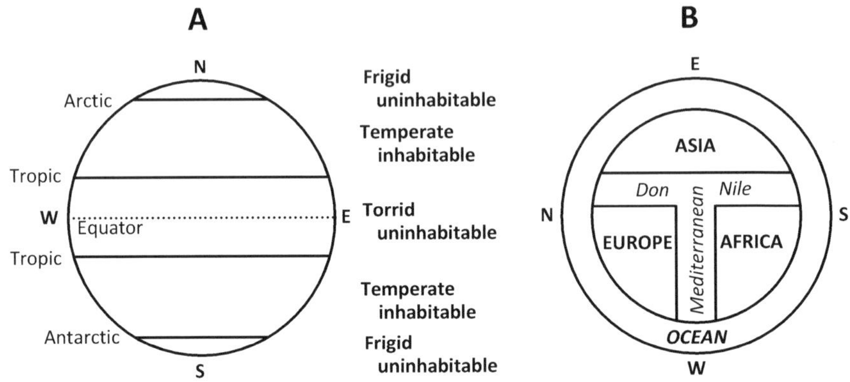

Figure 2.4 (A) Zonal map showing its five zones and their habitability status. (B) T-O map with the three continents of the oikoumene surrounded by the ocean. Note the different orientations in each map (modified from Felipe, 2012, 283, Figure 1, used with permission).

The older familiar Homeric 'flat' disc of land surrounded by the ocean of the ancient Greek maps survived by becoming just a map of the known inhabited 'oikoumene-island' (Latin *orbis terrarum*), covering less than half of the temperate band of the Northern Hemisphere, part of a much larger spherical Earth. Although several authors cautioned about the accuracy of these maps because of the severe constraints and distortions that their circular frames imposed on the shape of the three continents of the oikoumene, they retained their popularity, becoming the medieval T-O maps, as mentioned above (see Figures 2.2 and 2.4B).[16]

These theoretical reflections on the symmetry of the surface of the Earth led later to Crates of Mallus (2nd century BC), suggesting that the Mediterranean oikoumene was 'balanced' by another landmass on the other side of the Northern Hemisphere (whose habitants were called *perioecians*), and two more oikoumene-islands in the Temperate Zone of the Southern Hemisphere. This occupation of the surface of the sphere was the final consequence of the symmetrical arrangement of space already observed in Anaximander's map (Figure 2.1). To the south of the Mediterranean, beyond the Equator, lived the *antoecians*, while the *antipodeans*[17] were placed diametrically opposite to the oikoumene (Figure 2.5).

This view appears in Eratosthenes's poem *Hermes* (Gregorio, 2010) and in Cicero's *Dream of Scipio*: "You see that humans inhabit small and scattered portions of the earth, and that huge emptiness separates the blotches of human habitation" (*Republic* 6.20, 1999, 99). Cicero's text exerted a great influence in the Middle Ages, in particular through the commentary by Macrobius in the 5th century (see pp. 32, 55). This image has been termed *orbis quadratus* or *orbis quadrifariam* and was transmitted to the medieval authors (Stevens, 1980, 269–271; 2020). Of course, the existence of these additional three oikoumene-islands and their potential population with intelligent, 'human-like' individuals became a matter of speculation and contention in ancient times, as in Strabo, *Geography* 2.5.13.[18]

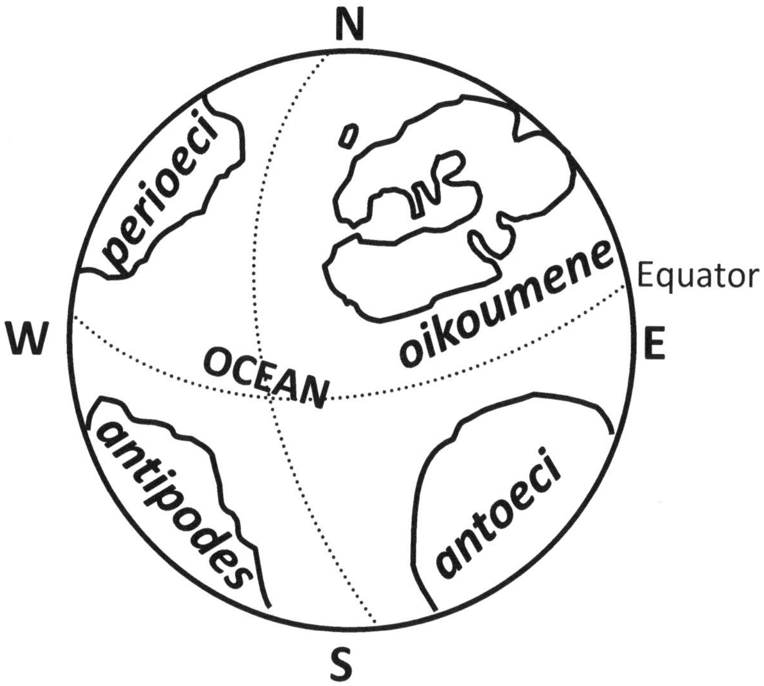

Figure 2.5 The Cratesian model with four symmetrical landmasses covering the Earth's surface. Note the oceanic and equatorial barriers separating them (from Felipe, 2012, 283, Figure 2, used with permission).

To make things more complex, it was usually assumed that communication between the four quarters of the Earth was impossible due to the large size of the oceans separating them and, even more pressing, the extreme heat at the Equator. However, Aristotle (writing before Crates) suggested that the distance by sea between Spain and the Far East was rather small (and therefore the oikoumene covered more than 180° west-east), an idea that proved influential for Christopher Columbus (c. 1451–1506).[19]

Aside from these speculative geographical ideas, Hellenistic scholars were more interested in an 'empirical' geography and astronomy, and did not feel so constrained by symmetrical and aesthetical concerns. For Strabo, "the geographer undertakes to describe the known parts of the inhabited world, but he leaves out of consideration the unknown parts of it—just as he does what is outside of it" (Strabo, *Geography* 2.5.5, 1917, vol. 1, 433). They centred their efforts in enlarging the oikoumene towards the Equator and Arctic Circle, as was also the case later in Roman times with Ptolemy (c. 100–c. 170). In his *Geography* (c. 160), Ptolemy built on the efforts initiated by Eratosthenes (c. 276–c. 195/194 BC) and described the oikoumene as a west-east rectangle in the Northern Temperate Zone, with Africa and Taprobane (Sri Lanka) extending towards the south to the Torrid

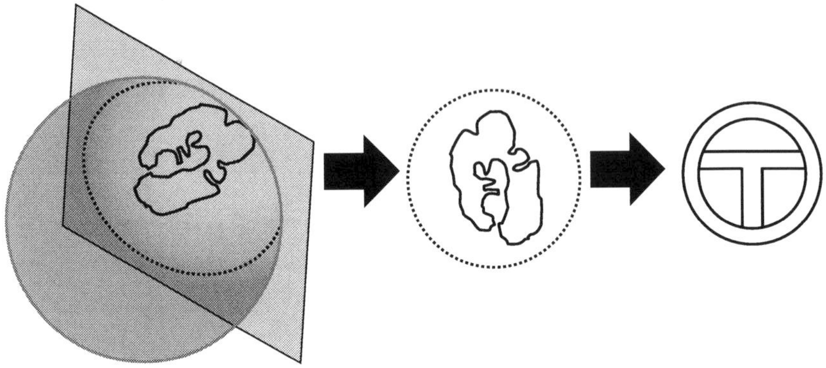

Figure 2.6 The ancient Mediterranean world captured in a small convex spherical cap produced by the intersection of a plane with the surface of the spherical Earth. This convex spherical cap can be easily visualised from above as a traditional T-O map.

Zone and even below the Equator.[20] These views were a clear contradiction of the ideas of Crates and his Equatorial Ocean and also of the fear of an uninhabitable Torrid Zone, which were outdated concepts in light of the knowledge of the interior of Africa by the time of the Roman Empire.[21] However, reports of ancient circumnavigations of Africa[22] were not considered by Ptolemy, who connected Africa with Asia and regarded the Indian Ocean as an enclosed sea.[23] Ptolemy's map was not surrounded by the ocean in all directions, showing large stretches of land on some of its sides, without a clear end towards the east (something that was in some way a reversal of Crates's model where water surrounded the four oikoumene-islands).[24]

However, the circular frame continued in use until the late 15th century. There were practical reasons for this preference, such as the visible horizon that appears circular and the fact that the ancient Mediterranean world could be captured in a small portion of the sphere (intersected by a plane) that would result in a small convex spherical cap with a circular edge at the bottom (visualised well with a circular map, Figure 2.6).

Discussions on the flat Earth and the antipodes in Roman times

The sphericity of the Earth was not immediately accepted in Greek culture, and some debate seems to have lingered for a while. Aristotle recounts resistance to the globe, though he considered it to be outdated. Key Roman authors from the time of the beginning of the Roman Empire, such as Geminus (1st century BC), Strabo (c. 64 BC–c. 24 AD), Pomponius Mela (died c. 45 AD)[25] or Pliny the Elder (c. 23–79 AD), supported the sphericity of the Earth, which at the time was already the general view, particularly among scholars.

However, opposition to the sphericity of the Earth survived during the Roman period. The Epicurean poet and philosopher Lucretius (c. 99 BC–c. 55 BC), with his famous Latin poem, *On the Nature of Things* 1.1052–82, is often mentioned as an example of those who denied the Earth's sphericity. Nevertheless, his remarks are quite unclear, and, in fact, it has been recently argued that the Epicureans were not flat-earthers (Bakker, 2016, 162–263). Epicureans believed that the universe was infinite and eternal, that there was no centre of gravity in the universe, and that all matter was eternally falling; therefore, for them, the spherical symmetry of the Earth was absurd. Historian of Epicurean philosophy Frederik A. Bakker indicates that "nowhere in Epicurus" nor "in Lucretius' *De rerum natura* do we find any explicit statement about the shape of the earth" (Bakker, 2016, 163). Referring to the possibility of the Epicureans being flat-earthers, Bakker notes that "even their most ardent ancient critics never accused them of saying so" (Bakker, 2016, 262). His conclusion is that they were uncommitted: "Epicureans had no firm conviction as to the shape of the earth at all: to them, as Woltjer (1877) already suggested, the earth like the cosmos (or the sun), might have any shape" (Bakker, 2016, 262).

In the Roman orator, politician and sceptical philosopher Cicero (106 BC–43 BC), we can see how the sceptical philosophy could be used to blur the issue of the antipodeans, although not to cast doubts on the sphericity of the Earth. In the context of a long discussion regarding the disagreement among philosophers in his *On Academic Scepticism* (*Lucullus)*, dated 45 BC, Cicero wrote about the difficulty of knowing things that are hidden to our senses, mentioning what is inside our bodies, and then moving to cosmological matters:

> Xenophanes says that there's life on the moon and that it's a land with many cities and mountains. These claims seem marvellous, but their author could no more swear that it is so than I that it isn't. You even claim that there are people directly opposite us on the other side of the earth, their footsteps standing opposite ours (the people you call 'Antipodeans')! So why are you more angry with me, though I don't reject such views, than with those who think you're crazy when they hear them?
>
> (Cicero, *Academic Questions* 2.122–123 [or 2.39 in other editions], 2006, 71–72)

In his *Natural History*, dated to the AD 70s, Pliny clearly stated the sphericity of the Earth and mentioned evidence for it.[26] However, after saying in chapter 2.64.160, "But her shape [the Earth's] is the first fact about which people's judgement agrees," he recognised in 2.65.161–165 that, in fact, not all agreed:

> Here there is a mighty battle between learning on one side and the common herd on the other: the theory being that human beings are distributed all round the earth and stand with their feet pointing towards each other, and that the top of the sky is alike for them all and the earth trodden under

> foot at the centre in the same way from any direction, while ordinary people enquire why the persons on the opposite side don't fall off - just as if it were not reasonable that the people on the other side wonder that we do not fall off. There is an intermediate theory that is acceptable even to the unlearned crowd - that the earth is of the shape of an irregular globe, resembling a pine cone, yet nevertheless is inhabited all round. [...].
> (Pliny, *Natural History* 2.65.161, 1938[1949], vol. 1, 296 [Latin], 297 [English])

Pliny also found that it was difficult for many common people to understand the curvature of the water: "But what the crowd most debates is if it must believe that the conformation of the waters also rises in a curve. Nevertheless, nothing else in the natural world is more visibly manifest" (Pliny, *Natural History*, 2.65.163, 1938[1949], 299). He insisted later: "The cause of the remaining facts that surprise us is found in the shape of the earth itself, which together with the waters also the same arguments prove to resemble a globe" (Pliny, *Natural History*, 2.71.177, 1938[1949], 311). That water and earth formed a single globe was also mentioned by his contemporary Seneca in *Natural Questions* 3.28.4-5 (written c. AD 65) or in Strabo's *Geography* 2.5.5 (early 1st century AD).

However, opposition to the Earth's sphericity was not to be found only among ignorant people. Apart from the Epicurean cosmology (if the traditional view of them as flat-earthers were true, see above), it was seriously discussed by Plutarch (c. 45–120), a Greek author of the Roman Empire, who defended Platonic philosophy and attacked particularly Stoics and Epicureans. He referred to the antipodes in non-committal terms in *On the Malice of Herodotus* 37 and *On the Stoic Self-Contradictions* 34. However, the longest discussion of both the antipodeans and the shape of the Earth appears in his well-known *On the Face Which Appears in the Orb of the Moon* 7, dated c. AD 100 (see Plutarch, 1957, 60–67). Lamprias, the main character in the dialogue, is very critical of the Stoics, represented by Pharnaces, and in this context launches an attack on the sphericity of the Earth. Harold Cherniss, the translator of the edition here consulted, indicates that: "Plutarch [...] probably did not intend this or the subsequent paradoxes to be taken too seriously. Lamprias is simply riding Pharnaces as hard as he can, using any argument good or bad, to make him appear ridiculous" (see in Plutarch, 1957, 63, note c). However, as with Pliny, such words kept alive a debate already closed for astronomers and geographers centuries ago.

That the issue of the antipodes was difficult to accept for many people in Antiquity can be seen in a later example. Discussing in the early 5th century the above-mentioned Cicero's *Dream of Scipio*, the Roman author Macrobius affirmed in his *Commentary on the Dream of Scipio*:

> [...]. I can assure you that the uninformed among them [the antipodeans] think the same thing about us and believe that it is impossible for us to be where we are; they, too, feel that anyone who tried to stand in the region beneath them would fall. [...].
> (Macrobius, *Commentary* 2.5.26, 1952[1990], 204)

This resembles the words by Pliny discussed above. It is therefore not surprising that sooner or later criticisms regarding the existence of the antipodes could spill over into attacks on the sphericity of the Earth, as this was the origin of the concept of the antipodes itself. Lactantius was one such example.

The flat Earth and the antipodes in the apologetic strategy of early Christian writers

From indifference to criticism

Although there are references in the Bible reflecting a flat Earth view, particularly in the Old Testament at a time when this was commonly accepted (e.g., Isaiah 40:22), they are not a constitutive part of the religious doctrines explained or defended by the biblical authors. After encountering Hellenistic culture, it was not uncommon for Hellenised Jews, such as the Alexandrian Philo, to develop sophisticated allegorical interpretations of the Bible (following pagan literary methodologies used to deal with ancient literature, in particular, Homer). In this way, the clash with contemporary natural philosophy could be avoided. Dated terminology (such as the four corners of the Earth) could continue to be used in religious contexts without being expected to be scientific descriptions of the cosmos. However, when confronted with some texts, in particular apocalyptic texts, like the Jewish *Book of Jubilees* or the Christian *Revelation*, it is not easy to discern whether their cosmological references were intended literally, allegorically, or even both. In addition, not all Jewish authors were as Hellenised as the Alexandrians, and views could be different in other places less exposed to Greco-Roman natural philosophy.

Among Christian writers, an early veiled positive reference to the antipodes (but without voicing an opinion in favour of or against antipodeans and not mentioning the shape of the Earth) has been usually identified in a passing reference to "worlds beyond [the ocean]" in Clement of Rome's *1 Clement* 20.8, late in the first century: "The ocean, which human beings cannot pass, and the worlds beyond it, are ruled by the same injunctions of the Master" (Clement, 1912[1965], 45, slightly modified).[27] In *De Principiis* 2.3.6 (dated in 220s), Origen quoted this small text as supporting the *antichthones*. Origen's work was known to the west thanks to the Latin translation by Rufinus in 397.[28]

From the late 170s, we have in Athenagoras of Athens's *A Plea for the Christians* several positive references to the sphericity of the world. He affirmed the sphericity of the 'sphere of the fixed stars' (Athenagoras, *Plea* 6.3) and the universe (*cosmos*) in general (Athenagoras, *Plea* 8.4; 16.1), also stating: "The best sacrifice to him [God] is for us to know who stretched out the heavens and gave them their spherical form and established the earth as a centre, [...]" (Athenagoras, *Plea* 13.2, 1972, 26, 28 [Greek] and 27, 29 [English]).

How did these topics come to be mentioned with contempt by Lactantius in the early 4th century? An important clue can be found in the specific literary genre of Lactantius's *Divine Institutes*. It is not a treatise of astronomy or geography; it is instead an apologetic treatise to defend Christian views against pagan criticisms.

Christian apologists had to confront many criticisms against Christianity, but they also launched their counterattacks against paganism. One of their targets was 'popular' pagan religion, but they also engaged with the philosophic-religious ideas of the ancient Greek, Hellenistic and Roman philosophers. For this latter debate, Christians used the strategy of a particular school of philosophy, scepticism.

Attacking philosophers with their own weapons

During the Hellenistic era, following the conquests of Alexander the Great, the new Epicurean and Stoic philosophical schools had a 'dogmatic' belief in the validity of sense perception. An additional strand of ancient philosophy which reacted critically to the previous dogmatic schools, and which is often overlooked, is 'ancient scepticism'.

One strategy of the sceptics against dogmatism was to highlight the disagreements among different dogmatists, not only to undermine their ideas but also to illustrate the impossibility of being sure of anything. That was a gold mine of argumentations to use against philosophers when early Christian thinkers wanted to bundle them all together for ridicule already in the late 2nd century with Tatian's *Address to the Greeks* (chapter 25) and Hermias's *Derision of Gentile Philosophers*, who explicitly mentioned the Academic sceptics (chapter 15). While the sceptics used these contradictions to undermine dogmatism, Christians used them to destroy all philosophy. The idea of pitting philosophers against each other in a humorous way that we find in Hermias is also found in contemporaneous pagan literature, such as that of the 2nd-century satirist Lucian of Samosata. In a more serious tone, but still with some irony, we saw how some cosmological topics were treated by Cicero, who was a direct source for Lactantius's criticism regarding the sphericity of the Earth and antipodes. People without a technical background in natural philosophy, astronomy and mathematics, and without a clear view of the historical progress of these disciplines, were not able to appreciate the fact that in some cases, such as on the sphericity of the Earth, a consensus was achieved many centuries earlier.

A second point of interest is that radical scepticism as well as other extreme positions of philosophical schools gave plenty of colourful daily-life anecdotes to caricature the philosophers,[29] which became a source of ammunition for Christian apologists to show how absurd these 'pagan' sages were and further to undermine their authority.[30] Some anecdotes would reveal contradictions between the teachings and the daily life of philosophers, while in other cases they would illustrate the absurdity of some of their teachings against 'common sense'. The mockery of the philosophers for their bizarre interests was as old as the philosophers themselves. Already in the late 5th century BC, Aristophanes made fun of what he considered absurd topics of interest, including astronomical and cartographical topics (*Nubes* 133–235).

Pagan philosophers were attacked in various fields (and with different methodologies), including natural philosophy and cosmological topics that, from a Christian point of view, either disagreed with the Bible and Christian theology and needed to be criticised or that were against common sense and therefore offered an easy target for discrediting pagan philosophy in an acceptable way to a non-Christian audience. Christians exploited these topics to undermine the pagan philosophers' authority using the techniques of the sceptical philosophy. Such was the strategy of Lactantius, who considered the sphericity of the Earth and the antipodeans as the ultimate example of pagan philosophers' stupidity. Such topics had rarely been used before Lactantius by Christians against pagan philosophers.

In fact, there was an ambivalent value in the philosophical approach of scepticism. On the one hand, its challenge of dogmatism could undermine the strength of Christian ideas; on the other hand, its self-critical attitude towards its philosophical context could be used as a first apologetic step to undermine pagan philosophy and vindicate religious faith as the only guide (Floridi, 2010, 267–287). Augustine saw scepticism in this way, as pointing beyond itself towards a revived Platonism that, in his view, led towards Christ.[31]

The precedent of Theophilus of Antioch against the sphericity of the world

In the late 2nd century, Patriarch Theophilus of Antioch's apologetic book *Ad Autolycum* criticised the sphericity of the world (*cosmos*), although in a tangential way. Interestingly, Antioch was where the anti-sphericity took later stronger roots (4th–6th centuries), arriving at the fully fledged flat Earth model that culminated in the work of the Cosmas Indicopleustes in the 6th century (see pp. 48–49).

Although Theophilus dealt with the sphericity of the Earth in two texts (*Ad Autolycum* 2.32.4 and 3.2.3), his criticism of the sphericity is only apparent in the first one. Theophilus criticises the contradictions between various pagan accounts of poets and philosophers regarding the gods, providence and origins in contrast with the accounts of the biblical prophets:

> Writers [pagans] who do not know these things want to call the world spherical or to compare it with a cube. How can they speak truthfully in these matters when they do not know how the world was created or how it was inhabited?
>
> (Theophilus, *Ad Autolycum* 2.32.4, 1970, 82 [Greek], 83 [English])

Theophilus does not develop this criticism further, but it seems clear that he disagrees with the sphericity of the Earth (the reference to population and inhabitants after the flood suggests that he uses world [*cosmos*] for the Earth). The reference to a cube could be a misunderstanding of the Platonic attribution of the cubical shape to element earth (not to the planet Earth!).[32]

Although Lactantius knew Theophilus of Antioch's *Ad Autolycum*, which he used in *Divine Institutes* (Nicholson, 1985), we will see that the shape of Lactantius's argument has little to do with Theophilus, despite both texts being apologetic books and denying the sphericity to mock the philosophers. It seems more likely that Lactantius was inspired by Cicero (see above).[33] That is the reason why we will not delve into a detailed study of Late Antiquity Antiochian flat-earthers.

Lactantius's denial: the flat Earth and antipodes debate in Western Christianity

Lactantius's criticism of philosophy in the apologetic strategy of Divine Institutes

Lactantius (c. 250–c. 325) was born in the Roman territory of North Africa. He had a distinguished career teaching rhetoric, and in the 290s he was brought by Emperor Diocletian to his new eastern capital, Nicomedia, in order to teach Latin rhetoric. It is thought that it was there that Lactantius converted to Christianity about 300, putting his rhetorical skills at the service of his new faith at the time of the fiercest persecution against Christians. To this effect, he wrote the *Divinae Institutiones adversus gentes (Divine Institutes, DI)*, usually dated between 304/305, during the last great persecution under Diocletian, and 310/311 (Digeser, 1994; 2019; Schott, 2008, 106–107). After the persecution, Lactantius seems to have been a significant influence at the imperial court, with Timothy Barnes calling him Constantine's "intellectual mentor" (Barnes, 2011, 118). Nothing is known of his last years; for a biographical update, see Bowen and Garnsey, 'Introduction' in Lactantius (2003), and the first pages of Rutherford (2015).

The *DI* is one of the last books of Christian apologetics composed before religious freedom was granted to Christianity in the early 4th century. In *DI* 5.4 Lactantius criticised Tertullian's *Apologeticum* – which only criticised the pagans and did not give a positive exposition of Christianity – and Cyprian's *Ad Demetrianum* – which defended Christianity merely with scriptural arguments, unlikely to convince non-Christians – while praising Minucius Felix's *Octavius*, which sought to attract educated pagans (Alban, 1943).

Lactantius begins *DI* attacking the pagan gods in books 1 and 2 (*On the False Religion* and *On the Origin of Error*): "At the close of book 2 [*DI* 2.19.2] he previewed the 'greater and more difficult struggle' that faced him in book 3" (Rutherford, 2015, 489). In book 3, Lactantius criticises the philosophers and their vain search for wisdom (*On False Wisdom*). Finally, starting with book 4 (*On True Wisdom and Religion*), he explains Christianity as the right alternative to the previously criticised errors. On the carefully structured concentric design of the whole book (books 1–3 are paralleled in 4–6), see Ingremeau (1993).

Although Lactantius did not hide his criticisms of paganism, in line with Minucius Felix, he chose a strategy sympathetic to pagan learning and tried to establish bridges with the culture of educated pagans, particularly philosophical monotheists (such as followers of Neoplatonism and Hermetism). Lactantius offered a sophisticated apologetic (that prefigured the stance taken by Augustine in *City of God*), which required a radical step away from the more negative attitude towards the pagan tradition in Christian writers before him. In what historian Elizabeth DePalma Digeser has described as an "inclusive Christianity" (Digeser, 2000, 84), Lactantius drew parallels between Christian teaching and ideas in the revered classical literature of poets such as Vergil and Ovid, as well as in the Delphic and Sibylline Oracles, and was willing to praise key pagan authors as being rather close to Christian views, Lactantius considered Plato the wisest philosopher (*DI* 1.5.23), saw Cicero as inspired by God (*DI* 6.8.6–12) and considered that Seneca could have been a Christian if someone had guided him (*DI* 6.24.12–14). More on the apologetic strategy of Lactantius can be found in Bryce (1973[1990], 11–15), Faes de Mottoni (1982), Garnsey (2002[2004], 157–162) and Schott (2008, 79–109).

Lactantius and the role of the flat Earth and the antipodes in his criticism of philosophy

Lactantius's criticism of philosophy in *DI* appears in book 3, arguing that the so-called wisdom of the pagan philosophers is not true wisdom. However, as we saw, Lactantius rejected previous Christian apologists that used upfront Christian texts as evidence in their debates with pagans. Instead, Lactantius uses the forensic strategies of the Academic sceptics, created by the Sophists, which he borrows from Cicero, to cancel out all the philosophical schools and undermine confidence in philosophy. Finally, from that *tabula rasa,* he proceeds to expound the Christian alternative, the *divine wisdom*, now finally using revelation as a source of evidence beyond the sceptical attacks.[34] That is the real explanation of Lactantius's strategy of setting aside the Bible at the beginning of his apologetic book, using instead the pagans' own weaponry turned against themselves – in order that their own authorities might refute them, *DI* 3.1.2, 3.4.8–10, 5.4.6 – so that they "perish by their own sword" (Lactantius, *DI* 3.28.20, 1871, 205).

Lactantius deploys various strategies to carry out his criticism of pagan philosophers (see pp. 34–35). For the relevant text, see Appendix, no. 12 (available at www.routledge.com/9781032893068), taken from Lactantius (1871), slightly modified, which is quoted below (for a more recent translation, see Lactantius, 2003).

Among his criticisms, Lactantius points to the defence by the philosophers of ideas that were considered ridiculous (in some cases just because they were against 'common sense') to undermine their authority. After discussing other moral, social and economic topics, Lactantius picks as the last set of ridiculous ideas those referring to issues of natural philosophy.

Lactantius ends his list in chapter 23 referring to such odd examples as someone who "asserted that snow was black" and "that pitch was white" (*DI* 3.23.11) or others who said that other human beings were living on the Moon (Xenophanes, *DI* 3.23.12–13) or the Sun (an unidentified Stoic mentioned by Seneca, *DI* 3.23.14). And it is here, at this climax, when the reader believes that nothing more ridiculous could be sustained by natural philosophers that Lactantius starts chapter 24, entirely devoted to what he seems to consider the most absurd of all those ideas related to issues of natural philosophy: the idea of people living upside down.

Analysis of Lactantius, Divine Institutes 3.24

As I noted in Felipe (2012), the argumentation of Lactantius uses two strategies: one is a direct ridiculing of the idea of the antipodes/antipodeans, that is seen as an *absurdity* for common-sense perception, which is found right at the beginning: "is any one so senseless as to believe that there are people whose footsteps are higher than their heads?" (*DI* 3.24.1). However, Lactantius goes further and launches a second attack as he intends to find out the origin of this error to which he then devotes the main part of his chapter 24, starting with these words: "The origin of this error must also be set forth by us," echoing the title of book 2, *On the Origin of Error*. This second attack effectively turns into a deconstruction of the idea of the antipodeans, which is considered to be based on *a purely deductive method of reasoning* from a false first premise that no one is open to reconsider, and not on real hard evidence (*DI* 3.24.2-10): sphericity of the heaven → sphericity of the Earth → symmetry of the surface of the Earth → antipodes → antipodeans. The discussion returns to the argument of absurdity in *DI* 3.24.11, bringing his argument to an end.

Remarkably, Lactantius's argumentation against the antipodes/antipodeans and the sphericity of the Earth does not contain any theological or biblical references, although sometimes the opposite is wrongly stated.[35] For Lactantius, religious reasoning is not needed here; these ideas can be demolished without reference to the Bible, as they are absurd. In addition, in the general scheme of the book, the usage of the Bible is restricted to the second part of the work, particularly from book 4 onwards, once the demolishing work of the pagan religion and philosophy has been done with purely 'secular' arguments.

What was the problem with the argumentation put forward by Lactantius?

Sadly, although Lactantius was right in criticising the abuse of deductive methods, he failed to separate the issue of the sphericity of the Earth from that of the speculations on what was on its other side, even though he was able to explain step by step that the Earth's sphericity came first and the speculation on the antipodes came later. If he had been able to discriminate better between both issues, he could have realised that the sphericity of the Earth was not just the result of speculation, but that there were solid arguments, as we saw in sections above, that he did not even mention. For anyone familiar with the philosophers' and even more with the astronomers' discussions of the shape of the Earth, the vast effort of Lactantius in book 3.24 would have seemed like a futile rhetorical exercise closer to the sophistic argumentation, at least on his failed attempt to discredit the sphericity of the sky and the Earth.

But what about the antipodes/antipodeans? Was Lactantius right in claiming that the antipodes/antipodeans were just the "last consequence" of the "rotundity of the heaven" (*DI* 3.24.8)? Later generations of Christians, starting with Augustine, rightly thought so, and they carefully separated the issues of the sphericity and the antipodes, opening a debate that spanned more than a millennium. That was a good strategy, as indeed there was less evidence for the antipodes/antipodeans than for the sphericity of the Earth. Lactantius was right in his observation that it was a geometrical and symmetrical/analogical argumentation that was behind the claims about the antipodes and their potential inhabitants, a pure inference based on theoretical reasoning. That was also derived from the Aristotelian principle that nature does nothing in vain, that it abhors vacuum (Thomson, 1948, 214). At around 200, the Greek astronomer Cleomedes, based on the teachings of Posidonius in the 2nd/1st century BC, justified the existence of antipodeans by a plenitude principle:

> [...]. Yet the regions of the Earth that are equally temperate are necessarily inhabited to an equal extent, given that Nature loves Life, and Reason requires that all [parts] of the Earth, where possible, be filled with animal life, both rational and irrational.
>
> (Cleomedes, *Heavens* 1.1.262–269, 2004, 37)

Lactantius's flat Earth in the context of his time

In light of the above, we can see how differently the topics of the sphericity of the Earth and the antipodes appear in Lactantius in comparison with the Antiochian real flat-earthers. Lactantius argued rhetorically against the antipodeans (and in that process also against the sphericity of the Earth) as part of a strategy to undermine the credibility of pagan philosophy at large. However, the Antiochians focused on developing a clear alternative cosmological proposal (congruent with their interpretation of the Bible) against the received 'pagan' cosmology. Even if Lactantius was a true flat-earther, his *Divine Institutes* was never intended to be a direct defence of the flat Earth.

This is probably why scholars fail to relate Lactantius with the Antiochians flat-earthers. As an example, Leone Montagnini, a multidisciplinar scholar interested in the history of ancient science, gives three 'traditions' among early Christians to deal with the sphericity of the Earth: the Alexandrians accepting the sphericity and using allegory to interpret the Old Testament; the 'conciliatory' position between the Old Testament and the sphericity defended by Basil, Ambrose and the late Augustine; and the Syrian-Palestinian rejection of the sphericity that used the typological interpretation of the Old Testament. However, Montagnini fails to associate Lactantius with any of these traditions and just links him with fellow North African Tertullian (Montagnini, 2009, 49). However, Lactantius was critical of Tertullian (see p. 36), and in any case, the topic here was not discussed by Tertullian, who just mentioned once the antipodes in passing (although apparently taking them as real) in *Adversus nationes* 1.8.1 (Hiatt, 2008, 55).

Augustine's scepticism: The antipodes debate becomes part of Western Christian theology

Augustine's position in the flat Earth and antipodeans debates

In the above classification of the early Christian reactions to the sphericity of the Earth, the late Augustine is placed among Christians who offered a way of approaching the developments of ancient science between the positive, allegorical approach of the Alexandrians and the negative, typological approach of the Antiochians. In such a concordist way of relating science and Christianity, it is difficult to envisage Augustine rejecting up-front well-known, consolidated results of ancient learning, such as the sphericity of the Earth. In fact, the young Augustine (before his conversion to Christianity) had interest in and knowledge of astronomy and astrology (Macías, 2004). However, a different matter was the speculative and uncertain idea of the existence of antipodes, let alone antipodean inhabitants.

Although the interpretation of Augustine's statements as being favourable to the sphericity of the Earth has been the most common view through history, in 1834 Letronne classified Augustine as a flat-earther (Letronne, 1834, 604). Other authors from the 19th century onwards considered him as a doubter or ambiguous on this matter. Recently, Augustine scholar Leo Ferrari described Augustine again as a flat-earther (Ferrari, 1996; 1999). Historian of science C.P.E. Nothaft has vigorously contested this view (Nothaft, 2011).[36] My view sides more with Nothaft, as whatever the interpretation of some obscure passages in the large Augustine corpus, he was very far from defending or holding a position such as the Antiochian flat-earthers on this topic.

For our purposes here, the only piece of information we need to study is a single passage from *City of God* (thereafter, *CG*), whose reception in the Middle Ages was influential in the denial of the antipodes. However, its (mis)reception in modern times became vital for creating the flat error, although the text itself deals – even more precisely than the one from Lactantius's *DI* – with the antipodes/antipodeans and not primarily with the shape of the Earth. The sophisticated discussion of the sphericity of the Earth in Augustine, also looking at other passages in his work, such as can be found in Ferrari and Nothaft, is a recent debate that did not take place until the end of the 16th century (see Chapters 4 and 5). Medieval and Early Renaissance authors repeatedly pointed to Augustine as denying the antipodeans but never considered him a flat-earther (see next sections below and Chapter 3).

Augustine's criticism of the idea of the antipodeans in the City of God

Once Christianity triumphed in the Roman Empire, Lactantius's 'defensive' apologetic, written during the last Christian persecution, was no longer up to date with the current situation of the Church. A century later, the situation was very different, and one of the leading figures in the Latin Christian Church, Augustine of Hippo (354-430), conceived a new apologetic text, *De Civitate Dei contra Paganos* (*City of God, CG*). Composed c. 412-426, the books 15 and 16 discussed here were written after 419 (O'Daly, 1999, 35). Augustine was aware of the work of Lactantius and mentioned it (on the influence of *DI* on *CG*, see Garnsey, 2002[2004]). We will discuss the parallels in pp. 42–43; here, we will explain the place of this passage in the context of the whole *CG* (on that see also Felipe, 2012).

In Augustine's times, even if it was on the retreat, paganism continued to pose a challenge for the Christian intelligentsia. As in the case of Lactantius's *Divine Institutes*, Augustine divides his apologetic book into two parts. The first part, with ten books, attacks pagan religion and philosophy. The second part, comprising twelve books, explains Augustine's Christian view of history, starting with the Genesis narratives and detailing his notion of the two cities (i.e., the city of God representing those who follow God and the earthly city those who follow a worldly life in opposition to God).

Book 16 of the *CG* is a retelling of the biblical narrative of Noah's sons after they leave the ark where they survived the flood (for the structure of this book, see Mazzucco, 2009). Augustine considers three digressions to ponder on small groups of living beings located in faraway places on the surface of our planet: in chapter 7, the population of remote islands by animals, and in chapter 8, the presence of monstrous human beings believed at that time to populate the distant borders of the oikoumene.[37] Then, in chapter 9, he wonders about an even remoter place beyond the oikoumene: the antipodes, pondering about the possible existence of such a place and the existence of humans there (antipodeans). The Patristic scholar Clementina Mazzucco points to an interesting detail regarding these three groups; while the first two are introduced as *quaestiones*, the third one is introduced with the more controversial words: *Quod vero et antipodas esse fabulantur*... (Mazzucco, 2009, 193, note 91).

If we compare Lactantius's *DI* and Augustine's *CG*, it is clear that they share the same structure of, first, criticising pagan culture (even preserving the twin structure of first criticising popular/official Roman pagan gods and then philosophical religion) and, after that, presenting the Christian alternative. Seen in this way, it is clear that while *DI* 3.24 is part of the attack in the first 'destructive' part of Lactantius apologetics, the *CG* 16.9 belongs to the second 'constructive' part in the work of Augustine. This is a significant difference that I believe explains the different content and tone of *DI* and *CG*, as we will see in the following sections. Curiously, this placement of the criticism of Augustine to the antipodes is similar to the location of the criticism of the sphericity of the world in Theophilus's *Ad Autolycum* (the spread of humankind over the Earth after the flood, see p. 35). However, I do not have any justification to speculate for a particular reason or connection for this similarity, as both texts are not really parallels.

Analysis of Augustine, City of God *16.9*

As I indicated in Felipe (2012), Augustine's objective in this chapter is to deny the existence of human antipodeans, deploying at least three strategies. For the relevant text, see Appendix, no. 1, taken from Augustine (1872), with several modifications, which is quoted below (for a more recent translation, see Augustine, 1998).

First, *Augustine calls into question the reality of the existence of antipodeans*, "the fable that there are Antipodes," due to the lack of evidence ("historical knowledge"). Scepticism about speculations regarding the inhabitants of the south of Asia and Africa can be found in the 2nd century BC Greek historian Polybius, *Histories* 3.38. In the 1st century BC, the Greek astronomer Geminus, in his *Introduction to Phaenomena* 16.19–20, indicated that there is no historical evidence for the antipodes and that all that could be affirmed was that a potentially inhabitable temperate Southern Zone was potentially possible. In the Imperial Roman era, Cleomedes referred to the inhabitants of the Mediterranean as "we humans, of whom there are direct reports, inhabit one of these [habitable zones]" (Cleomedes, *Heavens* 1.1.209, 2004, 34) and indicated that regarding the inhabitants of the other three Cratesian landmasses, "none of these are described by direct reports" (Cleomedes, *Heavens*, 1.1.262, 2004, 37), although he was convinced that they should exist on philosophical grounds (the plenitude principle mentioned in pp. 38–39). However, Augustine's use of the term "fable" (*fabulantur*) could also be seen as attributing the idea of antipodes to poets rather than rigorous historians or geographers. The criticism of the poets' geographical ideas had an older genealogy, going back at least to Herodotus's attacks on the imaginary geographical views of Homer (*History* 2.23; see p. 23) and his project of replacing them with a more rigorous geography (Romm, 1992, 33–41). In fact, the flat Earth geography of Homer (*Iliad* 18.478–608, 14.200–201, 21.195–96; *Odyssey* 9.13, 24.11–12) raised intense discussions in the ancient world, as Crates and other Hellenistic authors attempted to allegorise it, while Strabo's *Geography* 1.1.6–7 and Geminus's *Introduction to Phaenomena* 16.27–29 criticised it as outdated. In light of this context, Augustine, in pointing to the lack of historical evidence regarding the antipodes, simply reflects the state of the question at that time. Criticisms of poets' geographical views also appear in Isidore of Seville's denial of the antipodes, as we will see below.

In the face of such a lack of evidence, it is an obvious question to ask why a discussion on the existence of the antipodes became nevertheless so common. Here is where we can see Augustine using the explanation put forward by Lactantius a century before, which we have discussed above. *Augustine also attacks the deductive method used to demonstrate the existence of antipodeans*. This "reasoned conjecture" follows the familiar path of sphericity of the heaven → sphericity of the Earth → symmetry of the surface of the Earth → antipodes → antipodeans. Following this dissection of the reasoning behind the idea of antipodeans, while Lactantius then ridicules the concept of sphericity (first step of the "reasoned conjecture"), Augustine criticises the next point in the process: the idea of symmetry (see for more details pp. 46–47). He first attacks the inference of a symmetrical dry land on the other side of the Earth (because it could be covered by water). Then he points out that even if dry land existed in the antipodes, it does not follow that it must be inhabited by humans (against the plenitude principle). Augustine has no problem with an uninhabited antipodean continent or one inhabited by animals; his sole concern is the antipodean humans.

Finally, *Augustine points to a theological argument, as the existence of the antipodeans could imply in some way a falsity in the Bible*. The composition of this sentence in the final section of his argumentation in chapter 9 is not entirely clear: "For Scripture, which proves the truth of its historical statements by the accomplishment of its prophecies, gives no false information." It seems that Augustine wants to start his theological argument by highlighting the historical reliability of the Bible, which he considers confirmed by the fulfilment of its prophecies in the past. Then, his theological point is the common origin of all humans in a single individual (i.e. Adam): "and that thus even the inhabitants of that distant region are descended from that one first human." In fact, this is the same problem that he ponders concerning the existence of monsters in chapter 8, where he prefers to consider them, if they exist as human beings, as descendants of Adam: "either these things which have been told of some races have no existence at all; or if they do exist, they are not human races; or if they are human, they are descended from Adam" (Augustine, *CG* 16.8, 1872, 118).

Augustine may have had in mind several ideas: (i) the lack of any biblical mention of antipodes/antipodeans,[38] (ii) the biblical narrative of the common origin of all humans,[39] and (iii) the theological need for this common origin to encompass all real 'humans' due to Augustine views of transmission of Adam's original sin that had reached all people, and symmetrically have all of them available to receive the redemption through a single individual, a second Adam: Jesus.[40] In this context, to accept the existence of human antipodeans could imply for Augustine doubting the idea that "Scripture [...] gives no false information," with which he introduces this argumentation.

However, in the same way that he considered the possibility of humans travelling to distant islands in chapter 7, it could be argued that they could also have travelled to the farther antipodes. Augustine considers that caveat, showing that he had not a scientific problem with the sphericity of the Earth or even the possibility of the existence of antipodeans upside down. That solution could have reconciled even Augustine's theological concerns with the existence of antipodeans. Nevertheless, Augustine rejects this solution just on a very practical issue: the immensity of the ocean. This seems reminiscent of the Cratesian model where large oceans separated four landmasses. However, a few lines before, Augustine also suggested a model where only the Euro-Asian-African landmass would emerge from a planet covered everywhere else by a vast ocean ("the gathering together of the waters"). Curiously, he did not mention other reasons to impede the travel accepted at that time, such as the heat of the Torrid Zone.

Although finally, Augustine opted for the "simplest solution: one landmass, one origin for humanity (Adam/Noah), one act of "original sin," one sinner, one redeemer (Christ)," (Felipe, 2012, 291), the fact that he considered the possibility of human colonisation of the antipodes indicates that Augustine was only concerned with the theological problem of the unity of the humanity, rather than with philosophical or scientific issues as Lactantius. Therefore, the "absurd" that Augustine mentions (connected with the practical impossibility of crossing the ocean) is very different from the one denounced by Lactantius, who considered as "senseless," "folly," etc., the existence of humans upside down, a point not even mentioned by Augustine. This also explains the reference to the lack of historical evidence for the antipodes/antipodeans in Augustine: again, a case of practical knowledge, not mentioned by Lactantius. We can see how different Augustine is from Lactantius in tackling this topic, despite Augustine's similarities and borrowings.

What was the problem with the argumentation put forward by Augustine?

Augustine in light of Cratesian geography

The caveat discussed above shows that the rejection of the antipodeans in Augustine was not derived from a rejection of ancient geography but rather from his acceptance of a particular point of it: other potential landmasses would be separated by immense oceans impossible to cross. In some way, Augustine's denial of the antipodeans was the consequence of taking to the ultimate consequence the zonal model (Figure 4A) or even Crates's model (Figure 2.5).

Within this framework, some people believed that all of these landmasses were inhabited, but natural barriers made them sceptical about the possibility of reaching those populations. An essential text in this context was Cicero's final part of his *Republic*, where in the *Dream of Scipio*, which was well known in Late Antiquity, he described first the Cratesian model and then the zonal model, indicating the impossibility to reach the inhabitants of the Southern Zone (Cicero, *Republic* 6.20, 1999, 99–100). Pliny indicated that the seas blocked access to half of the world (Pliny, *Natural History* 2.67.170, 1938[1949], vol. 1, 305). He then described the five zones and referred to the heat of the Torrid Zone as a barrier (Pliny, *Natural History*, 2.68.172, 1938[1949], vol. 1, 307). Augustine's contemporary, Macrobius, indicated that the antipodeans were an inference from reason that could never be reached due to the Torrid Zone (Macrobius, *Commentary on the Dream of Scipio* 2.5.17, 1952[1990], 203).[41] Augustine was no fool in pointing to such difficulty.

Augustine in light of Ptolemaic geography

The problem Augustine faced in taking a position on the antipodes was complex because it was possible to dispute the claims of a lack of historical evidence and an Equatorial Ocean/Torrid Equatorial Zone that could not be crossed with the claim that sailors travelling around Africa had been in the antipodes (at least in the southern portion of the Earth, the *antoeci*). However, many ancient authorities disbelieved those stories, for example, Herodotus 4.42 and Strabo 2.3.4–5 (see more references in endnote 22). Furthermore, as we saw, Ptolemy considered in his *Geography* that Africa and Asia were united by a bridge of land, converting the Indian Ocean into an interior sea and making travel around Africa impossible.

Nevertheless, some ancient authors accepted the existence of equatorial inhabitants. Geminus disbelieved the Equatorial Ocean of Crates (which he saw as an old and discredited opinion) and the inhabitability of the equatorial region, based on the evidence of Egyptian explorations to the south (quoting a lost book from Polybius). However, such 'official' explorations had not yet reached the equatorial lands and even less the antipodes, as noted by Ptolemy, who accepted that Africa extended to both sides of the Equator but remarked we know nothing about the "inhabited regions" of the Equator: "For up to now they are unexplored by people from our part of the inhabited world, and what people say about them must be considered guesswork rather than report" (Ptolemy, *Almagest* 2.6, 1984, 83, with slight modifications). Indeed, when in his *Geography* Ptolemy mentions the southern land connecting Africa and Asia below the Equator, he recognises that this is an "unknown land that encloses the Sea of India and surrounds Aithiopia south of Libyē" (Ptolemy, *Geography* 7.5, 2000, 108). As late as the early 5th century, a compiler such as Capella, who was still reproducing the already discarded Cratesian views, was so careless that he could refer to the uninhabitability of the Equatorial Zone (*The Marriage of Philology and Mercury* 6.602–605) and without hesitation mention in the next lines the inhabitants of the Equator (*The Marriage of Philology and Mercury*, 6.607).

In any case, if accepted, these antipodes (*antoeci*) would not have lived on a landmass disconnected from the landmass where Augustine lived. The above stories would have shown just that Africa extended to the southern side of the Earth, and any human inhabitant there could have arrived walking and, would, therefore, not have posed a problem for Augustine. That is probably why Augustine does not mention the fiery Equator as a problem, only the immensity of the ocean. In fact, southern Africa was not the real problematic antipodes to the Mediterranean that Augustine had in mind, just the *antoeci*.

We can see that Augustine was not discussing lands in the same meridian as he was, but on the other half of the Earth (the Western Hemisphere, where America is placed)[42] as he talked about "people on the opposite side of the Earth, where the Sun rises when it sets on us, people who walk with their feet opposite ours." Furthermore, a few lines below, he identified his problem with the conclusion "that the other part of the Earth, which is beneath, must also be inhabited."

In fact, Augustine was a North African familiar with regions further south of the Roman territory of North Africa. Right before the composition of the *CG* 15 and 16, in an epistolary exchange with Hesychius (Bishop of Salona, near modern Split) in 418, Augustine discussed the Apocalyptic text from Matthew 24:14. There he entered into considerations about the extension of the evangelisation, the existence of peoples outside Roman jurisdiction, and a connection between the fulfilment of the evangelisation and the second coming of Christ in light of that Gospel verse.[43] In his reply, Hesychius also threw into the discussion the texts of Acts 1:8 and Romans 10:18 (containing a quotation of Psalm 18/19:4/4-5), suggesting that the Apostles had already fulfilled the Gospel's prophecy.[44] This last text would be the object of intense discussion regarding the reaching of the Gospel to all the Earth when new lands were found in the 15th 16th centuries (see Chapters 3 and 4). Augustine replied the following year with the long letter nº 199, containing a treatise *On the End of the World* (*De fine saeculi*), where he rejected the idea that the whole world had been already evangelised, pointing precisely towards unreached populations in the south: "For there are among us, that is, in Africa, innumerable barbarian tribes among whom the Gospel has not yet been preached. [...], yet it cannot rightly be said that the promise of God does not concern them" (Augustine, *On the End of the World* 46, 1955, 394).

In the face of all the uncertainties associated with the idea of the antipodes, it is then not surprising that Augustine considered the possibility that all the known emerged oikoumene was a single landmass: "the whole world, which is, in a sense, the greatest island of all because the Ocean also girds it about" (Augustine, *On the End of the World* 47, 1955, 395).

Augustine and a single landmass emerging from the "gathering together of the waters"

Professional geographers were not very fond of Crates's speculations,[45] as Ptolemy who described a single landmass composed of Europe, Asia and Africa (see pp. 29–30). In light of these varying geographical theories, it was not impossible for Augustine to reject the Cratesian model, as Geminus did, and to imagine a single landmass, as Ptolemy did, considering the rest of the planet as covered by water. That last point could even be read back into Genesis 1:9. Augustine's mention of "the gathering together of the waters" could be considered as an echo of this biblical text and not a scientifically triggered proposal, as Augustine does not seem to be familiar with those authors mentioned above.[46] In any case, Augustine's sceptical attitude, contrasting with the aggressive denial in Lactantius, was a rather appropriate approach at that time.

Augustine and the sphericity of the Earth in City of God *16.9*

The difference of approach between Lactantius and Augustine mentioned above (philosophical-scientific vs historical-practical) also helps to understand the key issue that separates both: Augustine avoids a direct attack on the sphericity of the Earth such as made by Lactantius. As indicated in p. 37, while this attack formed the climax of Lactantius's polemic arguments against pagan philosophy, Augustine's discussion is not part of an argument against any 'enemy' but just a reflection on a particular point of the biblical narrative about the population of the Earth. Therefore, Augustine engages in discussing several alternative scenarios and caveats not envisioned by Lactantius.

Common in both texts is the criticism of the deductive process concluding from the sphericity of heaven to the existence of antipodeans. As explained in pp. 41–43, Augustine only uses this criticism to expose the speculative nature of the various discussions of the antipodeans but does not directly deny its starting point, i.e. the assumption of the sphericity of the heaven and Earth. He instead attacked a later point in the deductive chain, namely, the key idea of Crates: the symmetry of landmasses:

> [...]. But they [i.e. the defenders of the existence of antipodes] do not remark that, although it be believed or demonstrated by reason that the world is of a round and spherical form, yet it does not follow that the other side of the Earth is bare of "the gathering together of the waters." [...].
>
> (Augustine, *CG* 16.9; Appendix, no. 1)

In modern times (Ferrari, 1996, 145–148) this has been regarded sometimes as a denial of the sphericity of the Earth or as an indication that Augustine had doubts about the sphericity of the "world" (*mundus*, used in this text as synonymous with Earth, *terra* – both terms are used). Nevertheless, no readers of Augustine seem to have understood that he was calling into question the sphericity of the world for centuries; this is a recent reinterpretation linked to the flat error (see Chapters 4 and 5). Augustine's claims can be read more as rhetorical statements than as scientific ones. Could it be that Augustine was just using the rhetorical device of concession (*concessio*, see Quintilian, *Institutio oratoria* 9.2.51), trying to demonstrate that starting from the sphericity of the Earth there is no logical need to believe in the antipodes? In this case, could we read this text not as taking any particular position about the actual reality of the sphericity of the Earth but just as saying that the belief in the sphericity of the world, regardless of how well it can be proven, as a premise does not imply the existence of the antipodes, without Augustine committing himself to accept or reject the premise? If so, we should not conclude that Augustine was denying or doubting the premise (the sphericity of the world), only that the construction of the sentence was just a rhetorical device to show the invalidity of this inference. In fact, Nothaft defends this interpretation:

> [...]. Augustine's countered this argument [that of the antipodes inferred from the sphericity of the Earth] by showing that it contained a non-sequitur [...].
>
> There are basically two ways of refuting an argument based on conditional reasoning: by denying its *antecedens* or by showing how the *consequens* fails to follow from the former. What Augustine wants to stress is that *even if* the *antecedens* is accepted in the above case, the *consequens* (namely the existence of Antipodes) can still be denied. [...].
>
> (Nothaft, 2011, 47)

In any case, from the rest of his work and close comparison with the parallel text from *DI*, it is clear that Augustine did not intend to criticise the sphericity of the Earth. We said the same regarding Lactantius, but he did take the strategy of criticising the sphericity of the Earth once he started discussing the antipodes (i.e., Lactantius tried to deny the *antecedens*). However, Augustine preferred to base his criticism – apart from disarming the non-sequitur – on historic-geographical argumentations.

Indeed, the series of alternatives that Augustine proposes to the existence of the antipodeans only makes sense if a spherical Earth is accepted. Even the caveat that "some people might have taken ship and traversed the whole wide ocean, and crossed from this side of the world to the other" (Augustine, *CG* 16.9; Appendix, no. 1) is refuted without a hint of flat Earth reasoning. That is why these alternatives, not mentioned by Lactantius, who could not even consider them, became intensely disputed during the Middle Ages in the context of a clear view of the Earth as a sphere (see below).

Flat-earthers in the Middle Ages?

As discussed in Chapter 1, the popular view of the medieval period is that all, or most, people of that age were flat-earthers under the influence of early Christians, such as Lactantius and Cosmas. Although, as we saw, this view has long been discredited, even some well-known historians still assume, from time to time, the existence of such influence. Consider, for example, the words of historian of medieval science Edward Grant: "There was, however, a medieval tradition of a flat earth derived from Lactantius [...]. When pious Christians sometimes placed Jerusalem at the centre of the earth, they seemed to imply, if they did not explicitly proclaim, a flat earth" (Grant, 1994, 628). Four decades before, historian and philosopher of science Thomas Kuhn affirmed: "The cosmologies of men like Lactantius and Kosmas never became official Church doctrine. [...]. But though these cosmologies [...] were never official, they are representative. They illustrate the decadence of secular learning that characterized the Dark Ages, [...]" (Kuhn, 1957[1970], 107).

However, current scholarship shows: *first*, medieval T-O maps were not a representation of a medieval circular flat Earth, but only of the emerged landmass covering a portion of the Northern Hemisphere (Figure 2.6), rooted in pre-Christian times, as mentioned above in pp. 23, 28; *second*, the centrality of Jerusalem in some medieval maps was a late development, not connected with an earlier Christian tradition and not implying a flat Earth (see Chapter 2, endnote 16), and *third*, there are no references in medieval maps or geographical texts to Lactantius or a flat Earth, Lactantius's primary influence could be the reference to the antipodes by Augustine in his *City of God*, as previously discussed.

What medieval maps tell us

No maps have survived from the Greek-Hellenistic period, and we only have very few traces of Roman mapping. The study of early medieval maps has been used to attempt the reconstruction of Roman maps, leading during the 20th century to several proposed classifications that have been less productive than expected (Edson, 2008; Arnaud, 2014, 31–57).

Among those maps are the Albi map (8th century), the 'Isidore Map' in the Vatican (an 8th-century copy of Isidore of Seville's 7th-century *Etymologiae* with a map not related to the Isidorean text), and Beatus of Liébana's map (found in his 8th-century *Commentary on Revelation*, of which the earlier known copies are dated to the 10th century). There are other maps that, although of later date, are believed to represent older Roman originals: the 10th century Cotton or 'Anglo-Saxon' map and the 12th-century 'Jerome' map of Asia (contained in a copy of St Jerome's *De locis sanctis*). In addition to those maps of the known inhabited lands, there is a late medieval Byzantine zonal map derived from an early Roman imperial model (Hiatt, 2008, 42–44). Zonal maps are key for the debate on the sphericity of the Earth, as they depict not just the oikoumene but the whole planet as a sphere (Figure 2.4A).

Also dating from Late Antiquity are the maps contained in the 6th-century *Christian Topography*, known from later extant copies (9th and 11th centuries), by the retired merchant and amateur geographer Cosmas Indicopleustes. Although a follower of the Antiochian flat Earth cosmology, Cosmas wrote in Alexandria, home to a rival Christian school of theology, and where he was confronted by the Christian philosopher John Philoponus (6th century). Philoponus defended the sphericity of the Earth and vigorously attacked Antiochian flat-earthers (without naming Cosmas) in his commentary on Genesis (*De opificio mundi* 3.10). The patriarch of Constantinople, Photius (9th century), read Cosmas and criticised him in the *Bibliotheca* 223 (for that debate, see Nicolaidis, 2011, 24–39).

Cosmas attributed a schematic map that he reproduced to Ephorus from the 4th century BC (Figure 2.3). It is a rectangular oikoumene that Cosmas appropriated for his flat Earth geography (Kominko, 2013, 35–41). It is important to mention that by the time of Ephorus the Earth was already considered to be spherical, and this frame (see pp. 24–26) was probably not intended to support the claim that the Earth was flat, just delineate the rectangular oikoumene. This could easily lead to confusion between the representation of the whole planet and just the portion of the oikoumene, which once in the distant past had been considered to be all that existed (Kominko, 2013, 54). In any case, Cosmas's map had no influence on Western Latin medieval culture. The situation was different in the east, and similar rectangular frames appear in Byzantine religious culture (Kominko, 2013, 54–65; Caudano, 2016).

In general and practical terms, we can distinguish two different kinds of maps that were transmitted from Late Antiquity to medieval times. First, *ecumenical maps* depicting the three known continents of the oikoumene. T-O maps (see Figures 2.2, 2.4B and 2.6), are extreme simplifications of ecumenical maps, while mappaemundi (such as the Ebstorf or Hereford maps) are complex representations, usually incorporating religious and secular historical information. Both maps could display more or less additional unknown surrounding areas, such as the Paradise in the east or regions with monsters in distant Asia and Africa. The stories of monsters already appeared in Herodotus (5th century BC) and became very popular in the Late Roman period, being transmitted to the Middle Ages. Second, we find *zonal maps* depicting the whole globe, including the unknown regions below the Equator. They portray an astronomical point of view where the antipodes have a clear space in the Southern Temperate Zone (Figure 2.4A). After the long history of multiple classifications, such a simplification seems to be the most realistic option, giving wide scope for variation in each category (Edson, 1997, 164). Clearly, the zonal maps could only be as old as the theory of the zones in the spherical Earth. However, ecumenical maps could have, at least theoretically, earlier precedents (see pp. 23–25).

Both kinds of maps appear together in some medieval and Renaissance books (even side by side, as in Figure 2.4), showing the compatibility of the two geographical diagrams: "Examples would be the two maps in the British Library (Cotton MS, Tiberius B. V. 1, mid-11th century), one of which is the famous rectangular world map, and the other a zonal map" (Edson, 2008, 220, note 5; Tedford, 2020). A 14th-century copy of Gautier de Metz, *L'Image du Monde* (composed in 1245) contains a famous diagram illustrating the sphericity of the Earth, and a few pages later T-O maps (see Chapter 3, endnote 68). Both maps appear in Lilio (1493), fol. a.iiii[v] (on this book, see pp. 119–121). Another interesting late example appears in the *Novissime hystoriarum omnium repercussiones* by Foresti (1483), which in a printed edition of 1503, fol. 7[r], shows a zonal map and a T-O map side by side in identical disposition as Figure 2.4 above (available at: https://en.wikipedia.org/wiki/Giacomo_Filippo_Foresti#/media/File:Bergomensis_T-O_World_Map_1503_Cornell_CUL_PJM_1003_01.jpg).

Furthermore, mixed maps have also been found, i.e. zonal maps that incorporate the tripartite continental representation from T-O maps in the Northern Temperate Zone (as in Figure 2.7), clearly showing that for medieval people the ecumenical maps (both the T-O maps and mappaemundi) were not representations of a disc-shaped flat Earth (even though they displayed similarities with older, genuine flat Earth representations). See for example a 1055 map mentioned by Evelyn Edson: "The Ripoll map in the Vatican Library (MS Reg. Lat. 123, fols 143v.-144r., 1055) has an ecumenical map drawn in the northern half of a hemispheric map" (Edson, 2008, 220, note 5).[47] The maps of the early 12th-century *Liber floridus* of Lambert of Saint-Omer also include an elaborate map of this kind where the Northern Temperate Zone is drawn in much detail, reflecting more a mappaemundi than a T-O map (see p. 55). Other well-known examples of zonal maps with a tripartite division in the Northern Temperate Zone are found in the 12th-century Bible Arnstein (British Library Ms Harley 2799, fols 241ᵛ, 242ʳ and 242ᵛ).[48] One of these mixed maps with a clear link to the Late Roman period appears in Macrobius's *Commentary on the Dream of Scipio* (early 5th century) which transmitted Crates's views. The text has instructions for drawing four diagrams, including a zonal map. An additional fifth diagram is a zonal map with an ecumenical map in the Northern Temperate Zone (Hiatt, 2008, 48–52).[49] Together with the explanations in the text,

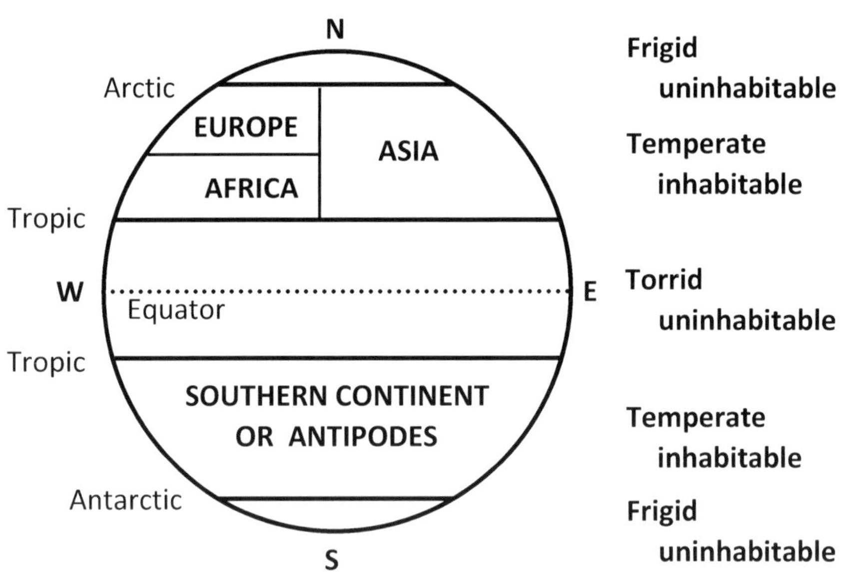

Figure 2.7 Zonal map with the three known continents of the oikoumene in the Northern Temperate Zone. The Southern Temperate Zone is typically the place for the hypothetical antipodes in a continent further south from Africa (from Felipe, 2012, 283, Figure 1A, used with permission).

Macrobius transmitted to the Middle Ages the sphericity of the Earth and the theory of the antipodes.

A further indication of how the T-O maps could be compatible with a spherical Earth is found in some medieval T-O maps which have curved one or both of the lines forming the T to better create the perception of curvature, as in the 14th-century *Traité de la sphère* by Oresme (Paris, Bibliothéque Nationale, Ms. Fr. 7065 (565), fol. 23[r] mentioned in pp. 82–83). For more examples, see Sáenz-López (2013, 544–554). In some cases, these tripartite images represented the elements (earth, water and air) instead of the three continents (as in the case of the mentioned example of Oresme).

The eclipse of Lactantius in Late Antiquity

In his version of Eusebius's *Chronicle* (year 2333/317 AD), Jerome famously called Lactantius "the most eloquent man of his time" (Bowen and Garnsey, in Lactantius, 2003, 4) and in a letter to Paulinus of Nola praised his "stream of Ciceronian eloquence" (*Epistle* 58.10, quoted from Bowen and Garnsey, in Lactantius, 2003, 4). Jerome included a sketch of Lactantius's life in *Lives of Illustrious Men* 80 (and he was mentioned in 18, 58 and 111). Jerome also referred to Lactantius in several letters and other of his books. However, Lactantius soon became almost forgotten during the medieval period (Galynina, 2017; Digeser, 2019, 249). In fact, Augustine and Jerome criticised him on theological grounds. Some of his views made him suspect of heterodoxy over the centuries. Not only his friendly attitude to some pagan texts, such as the Hermetic corpus, and his scant use of the Bible, but also theological points, particularly certain dualism, and his unclear view of the Trinity, as he lived before the theological debates against the Manicheans and Arianism. In the above quoted letter to Paulinus of Nola, Jerome said of Lactantius: "Would that he had been as good at affirming our doctrines as he was at demolishing those of others' (*utinam tam nostra adfirmare potuisset quam facile aliena destruxit*)" (*Epistle* 58.10, quoted from Bowen and Garnsey, in Lactantius, 2003, 4).

By the 12th century, such criticisms became part of the *testimonia* appearing in the manuscripts of his works, at the same time praising and criticising Lactantius: "The earliest example known to me is Arras: Bibliothèque de la Ville, MS 549, s.XII" (Rutherford, 1998, 174, note 13). A direct criticism of Lactantius (concerned with his eschatology) is found in letter 147 from Abbot Wibald of Corvey to Manegold of Paderborn in 1149 (Rutherford, 2015, 471, note 96). Historian David Rutherford, an expert in Lactantius's reception, has also noted that two centuries later the *testimonia* grew, and in the 15th century they were complemented by marginal caveats (Rutherford, 1998, 174; 2017). These *testimonia* were continuously enlarged in the printed editions (see p. 73).

In any case, as we will see, only with Cardinal Fillastre in the 15th century (p. 70) can we find a reflection (in this case critical) about the famous passage in *DI* 3.24. As far as I know, no one before made any comments on Lactantius's attack on the antipodeans and the sphericity of the Earth. Therefore, I have found no justification for the claim, as it appears in authors quoted in p. 48, that the flat Earth views of Lactantius influenced medieval cosmography.

The eclipse of Lactantius was paralleled by the extinction of the flat Earth cosmology from the Western European culture during the Middle Ages (the situation in the east was more complex due to the presence of the school of Antioch). The real medieval debate was that of the antipodes and the antipodeans, and Lactantius was forgotten in it.

The medieval debate on the antipodes sparked by Augustine

The antipodes debate in the Western Church of the Early Middle Ages (5th to 9th centuries)

Augustine's *City of God* (*CG*) was a popular book throughout the Middle Ages (Sloan, 2013). It was reported to be a favourite of Charlemagne (Einhard, *Life of Charlemagne* 24). References to *CG*'s discussion of the antipodes appeared not only in texts quoting and/or discussing this topic but also in maps and illustrations. Augustine's words can be found as direct quotations, but also often abridged or paraphrased. We could summarise Augustine's influence on this topic during the subsequent millennium with historian Alfred Hiatt's words: "For any medieval audience, the classical and late antique tradition of antipodal reference could not be read without the condemnation of the theory of human inhabitation on the underside of the earth in Book 16 of *De Civitate Dei*" (Hiatt, 2005, 12). This was in line with the consideration of Augustine, particularly in the Early Middle Ages, as the "omnipresent fount of knowledge" (Otten, 2013, 30), implying that "having Augustine on one's side was considered key to having one's argument win the day" (Otten, 2013, 30).

Augustine's criticisms were echoed by the major ecclesiastical authors of the Early Middle Ages; among them, the most influential was the scholar and Bishop Isidore of Seville (c. 560–636). His influence during the Early Middle Ages extended to Geography (Lozovsky, 2000, 107–113). His ideas about the natural world appeared in *De Natura Rerum* (*On the Nature of Things*, *ONT*, 613) and the encyclopaedic *Etymologiae* (*Etymologies*, between 600 and 636). His descriptions of the Earth belong to the classical astronomical 'spherical' tradition. Sadly, his unhelpful diagrams and his unclear explanations of some topics, such as references to the Earth as having the shape of a wheel (*Etymologies* 14.2.1), led modern commentators (not medieval ones!) to misrepresent Isidore as a flat-earther.

Isidore was referring to the oikoumene, not the planet, but he was betrayed by the term *orbis terrarum*, which could be used for a two- or a tri-dimensional figure (see p. 17). The debate whether Isidore was a flat-earther still rages.[50] In any case, Isidore's explanations (as Augustine's) have nothing in common with true flat-earthers such as Lactantius or Cosmas Indicopleustes.

In *ONT* 40.3 Isidore quoted Clement (the text discussed in p. 33) to deny the possibility of crossing the ocean. The antipodes were rejected, betraying a clear influence from Augustine in *Etymologies*:

> Now indeed the people called Antipodes (i.e. "opposite-footed") – because they are thought to be contrary to our footprints, as if from under the earth

> they make footprints upside-down from ours – are on no account to be believed in, because neither the solidity nor the central space of the earth allows this. Indeed this is not confirmed by any knowledge of history, but poets conjecture it as it were by sheer inference.
>
> (Isidore, *Etymologies* 9.2.133, 2006, 199)

> Apart from these three parts of the world there exists a fourth part, beyond the Ocean, further inland toward the south, which is unknown to us because of the burning heat of the sun; within its borders are said to live the legendary Antipodes. [...].
>
> (Isidore, *Etymologies* 14.5.17a, 2006, 293)

A third reference to the antipodeans is in fact a reference to 'monsters' like humans but with the feet back to front, derived from Pliny via Solinus, and mentioned in *CG* 16.8: "The Antipodes in Libya have the soles of their feet twisted behind their legs, and eight toes on each foot" (Isidore, *Etymologies* 11.3.24, 2006, 245).

The next principal author of this period was Bede (c. 672–735), heavily influenced by Isidore. In his *De Natura Rerum* (*On the Nature of Things*, c. 703), he clearly explained the spherical nature of the heavens and the Earth, the five zones and the division of the known lands into three continents (chapters 3, 5, 9, 45–47, 60). In chapter 32 of *De Temporum Ratione* (*On the Reckoning of Time*, c. 723), Bede reaffirmed the sphericity of the Earth:

> The reason why the same [calendar] days are of unequal length is the roundness of the Earth, for not without reason is it called "the orb of the world" on the pages of Holy Scripture and of ordinary literature. It is, in fact, a sphere set in the middle of the whole universe. It is not merely circular like a shield [or] spread out like a wheel, but resembles more a ball, being equally round in all directions, but not in a mass of equal magnitude – although I would believe that the enormous distance of mountains and valleys neither adds to it nor diminishes it any more than a finger would a playing ball.
>
> (Bede, 1999, 91)

Nevertheless, in chapter 34, Bede denied the antipodes in Augustinian terms:

> However, it can only be proved that one [zone] of them is inhabited; no credence whatever should be given to fables about the antipodes, nor does any writer claim to have seen, heard, or read of any who have followed the winter Sun to southern climes, so that, with this [region] behind their backs, and having traversed the scorching heat of Ethiopia, they have discovered, beyond these, temperate lands habitable by humans, with heat on this side and cold on the other. [...].
>
> (Bede, 1999, 99)

Although unimportant until unearthed in the 16th century (see pp. 170, 172), it is necessary to discuss the famous conflict between the English Boniface of Mainz and the Irish Virgilius of Salzburg – both churchmen active in Germany in the 8th century. In fact, we know very little about it, just a letter of 748 from Pope Zacharias I to Archbishop Boniface. The letter contains a reference to Boniface's complaint regarding the teachings of Virgilius, which are condemned and that seem to relate to a belief in the existence of antipodeans.[51] Although in his letter the Pope sided with Boniface and was inclined to expel Virgilius from the church, the outcome of the conflict is unknown, but it was neither an obstacle for Virgilius to become Bishop of Salzburg by 767 (or even as early as 755) nor for his canonisation in 1233 (Hiatt, 2008, 82–83).[52]

Beatus of Liébana (c. 730–c. 800) wrote *Commentaria in Apocalypsin* (*Commentaries on Revelation*) in 776 (Edson, 1997, 149–159; Williams, 1997). It contains a fascinating map that partially resembles a tripartite T-O map, which also has a southern landmass near Africa. According to Hiatt, the additional 'continent' appears in 13 of the 14 Beatus manuscripts with maps (Hiatt, 2008, 84). Of those, four copies label this land as the fourth part of the world, following Isidore's *Etymologies* 14.5.17 (Williams, 1997, 17). Five copies just described it as a desert land unknown due to the heat of the Sun. In two later maps, there is no label, and in two other (Burgo de Osma, 1086; Lorvão, 1189), the antipodeans appear together with skiapods (Hiatt, 2008, 85).

These two maps, and a few others, contain miniatures of the heads of the apostles at the locations where they were supposed to have preached, showing the expansion of the Gospel throughout the oikoumene, which was regarded as a precondition for the Last Judgment, Matthew 24:14, Acts 1:6-8 (Edson, 1997, 153, 157; Williams, 1997, 24). Interestingly, in the southern continent of the Osma map, we can see, instead of an apostle's head, a skiapod and a red Sun-disc with a legend derived from Augustine and Isidore: "This region remains uninhabitable and unknown to us on account of the heat of the sun. It is said that the Scopodes live there [...]" (Edson, 1997, 154).[53] These two features point to a region close to the Torrid Zone rather than to the Temperate Zone in the Southern Hemisphere (Williams, 1997, 21 and Figure 8, p. 22).[54] The mention in some cases of antipodeans could be, in fact, a reference not to the inhabitants of the antipodes but to monsters with eight-toed feet facing backwards from Ethiopia (*Etymologies* 11.3.24), the same place where skiapods were imagined to live (*Etymologies* 11.3.23).[55] Both monsters were considered descendants from Adam by Augustine (*City of God* 16.8).

The revival of Platonism and the renewed debate about the antipodes (9th to 11th centuries)

In spite of the overwhelming ecclesiastical criticisms, the concept of the antipodes/antipodeans, and even the Cratesian model, were transmitted to the Middle Ages in a favourable way by several late 4th–5th century non-Christian Latin authors with Platonic or Neoplatonic inclinations, such as Maurus Servius Honoratus (late 4th to early 5th century) in his commentary on Vergil's *Aeneid*, Macrobius Ambrosius Theodosius (early 5th century) in his commentary on Cicero's *Dream of Scipio* and

Martianus Capella (late 5th century) in his *De nuptiis Philologiae et Mercurii*, an encyclopaedia of classical learning, key in early medieval education, particularly in the 'Carolingian Renaissance'.

Finally, when the ecclesiastical and the non-Christian traditions met, some authors played down the conflict, such as a gloss to the *Categories* of Aristotle attributed to Heiric of Auxerre (c. 841–c. 876): "Some say that there are antipodes in another world separated from us by the ocean ... That they can live under the Earth is not contrary to faith, because the nature of the Earth as spherical [*spheroides*] allows it" (Gautier Dalché, 2013, 217).[56] However, Hiatt notes that frictions also appeared in the glosses where "authorities could be played off against each other," and "Macrobius was read through the frame of Augustine" (Hiatt, 2008, 73). Hiatt mentions a late 9th-century gloss to Boethius – "if every person [is descended] from Adam alone, whence then the antipodeans?" (Hiatt, 2008, 83)[57]– and a copy of Macrobius, from c. 1000, with this negative gloss: "Augustine speaks against this in book 16, *De civitate dei*" (Hiatt, 2008, 73).[58]

A detailed criticism of Macrobius appears in the *Liber contra Wolfelmum* (1085/1086) by the Augustinian canon Manegold of Lautenbach (c. 1030–c. 1103) against the Benedictine abbot Wolfhelm of Brauweiler or Cologne (died 1091, beatified by the Catholic Church). Following Augustine, Manegold did not deny the sphericity of the Earth and criticised the idea of antipodes *only if* these people were isolated and unreachable by the Gospel.[59]

The antipodes debate in the 'medieval Renaissance' (12th and 13th centuries)

The revival of Platonism and the interest in Macrobius and Capella in the late 11th century led again to the question of the antipodes. Lambert of St Omer (c. 1061–c. 1125) composed the encyclopaedia *Liber floridus* in 1112–1115, and revised it up to 1121 (Hiatt, 2008, 107). It contains several T-O and zonal maps. The most detailed map says at the south of an Equatorial Ocean:

> Southern region, temperate but unknown to the children of Adam, extending in no way to our race. [...] the sea that lies between the lands, which flows from the east to the west and divides the world [...] repels the approach of humans, nor does it permit transit to this zone by any means. Philosophers assert the antipodeans inhabit this zone [...].
>
> (*Sp[ha]era geometrica marciani;* full-text in Latin and English, with slight modifications, from Hiatt, 2008, 108)

It deserves mention that attached to an island in the western margin of the Northern Hemisphere, a label can be found with a text derived from Capella stating: "Here our antipodeans live..." (Hiatt, 2008, 109), likely referring to the underside of the Northern Hemisphere (Edson, 1997, 110; Hiatt, 2008, 108 109). This island is surrounded by a red line (a feature that only occurs with the Paradise, the whole Earth itself and the ecliptic which is a straight line), and that has been interpreted as meaning an "inaccessible" place (Lecoq, 1987, 15).

William of Conches (c. 1090–c. 1154) discussed the topic of the antipodes in *De philosophia mundi* (c. 1125–1130), *Dragmaticon* (c. 1144–1149), and in his glosses on Macrobius's *Commentarii*: "Macrobius does not affirm that there are men there, but he says they should be believed to exist because of the similar climate" (Hiatt, 2008, 109–110). For him, only one Temperate Zone was inhabited, and the references to the antipodeans by the philosophers were considered worth discussing, although "we do not believe exist, for the purpose of understanding philosophical reading" (*Philosophia* 4.1.5; quoted from Hiatt, 2008, 109; see also *Dragmaticon* 6.3.6). They also had zonal maps (Hiatt, 2008, Figure 14 in p. 111).[60] This strategy secured continuity of the dialectic debate over this controversial issue without the need to accept it. The *Dragmaticon* stressed the proofs of the sphericity of the Earth against "certain bestial people, believing more in perception than reason, who maintain that the earth is flat" (*Dragmaticon* 6.2.2; quoted from Hiatt, 2008, 110).[61] This rhetorical device should not be interpreted as a medieval revival of flat-earthers.[62]

Of great interest is Hermann of Carinthia's (c. 1100–c. 1160) work *De essentiis* (1143), where he described the Earth in Cratesian terms, indicating that we only know "our quarter," concluding that "the people of each limit [i.e. the ends of India and Libya] are antipodal to each other" (Hiatt, 2008, 98). The idea of 'internal' antipodes within their own landmass would make it easier to accept both the concept of antipodes and the possibility of a western passage from Europe to Asia (Hiatt, 2008, 99) and had old roots going back to Aristotle (Vogel, 1995, 226–227).

In *Microcosmus* 1.46, c. 1180. Geoffroy of St Victor (c. 1125–c. 1195), using Genesis 1:9, defended a single landmass (Bartlett, 2008[2011], 45) against the Cratesian views: "For it follows that [if there were antipodeans] not all people descend from a single man [Adam] and that [eternal] life was not restored to the world through the merit of a sole individual [Christ]" (*Microcosmos* 1.47; quoted from Randles, 1994[2000], 21, with slight modifications).

This Augustinian concern also appeared in *De naturis rerum* (c. 1190) by Alexander Neckam (1157–1217): "But is it possible for the antipodeans to have descended from the first parents? According to Augustine there are no antipodeans, this is only said as fiction or belief" (Neckam, *De naturis rerum* 2.49, 1863, 160).[63] This author also defended that the "seas were higher than the coastline," which was argued "by sight" (*ut visus docet*) and a generic mention to the divine command to limit the seas – no reference is given, but see Proverbs 8:29 (Neckam, 1863, 159). As we will see, whether the earth was higher or lower than the water/ocean became an intensely contentious topic up to the 16th century. These ideas in Neckam became part of the new scholastic eccentric model of the Earth (see pp. 58–60).

In addition to the academic discussion in texts and the pictorial depiction in maps, the antipodes appeared in popular European literature (Tattersall, 1981), mixing the sphericity of the whole planet with the circularity of the oikoumene, rather than displaying a true flat Earth cosmology (Cormack, 1994, 366–367; Lecoq, 1996, 28–30). Key problems are the use of ambiguous Old French terminology such as "roond" or "roondece," "terre" or "monde," and the brief nature of those references.

Despite the opposition, the topic of the antipodes appeared in commentaries on the most successful astronomical textbook of the Middle Ages, still used until the 17th century: *Tractatus de sphaera* (usually known as *De sphera*, c. 1220; early printed editions used related titles such as *Sphericum opusculum*) by John of Sacrobosco (c. 1195–c. 1256). The sphericity of the Earth was explained in the text, but the antipodes question was not (original Latin text and English translation in Thorndike, 1949).

An early commentary to Sacrobosco (c. 1230–1235), attributed with little certainty to Michael Scot (c. 1175–c. 1232), detailed the Augustinian theological concerns, as he considered that if the antipodeans were immortal, "they would not be descended from Adam because all those who are descended from Adam are mortal because of the sin of our first parent." On the other hand, sinful, mortal antipodes did not exist

> because Christ would have been born twice and would have suffered twice, which is false. [...] this zone is not habitable by mortals and the reasons which prove this must be accepted because it would be contrary to the faith [...].

The reason is a premise established at the beginning of the previous discussion regarding the isolation of zones, "because of the torrid zone which lies between the two zones" (*The commentary ascribed to Michael Scot*, lectio XI, translated in Randles, 1994[2000], 25, from Thorndike, 1949, 321). In addition, references to Mark 16:15, Romans 10:18 and Psalm 18/19:4/4-5 supported that the Gospel had already reached all peoples (Thorndike, 1949, 321–322).

A commentary to Sacrobosco by Robertus Anglicus (c. 1271) accepted the habitability of the Equatorial Zone (Lecture XIII, Thorndike, 1949, 239) but was sceptical of antipodeans:

> As for the related question, whether the quarter opposite to that in which we live is habitable, I think so, unless water prevents, as is likely. We should not, however, believe that anyone lives there, since no scientific expert has ever mentioned this.
>
> (Thorndike, 1949, 242)

His ideas seem to have been influential on the Franciscan John Pecham (c. 1230–1292), who shortly before becoming Archbishop of Canterbury rejected in his *Tractatus de sphera* (1277–1279) the existence of antipodeans (Hiatt, 2008, 141, note 106). From c. 1250 is a map by John of Wallingford (?-1258),[64] with the following text in the Southern Hemisphere: "It is not known if this middle part which is beyond the torrid part is inhabited, for no-one ever crossed from that part to us, nor from us to them" (Hiatt, 2008, 123 (Figure 18), 124, 142 (note 128)).

From the 13th century, there is a more positive acceptance of the antipodeans. The *Liber introductorius* (c. 1230) attributed to Michael Scot (Hiatt, 2008, 117), includes an unusual and sophisticated zonal map which considered the Southern Zone as inhabited and mentioned the internal antipodes as Hermann of Carinthia and William of Conches (Hiatt, 2008, 120–121).

In his 1298 *Declaratio Raimundi*, Ramon Lull (c. 1232–c. 1315) used the antipodes to give an example of something that can be conceived rationally (*intellectus*) but not grasped by the senses (*imaginatio*), and in his *Tractatus novus de astronomia*, he indicated that antipodeans "would imagine that people in Paris were falling towards the heavens" (Hiatt, 2008, 125; see also pp. 126 and 142, notes 132 and 135). A similar point had been already made by Pliny, Macrobius and Albert the Great (see pp. 31–32, 32–33 and 60, respectively).

In the next century, the idea of the antipodes was introduced in the fictional popular French travel book *Travels of Mandeville*, describing the travels of an English knight to the east from 1322 to 1356. It was widely read in Europe from its composition c. 1360 until the 16th century, often considered as an authentic historical travel narrative. It is important to warn here about some mistaken interpretations of Mandeville as reflecting flat Earth views (see Grant, 1994, 628–629). The reason for that is a reference to Jerusalem as being in the middle of the word (in the prologue), which has to be read as the centre of the oikoumene, not the whole planet. In fact, the book accepts without discussion both the sphericity of the Earth and the existence of reachable antipodeans (chapter 20). That led to an opposite and equally mistaken interpretation, as Mandeville has been wrongly seen as reacting against an alleged contemporary school of flat-earthers (Allegro, 2017; Nothaft, 2017, has criticised this paper).

The beginning of the scholastic eccentric model of the Earth (13th century)

As Aristotelianism extended from the 12th century onwards, a paradox became apparent. The Aristotelian theory of elements assumed a vertical arrangement with the heaviest (earth) at the bottom and then a superposition of spherical layers of water, air and fire (Figure 2.8A). Therefore, why is there dry land emerged above the waters of the seas? On this problem of Aristotelianism, see endnote 13.

Already in ancient times, there were attempts to find, in the words of the historian William Graham Lister Randles (1994[2000], 22–23), "plausible" explanations as to why the ocean "encircles" (Manilius, *Astronomica* 4.595-596) or "surrounded" the emerged land (Pliny, *Historia Naturalis* 2.112.242). Randles indicates that the terms they used to explain it are that the earth "is afloat" (*natat*) or "swims" (*innatans*). Abelard (1079–1142), influenced by these authors, explained the relation of the water and the emerged oikoumene-island in this way in his *Expositio in Hexameron*:

> Just as when a globe is plunged in water in such a way that a part of it emerges, so the globe of the earth lies (*insedit*) in the waters so that one part of it is in contact with the sea.
>
> (Original Latin text published by Migne, 1855, vol. 178, col. 748 AB)[65]

In the late 12th century, Geoffroy of St Victor concluded that the only way to sustain the Earth "above the water" was by "surrender[ing] at God's command" (*Microcosmus* 1.49; quoted by Bartlett, 2008[2011], 47). Sacrobosco (c. 1220) also became aware of the problem, but he only explained the emergence of a portion of Earth in terms of final causes: "to protect the life of animate beings" (Sacrobosco, *Tractatus de sphaera* 1; quoted from Thorndike, 1949, 119 [original Latin in pp. 78–79: *ad vitam animantium tuendam*]).

The commentary to Sacrobosco attributed to Michael Scot (c. 1230–1235) retained Cratesian terminology and concepts (suggesting several oikoumene-islands, Randles, 1994[2000], 24–25), with a view similar to an eccentric model in biblical terms of earth above the water: "[Aristotle] says that [all] round the earth lies the water. [But] this [opinion] seems to be unfounded, since it rather appears that the earth lies above the water, like an island in the sea" (lectio III; Latin text in Thorndike, 1949, 274; translated in Randles, 1994[2000], 24). He quoted Psalms 135/136:6 and 23/24:2 in support of this view (lectio III; Latin text in Thorndike, 1949, 274). As indicated by historian Francesc Relaño, "[o]nly another authority of at least comparable weight was to be able to reconcile the Aristotelian theory with the current practice showing a part of the earth established on the waters: the Bible" (Relaño, 2002[2019], 8–9). Later, he offered Sacrobosco's final causal explanation (lectio VI; Latin text in Thorndike, 1949, 296; see Randles, 1994[2000], 24–25), also mentioned in a commentary on Sacrobosco (c. 1271) by Robertus Anglicus (Commentary of Robertus Anglicus, second lecture, in Thorndike, 1949, 205).

Over the next few centuries some scholastic authors viewed the planet as a small sphere of earth of which only a portion (their oikoumene-island) emerged out of a large sphere of water (Figure 2.8B). Indeed, since Late Antiquity, it was believed that the relation of elements water to earth was 10:1 (Randles, 1994[2000], 9–10, 30). This *eccentric* model was based on two separate spheres of earth and water, and some authors compared this model with a floating apple. This model of the Earth incorporated not only the denial of the antipodes but also the view that the waters were higher than the earth (precisely in the antipodean location) as required by the literalistic reading of several biblical texts, particularly in the Psalms. For more details and references, see pp. 86–90 (specifically in the context of the discussion of the ideas of Paul of Burgos).

In competition with this model, there was the *terraqueous* model (using a 17th-century terminology; see p. 108) of classical inspiration where the Earth was a large sphere of earth, covered on its surface by water, out of which emerged one (for Ptolemy, Figure 2.8C) or several (for Crates, Figure 2.8D) oikoumene-islands. While these two situations were possible in such a model, a vital implication of the scholastic eccentric model (Figure 2.8B) was that only one oikoumene-island was possible, and antipodes were absolutely excluded (Relaño, 2002[2019], 9).

Robert Grosseteste (c. 1175–1253), in *De sphaera* (c. 1220), considered only one landmass inhabited and explained the relationship between earth and water following Ptolemaic views: "the earth together with the waters contained in it are

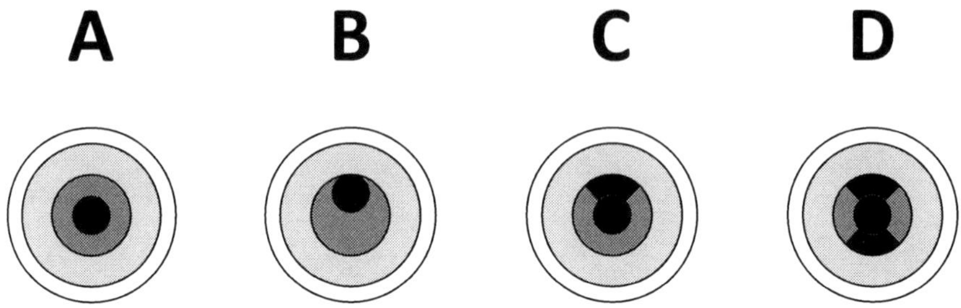

Figure 2.8 (A) Aristotelian 'theoretical' earth at the bottom of superposed elements. (B) Scholastic eccentric earth. (C) Ptolemaic single emerged landmass earth. (D) Cratesian multiple emerged landmasses earth (in this image, there are two landmasses, in reality, Crates defended four landmasses). The four elements are represented by concentric circles: Black (earth), dark grey (water), light grey (air) and white (fire).

nothing more than the sphere of the earth" (Grosseteste, *De sphaera* 1, printed in Sacrobosco, 1508, 87 [fol. 95^r]).[66] The uncovering of the dry land was also explained in a finalist way, with water retreating to underground caves that fitted the above Ptolemaic view (Grosseteste, *De sphaera* 1, printed in Sacrobosco, 1508, 87 [fol. 95^r]). Grosseteste had contradictory views regarding the habitability of the equatorial region, rejected in *De sphaera*, but accepted in *De natura locorum* (Relaño, 2002[2019], 26 and note 32 in p. 43).

Albert the Great (c. 1200–1280), in his *De Natura loci* (c. 1250), denied that the Southern Hemisphere was underwater (Randles, 1994[2000], 26–27) and defended the habitability of the southern antipodes and the possibility to reach them through habitable regions at the Equator (Hiatt, 2008, 100; Randles, 1994[2000], 27). Historian Klaus Anselm Vogel discusses Albertus's ideas on cosmography, quoting his scorn towards those who fear that people in the lower hemisphere could fall off the Earth as *vulgaris imperitia* (Vogel, 1995, 174–184).

His contemporary Roger Bacon (c. 1219/20–c. 1292), in his *Opus Maius* (1267), quoted Aristotle and Seneca to defend that the Western Ocean from Europe to Asia was 'small' and 'navigable' (see references in endnote 19). He also referred to the *Hexaemeron* of Ambrose and 4 Esdras 6.42, an ancient non-canonical Jewish text (Randles, 1994[2000], 27–28; Vogel, 1995, 171–174; see more details in Chapters 3 and 4).

Despite those early criticisms, the eccentric model continued gaining influence through the 13th century. In his *Tractatus de Sphaera* 5 (after 1268), Campanus of Novara (c. 1220–1296) provided a supernatural mechanism based on the separation of the waters from the dry land in Genesis 1:9 (lowering the sphere of water; see also Psalm 103/104:6–9), as implied already by Augustine (Randles, 1994[2000], 19–20 and here p. 46): "The whole of the dry land is like a very large island elevated up in the air above the surface of the water" (quoted in Randles, 1994[2000], 29).

Conclusions

After a complex generation process in classical Greece and general acceptance during the Hellenistic period, the sphericity of the Earth and the heavens became mainstream knowledge in the Roman Empire. Common people would have had notions of it, but only a select minority of Greek speakers in the Latin west had access to first-hand scientific information, while most Latin readers only knew secondary sources (such as Cicero or Pliny) and handbooks. Regardless of the language, a minority of intellectuals had reservations about the sphericity of the Earth, and Lactantius was one of them.

In addition, early in the Hellenistic period, speculation about other landmasses besides the European-Asian-African oikoumene became common. The term antipodes was used for these other landmasses, which some considered inhabited by human-like antipodeans. If uneducated masses doubted the sphericity of the planet, the existence of upside-down antipodes was seen as a dilettante's speculation in many cases. Ancient scholars also noted that there was no evidence for antipodeans, in line with Augustine.

That was the situation at the advent of Christianity, and it should not be so surprising – as it appears to be for some authors in the last two centuries – to find among some ancient Christian authors criticisms of ideas that are so removed from daily common sense. Historically, the more influential authors have been Lactantius and Augustine.

In both authors the target was not the shape of the Earth but the denunciation of the antipodes/antipodeans (as noted in Nothaft, 2011, 46–47). However, there are essential differences between both authors. Lactantius's discussion aimed, in the context of an apologetic text, to undermine the authority of pagan philosophers by exposing their absurdities against common sense. However, Augustine wanted to stress that all human beings must be descendent from Adam, and therefore linked by the original sin and by the opportunity of redemption in Christ –thus, a separate humankind was a challenge that he preferred to deny.

However, Lactantius went beyond the antipodes to deny the spherical Earth, a red line that Augustine never crossed. Contrary to common opinion, Lactantius, the more aggressive of both authors, never used biblical or theological arguments. Nonetheless, as was endlessly repeated by 15th- to 17th-century authors, Augustine was misled not only by his theology, but also by mistaken popular geographical ideas of his time and earlier non-scientific authors (such as Cicero) who considered impossible to cross the oceans.

Despite some of the 19th- and early 20th-century scholarship, it has been shown in the last decades that there was no debate on the sphericity of the Earth in Western Europe during the Middle Ages, and that the impact of Lactantius in this regard was non-existent. In contrast, there is plenty of evidence of an active debate enduring through a millennium about the existence of the antipodes and antipodeans inspired by Augustine. However, this debate never sparked a denial of the sphericity of the Earth.

Notes

1 Strabo, *Geography* 1.1.11, Agathemerus, *Outline of Geography* 1.1 and Diogenes Laertius, *Lives of Eminent Philosophers* 2,1.2, all texts derived from a lost text by Eratosthenes on the history of geography (Heidel, 1937[1976], 132). See Couprie (2011, 79).

2 In *Histories* 2.31–34, Herodotus defended a West African source for the Nile and the parallel course of the Ister and Nile.

3 In keeping with the principle of symmetry, and as a consequence of the legend about the Hyperboreans populating the extreme north, Herodotus suggested that there should be "also Hypernotians" in the south (Herodotus, *Histories* 4.36, 1890, vol. 1, 306). The discussions on the Ister/Nile and Hyperboreans/Hypernotians illustrate the importance of symmetrical reasoning in Greek culture (and in geography in particular, Priestley, 2014, 111–120). See also Figure 2.3 for an example of population filling the edges of the map.

4 Figure inspired in the first printed version of a medieval T-O map from Isidore's *Etymologies* (Augsburg, 1472); available at: https://upload.wikimedia.org/wikipedia/commons/7/70/T_and_O_map_Guntherus_Ziner_1472.jpg.

5 Redrawn from the manuscript found at Biblioteca Apostolica Vaticana, Vat.gr.699, fol. 19r, see http://digi.vatlib.it/view/MSS_Vat.gr.699/0045, legends from Heidel (1937[1976], 17).

6 For recent bibliography, see Couprie (2005; 2011; 2018; 2021 and forthcoming, 2025), Graham (2013; 2021).

7 In the later *Timaeus*, Plato would attribute to the whole universe the shape of a sphere (33b, 44d, 62d-63a) or a dodecahedron (55c), see Couprie (2005, 196; 2011, 205).

8 See also Couprie (2005, 198–200; 2018, 20–21, 148).

9 A discussion of Aristotle's arguments against the notion of a flat Earth can be found in Couprie (2011, 213–220; 2016; 2018, 241–260).

10 This only shows that the Earth is curved (Neugebauer, 1975, 576, 1093–1094; Couprie, 2016; 2018, 243). For a complete proof, it would be necessary to show that there is no other shape of the Earth and Moon able to give the shapes observed during the eclipses of the Moon. Anaxagoras, who remained a flat-earther, proposed additional unknown dark celestial bodies that could be responsible for some of the eclipses that were not explained by a flat Earth, while Pythagoreans could use the counter-Earth in the same way (Couprie, 2016; 2018, 167–194 [Anaxagoras], 244–245 [Pythagoreans]; for Anaxagoras, see also Graham, 2013, 123–124, and 192–198 for Pythagoreans). Ptolemy did not mention this argument in favour of the Earth's sphericity.

11 Given the small scope of the observations available at the time, this was compatible with a fully spherical Earth, but also with a convex Earth (Couprie, 2016; 2018, 246–248). Only long-range trips, particularly to the Southern Hemisphere, could settle the question by showing that the Earth was more than hemispherical and also, looking up to the sky, showing that the stars circled around a new South Pole. Herodotus mentioned these trips around Africa (see endnote 22); however neither he nor Aristotle, Ptolemy or other astronomers and geographers used this definitive argument for the sphericity of the Earth, as many ancient authors disbelieved these reports, because they considered the Equator to be an uncrossable barrier and the Indian Ocean a closed sea (Couprie, 2016; 2018, 251–252). With the enlargement of the oikoumene under Alexander the Great and the Roman Empire, the convexity of the Earth grew closer to a quarter of a sphere (see endnote 20). That could also push forward the idea of a full planetary sphere against a 'flat' convex Earth. In fact, Ptolemy argued the sphericity of the Earth in the *Almagest* on empirical/astronomical basis (without the misleading lunar eclipses) and only later introduced the idea of gravitation, central for Aristotle.

12 After Aristotle, other authors in favour of the sphericity of the Earth pointed to the fact that barely visible lights or shores in the distance became more visible when sailors

climbed up high of the masts of their ships or when these distant objects were elevated. Another observation was that a low shore became higher as one got closer (Strabo, *Geography* 1.1.20). Ptolemy compiled several astronomical observations in favour of the sphericity of the heavens: (i) the rising of the Sun, Mon and stars from one edge of the Earth to the opposite after traversing high in the sky in fixed periods of time; (ii) circular movement of the stars around the pole (*Almagest* 1.3). For the sphericity of the Earth he argued from: (i) different timing in the daily rising and setting of the stars from east to west, (ii) lunar eclipses that were recorded at different hours of the same day depending on the position in the east to west axis, (iii) different stars were visible in the sky when one moved from south to north, (iv) sailing towards the coast, it becomes higher the closer we get to it (*Almagest* 1.4). Another observation was that at noon, the shadows of the gnomon at the equinox are always towards the north, never towards the south (*Almagest* 2.1). An even more comprehensive list of arguments for the sphericity of the Earth is found slightly earlier in Pliny (see endnote 26). For a comparison of the arguments put forward from Aristotle to Cleomedes (c. 200), see Bakker (2016, 171–175) (including Table 4.2)

13 Aristotle, *Meteorology* 354b23b-25. Aristotle's model prompted the question of why the water element did not cover the whole earth element and why humans can live on emerged land. This was made even more problematic when after c. 250 BC Archimedes (c. 287–212 BC) demonstrated mathematically that the spheres of water and earth have to be concentric (Archimedes, *On Floating Bodies* 1.2). Plato and others imagined a water circulation cycle involving huge underground reservoirs of water, while Aristotle proposed a water cycle with the atmosphere by the opposite processes of rarefaction and condensation under the action of the Sun (Wey Gómez, 2008, 116–117).

14 Pseudo-Plutarch, *Placita Philosophorum* 3.11; Strabo, *Geography* 2.2.2; Achilles Tatius, *Isagoge in Arati phaenomena* 1.31.

15 According to William Arthur Heidel, Erich Frank "suggests that it was Eudoxus, the associate of Plato, who first transferred the celestial zones to the earth and thus introduced the notion of the 'clime' for the vague 'regions' (τόποι) of earlier thinkers" (Heidel, 1937[1976], 76, note 170).

16 Such popularity was probably reinforced by the Christian significance which was easily associated with them during the early medieval period: the identification of the "T" with the cross (Lanman, 1982) and the association of the three continents with the three sons of Noah (Williams, 1997, 13 and references in note 22; Relaño, 2002[2019], 11 and references in note 26, p. 19). Late maps were even slightly reorganised to have Jerusalem at the centre of the cross, such as in the 13th-century Ebstorf mappaemundi. This change can also be traced in texts, from Isidore, in the 7th century, where Jerusalem is at the centre of the "whole region [of Judea]" to Hrabanus Maurus, in the 9th century, where this text is slightly transformed into the centre of the "region and the whole earth" (Lozovsky, 2000, 112; on this topic, Woodward, 1985, 515–517 and Relaño, 2002[2019], 11 and 19 [note 27]).

17 This term was invented by Plato, according to Diogenes Laertius's *Lives* 3.24 (although in 8.1.25 it is attributed to Pythagoras), and it appears in *Timaeus* 63a (ἀντίπους, see discussion in Stevens, 2020, 116, note 6). Although strictly speaking *antipodes* was used for the quarter opposed to the Mediterranean across the centre of the Earth, the term *antipodes* was loosely used for any of the three separate landmasses. Similar ambiguous usage is found for *antichthones* (a term originally derived from the opposite or counter-Earth of the Pythagorean planetary system orbiting around a central fire, see endnote 10), which sometimes was used as a synonym for *antipodes*, but in other cases denoted any particular population 'opposed' to the Mediterranean oikoumene (see, for example, the discussion in endnote 25, and Hiatt, 2008, 17).

18 It is not difficult to appreciate the parallels with today's speculations about other inhabited planets. The existence of extrasolar planets, debated for centuries, has been finally

confirmed in the late 20th century, and now the debate is about the existence of life somewhere else in the universe.

19 Aristotle, *On the Heavens* 298a9-298a16; *Meteorology* 362b28-362b30. See Strabo *Geography* 1.1.13, 2.3.6 and Seneca *Natural Questions* 1.13. On that, see Couprie (2018, 248–249). For a recent review of scholarship on Seneca's texts related to Columbus, see Hine (2010, 147–150).

20 In the *Almagest* and *Tetrabiblos*, Ptolemy had the southern limit at 4¼° north of the Equator. However, in his *Geography*, where he considered the size of the Earth smaller, he increased the size of the oikoumene, which was extended south to 16° below the Equator with a total longitude from Spain to India of 180° (Berggren and Jones's note in Ptolemy, 2000, 20–22).

21 Several ancient authors assumed the existence of an inhabitable land right on the Equator that had a more temperate climate than the regions in the tropics (which in certain authors were also considered inhabitable, therefore accepting a continuous habitability between the Northern and the Southern Temperate Zones). The reason given was that the Sun should heat for a longer time in the tropics than in the Equator. The habitability of the equatorial lands was defended by Eratosthenes (Strabo, *Geography* 2.3.2), Polybius (Strabo, *Geography* 2.3.2 and Geminus, *Introduction to Phenomena* 16.32,33), Posidonius (Strabo, *Geography* 2.3.2 and Cleomedes, *Heavens* 1.4.90-146), Geminus (*Introduction to Phenomena* 16.25,26,30-38), and Ptolemy (*Almagest* 2.6). However, Strabo considered the existence of two separate inhabitable Temperate Zones, accepting at most an inhabitable area under the Equator that was separated from the Temperate Zones (*Geography* 2.3.2 and 2.5.34; for Strabo's views, Anastos, 1951[1979]). Although Ptolemy did not speculate regarding other oikoumenes, it could be deduced that he conceded their possible existence (Wey Gómez, 2008, 118).

22 Reports transmitted by Herodotus, *Histories* 4.42-43; Hanno, *Periplus or Circumnavigation of Africa*; Strabo, *Geography* 2.3.4-5; Pliny, *Natural History* 2.67.168-169; 6.34.175-176; Martianus Capella, *Marriage* 6.617-622.

23 This possibility was entertained before: "Hipparchus (cf. Strabo 1.1.8-9, Loeb 1:17-19) and Polybius (3.38, Loeb 2:89) had previously considered it possible that the Atlantic and Indian Oceans did not join South of Africa. Ptolemy does not actually provide coordinates for the coast of the unknown land linking Africa and Asia, but he refers to it verbally in [*Geography*] 7.3, 7.5, 7.7, and 8.1" (Berggren and Jones in Ptolemy, 2000, 22, note 23).

24 Ptolemy, *Geography* 7.5.2. In fact, the idea of earth, instead of water, encompassing all the oikoumene was very ancient, see in Herodotus (*Histories* 3.116; 4.45; see Romm, 1992, 33–41) and Hipparchus (according to Strabo, *Geography* 1.1.9).

25 Mela seems confused about landmasses outside the European-Asian-African. Rather than antipodes (or more precisely antoecians) he referred to the counter-Earth or counter-world (*antichthon*), opposite below in the Southern Hemisphere. However, while in his *Description of the World* or *The Chorography* (*De situ orbis* or *De chorographia*) 1.4 he seems to accept its existence as an unexplored but inhabited place, later in 1.54 he was less sure of that. Further, in 3.70 he confused the island of Taprobane (Sri Lanka) with an extreme of the *antichthon*, while at the time of Alexander it had already been circumnavigated and shown to be an island (see details on Taprobane in Wey Gómez, 2008, 477, note 93).

26 *Natural History* 2.2.5, 2.64.160, 2.65.164, 2.71.177-73.181, 2.74.182-76.185 and 2.77.186-187. See endnote 12.

27 ὠκεανὸς ἀπέραντος ἀνθρώποις καὶ οἱ μετ᾽ αὐτὸν κόσμοι ταῖς αὐταῖς ταγαῖς τοῦ δεσπότου διευθύνονται. (Clement, 1912[1965], 44). On this text, see my brief discussion in Felipe (2012, 284). Other early Christian thinkers seemingly accepting the antipodeans are Irenaeus, Tertullian and Hilary (Betten, 1923, 80; Moretti, 1994, 262; Hiatt, 2008, 55; Stevens, 2020, 127).

28 Only some fragments of the Greek original are known; this portion is only preserved in Rufinus's translation, preceded by a terminological discussion of Latin and Greek words which is clearly not by Origen (Origen, 2017, vol. 1, 168–169, see note 51 in p. 169).

29 Stories such as those of Thales falling into a well while looking at the sky, Pyrrho about to fall off a cliff or washing a pig, Diogenes dying, after a life of austerity, by eating an octopus, etc. Probably, the ultimate in this line of criticism was to point out how some philosophers carried their ideas to the extreme of committing suicide, as in Lactantius's *Divine Institutes* 3.18.5ff.

30 Tatian, *Address to the Greeks* 2, 3, 26, Tertullian, *Against Nations* 2.4 and *On the Soul* 1.8, or Lactantius, *Divine Institutes*, which will be discussed below.

31 Wills (1999), O'Daly (2001[2005]), and Lévy (2010, in particular pp. 98–100).

32 As explained, for example, in Plutarch, *Platonic Questions* 5. See Plato, *Timaeus* 53a-57d and Aristotle, *On the Heavens* 307a8.

33 On the influence of Cicero on Lactantius, see Schmitt (1972, 23–28, 29–33 in relation to Augustine) and Kendeffy (2015).

34 Rutherford noted that philosophers contemporaneous with Lactantius were also aware of the debilitating effects of scepticism during the previous centuries, and they "had at least tacitly accepted this premise and were resorting to divine oracles themselves" (Rutherford, 2015, 489).

35 For example: "Lactantius repudiates the sphere-shape of the earth, together with existence of the Antipodes and the concept of gravity on supposedly biblical grounds" (Simek, 1992[1996], 2). Russell (1991, 32–33) has a similar position, as well as Eco (2004, 276), who apparently confused the argumentation of Lactantius with that of Cosmas (see also Omodeo, 2014, 57 in p. 198 below). Here is another recent example that invents such alleged biblical grounds for Lactantius, confusing him with the Antiochian flat-earthers:

> "To be sure, Firmianus Lactantius (ca. 240–ca. 320) and Severianus, bishop of Gabala (fl. late fourth century, d. after 408), could not reconcile the idea of the spherical Earth with the literal teachings of the Bible (which contains the words "circle" and "four corners" but not "sphere" or "globe"), and Cosmas Indicopleustes (a sixth-century Nestorian Christian residing in Alexandria, who wrote a Christian Topography) described a flat, rectangular, four-cornered Earth with a vaulted heaven"
>
> (Woodward, 2013, 549).

36 Vogel (1995, 75–80) and Stevens (2020, 123–124) also reject to consider Augustine as a flat-earther. Ultimately, in *De genesi ad litteram* Augustine considered this issue of secondary importance for a Christian.

37 On the biased terminology of 'race' used in the common label 'monstrous races', see Mittman (2015), who suggests as an alternative 'monstrous peoples'. For 'monsters', see Friedman (2000), and Mittman and Dendle (2013).

38 As we will see in later chapters, this was not only a negative argument, as the Apostles were supposed to have already fulfilled Christ's commission to evangelise "all peoples" in their travels throughout the Old World (Matthew 28:19; Mark 16:15; Luke 24:47; Acts 1:8). Therefore, no inhabited land was expected outside it. Interestingly, Augustine had written extensively on this topic in his *Letter* 199 (entitled *On the End of the World,* i.e. *De fine saeculi*; see p. 45) in 418–419, just before he started chapters 15 and 16 of *CG* (see endnote 43). Augustine readily accepted the existence of more inhabited lands outside the Roman Empire, but not inhabited separate landmasses.

39 Not only does Genesis contain a monogenist view, it can also be derived from Paul in Acts 17:26. In contrast, Greco-Roman mythology felt comfortable accepting multiple human lineages (Wey Gómez, 2008, 119, 122 and particularly the long note 47 in pp. 473–474).

40 This was an essential point in Augustine's theology, which has been very influential in Western Christianity. Augustine supported his views particularly in his interpretation of Romans 5:12, a text with which he opens his *On Marriage and Concupiscence* (*De nuptiis et concupiscentia*) 1.1 (also 1.24). This book was composed in 418-419, at the same time as his *Letter* 199, just before he started chapters 15 and 16 of *CG* (see endnote 38), therefore Augustine had thought deeply on this topic before he wrote his famous passage on the antipodeans.

41 Similarly, the earlier Polybius (2nd century BC) considered that the far south of Asia and Africa were "still unknown territory" (Polybius, *Histories* 3.38, 2010, 159). Geminus (1st century BC) still considered the antipodeans as mere speculation (Geminus 16.21-38). Alexander of Aphrodisias (2nd–3rd centuries), emphasised in his commentary on Aristotle's *Meteorology* the immense size of the oceans, which made the antipodeans unreachable (Vogel, 1995, 57–58). Another sceptic of the possibility to reach the antipodeans due to the Torrid Zone and the oceans was Olympiodorus of Alexandria (6th century), in his commentary on Aristotle's *Meteorology* (Vogel, 1995, 59–61).

42 Ancient and medieval authors used to divide the Earth into two hemispheres, but not north-south as we usually do now, but east-west, assuming that the eastern one –where they lived– was on top, while referring to the other – Western – Hemisphere as being below or beneath.

43 *Letter* nº 197 from Augustine to Hesychius (end of 418); translated in Augustine (1955, vol. 4, 347–350). Such discussions, shortly before Augustine wrote chapters 15 and 16 of *City of God*, could be on his mind when stating vaguely: "For Scripture, which proves the truth of its historical statements by the accomplishment of its prophecies, gives no false information" (*CG* 16.9; Appendix, no. 1). Compare with *Letter* nº 199 from Augustine (418–419) where, in discussing if the prophecy of the weeks of Daniel 9 referred to the first or second coming of Christ, he affirmed: "For my part, I see that if the first coming did not fulfil them, the second must necessarily do so, since the prophecy cannot be falsified" (Augustine, *On the End of the World* [*De fine saeculi, Letter* 199] 21, 1955, 372; Latin text from Migne, 1845, vol. 33, col. 912: *Equidem video quia si primus eas non complevit adventus, necesse est ut secundus eas compleat; quoniam prophetia illa non potest esse falsa...*). Another parallel is: "If, then, as the prophecy of truth cannot be falsified, it must needs be that all nations, as many as God has made, should adore Him, [...]" (Augustine, *On the End of the World* [*De fine saeculi, Letter* 199] 48, 1955, 395; Latin text from Migne, 1845, vol. 33, col. 923: *Si ergo quia prophetia veritatis mentiri non potest, necesse est ut omnes gentes, quotquot fecit Deus, adorent eum...*).

44 *Letter* nº 198 from Hesychius to Augustine (end of 418); translated in Augustine (1955, 350–356).

45 Alfred Hiatt refers to Claude Nicolet's *L'inventaire du monde* as indicating that Crates theory was "more popular with philosophers than geographers" (quoted from Hiatt, 2008, 21).

46 The Latin expression used by Augustine *aquarum congerie* (also found in *CG* 11.29: "the gathering of the waters beneath, and the laying bare of the dry land" (Augustine, 1871, 473)) can be considered as a synonym of *congregatio aquarum* used in the Latin text of Genesis 1.9. As *congeries* is found several times elsewhere in the Old Testament, this could be the reason Augustine used this term here. In *De genesi ad litteram* (completed before, in 415) Augustine also discussed the gathering of the waters (1.12.26). There he suggested cavities to collect the waters that originally covered everything, without specifying a single gathering of waters and a single uncovered landmass (Vogel, 1995, 75–80).

47 Available at https://digi.vatlib.it/view/MSS_Reg.lat.123/0292 and https://digi.vatlib.it/view/MSS_Reg.lat.123/0293.

48 Available at: https://web.archive.org/web/20220125225320/https://www.bl.uk/catalogues/illuminatedmanuscripts/ILLUMIN.ASP?Size=mid&IllID=31187, https://web.archive.org/web/20211201004509/https://www.bl.uk/catalogues/illuminated-manuscripts/ILLUMIN.ASP?Size=mid&IllID=17378 and https://web.archive.org/web/20220125224258/https://www.bl.uk/catalogues/illuminatedmanuscripts/ILLUMIN.ASP?Size=mid&IllID=20319.

49 Hiatt's Figure 2 in p. 5 is a copy of Macrobius of c. 1000: Oxford, Bodleian Library, MS D'Orville 77, fol. 100r, available at: https://iiif.bodleian.ox.ac.uk/iiif/viewer/d76ac9b7-c1f5-4dc1-92d4-164029bb5704#?c=0&m=0&s=0&cv=1&r=0&xywh=-1197%2C133%2C4809%2C2585.

50 For the modern debate, see Stevens (1980) (against), McCready (1996) (in favour), and their letters in Stevens and McCready (1996). See also Lecoq (1996, 22–25) (against), and recently Fear (2016) and Hannam (2021; 2023, 207–211) who are in favour of considering Isidore a flat-earther without adding new argumentation.

51 For the earliest printed retelling of this incident in 1554, see pp. 170, 172.

52 This episode of Virgilius life was studied in detail by Mayaud (2005, 91–121), comparing all the early printed accounts of the letter of Pope Zacharias I from the late 16th and early 17th centuries, which used different medieval manuscripts –some as early as 9th century.

53 In the Osma and in the Lorvão maps this inscription lacks reference to the fourth part of the Earth. *Scopodes* are referred to in English as skiapods, sciapods or even skiapodes; here, the term has been left as in the original Latin legend. In the Osma map, near the Sun, there is a label, "this place is the sun" (*hic ortus est sol<is>*), see Relaño (2002[2019], 31 and note 67 in p. 46).

54 This interpretation is supported by Edson (1997, 33), and Scafi (2006, 110–111).

55 See p. 53. Interestingly, a map without the drawing of a skiapod, but with a text combining skiapods and antipodes, appears in the Lorvão manuscript (Edson, 1997, 154; Williams, 1997, 25).

56 His source is: "Paris, B.N., lat. 12949, fol. 30v°; voir Cappuyns (Maïeul), *Jean Scot Ériène, sa vie, son oeuvre, sa pensée*, Louvain-Paris, 1933, n. 4, p. 67 sq." The Latin text is available in Carey (1989, 3), from where I quote the term *spheroids*, which Carey describes as a rare word of Greek origin that suggests the influence of Servius. This text is also discussed by Hiatt (2008, 83), where he says that it is due to "an anonymous Carolingian commentator on pseudo-Augustine's *Decem categoriae.*"

57 Gloss to Boethius, *Liber contra Eutychen et Nestorium* 5, which used to be attributed to Eriugena.

58 Gloss to Macrobius's *Commentarii* 2.5.22.

59 The relevant Latin text can be found in Manegold, *Liber contra Wolfelmum* 4, in Migne (1854, vol. 155, col. 154–155); translated in English in Manegold, *Liber contra Wolfelmum* 4 (2002, 40).

60 A similar rhetorical strategy was used by the Benedictine Rupert of Deutz (c. 1075/1080–c. 1129) in his commentary of Genesis (Vogel, 1995, 101).

61 Also Elford (1983, 175–179). See Williams's diagrams in Hiatt (2008, 112, Figure 15).

62 An example of such misunderstanding is Lecoq (1996, 27–28). On the classical sources of the 12th century –or earlier– Pseudo-Bede, *De mundi celestis terrestrisque constitutione*, quoted by Lecoq, see Hine (1988, 121–122). A similar mistaken attempt to uncover supposedly medieval flat-earthers from the critical references to the flat Earth in medieval texts is Allegro (2017) (for a refutation, Nothaft, 2017). Equally wrong is the accusation to Isidore by Fear of having contributed "to a continuing and widespread 'semi-educated' view that the world was flat" (Fear, 2016, 77). There is no evidence for the supposition that flat-earthism was widespread after Isidore, which instead suggests that most likely Isidore was no flat-earther.

63 *Sed numquid de primis parentibus descenderunt antipodes? Secundum Augustinum, non sunt antipodes, sed doctrinae causa aut figmenti ita dici solet.* I thank Dr Miguel Marañón (Instituto Cervantes, Madrid, Spain) for this translation.

64 British Library Cottom MS Julios D.VII, fol. 46[r] (the manuscript is available at: https://web.archive.org/web/20221207105940/http://www.bl.uk/manuscripts/FullDisplay.aspx?ref=Cotton_MS_Julius_D_VII).

65 *Quasi enim aliquis globus ita in aqua constituatur, ut una pars eius superemineat, ita ille globus terræ in aquis insedit, ut ex una parte eum mare contingeret, [...]* (Translated by Randles, 1994[2000], 22).

66 *Eſtqʒ terra cũ aꝗs ĩ ſe cõtentis ſicut ſꝑa terræ* ſolũ (Translated by Randles, 1994 [2000], 22).

3 Revival and criticism of early Christians's geographical views in the Early Renaissance (14th and 15th centuries)

Renewed attacks on Lactantius, including his cosmological views, in the Early Renaissance

The rediscovery of Lactantius, the "Christian Cicero"

As we approach the 15th century, the increased interest in Lactantius can be seen by the number of manuscripts preserved (including texts mistakenly attributed to him, Rutherford, 2015, 475, note 106) and by the secondary references to him, which increased from the 14th century onwards. Among the admirers of Lactantius in the mid-14th century were the best-known promotors of the Renaissance in Italy: Petrarch (1304–1374)[1] and Boccaccio (1313–1375)[2]. However, other intellectuals as Giovanni Dominic (1356–1419) or Jean Gerson (1363–1429) also recovered Lactantius (Rutherford, 1998, 173–174; 2015, 476).

By the 15th century, several important authors referred favourably to the eloquence of Lactantius (Panizza, 1978). David Rutherford, an historian specialised in the reception of Lactantius, refers to "[t]he century between the 1340s and the 1440s, which could justly be called the century of *Lactantius redivivus*" (Rutherford, 2017, 155; also p. 157). The rhetorical "Academic" method of *Divine Institutes*, particularly his third book (and the data on several ancient philosophical schools contained in it), inspired Lorenzo Valla for his *De vero falsoque bono* in 1431 (Panizza, 1978) In his famous *De elegantia linguae latinae* (1441), Valla considered Lactantius together with other great theologians of the early church who valued Christian and pagan learning: "Hilary, Ambrose, Augustine, Lactantius, Basil, Gregory [Nazianzen], Chrysostom and many others" (Valla, 1952, 620). Marsilio Ficino recalled Lactantius in *Pimander* (1471; now known as *Corpus Hermeticum*).[3] The famous epithet of Lactantius as the "Christian Cicero," inspired by Jerome, is usually attributed to a work by Giovanni Francesco Pico della Mirandola (*De studio divinae et humanae philosophiae* 1.7, 1497, fol. h.ii^r).[4]

Finally, the early printing date of his books in 1465 (including the *Divine Institutes*)[5] and their frequent reprinting, again with a particular interest in *Divine Institutes*, are another evidence of the growing interest in Lactantius during the Renaissance. There were 15 editions from the *editio princeps* to the end of the 15th century and 36 editions in the 16th century (Panizza, 1978, 83, note 30).

DOI: 10.4324/9781003545682-4

In my estimation, and quite contrary to the enduring medieval influence of Augustine's arguments against the antipodeans, there seem to be no cases of medieval authors following Lactantius anti-sphericity stand. As we noted, starting with the Augustine's *City of God*, a century after the *Divine Institutes*, this part of Lactantius's anti-antipodean argumentation was omitted. In fact, the anti-antipodean polemic in the Middle Ages followed Augustine's argumentation instead of Lactantius's (see pp. 52–60). As we will see, even those Renaissance authors who rediscovered Lactantius with enthusiasm, such as Petrarch, when discussing the antipodeans referred to Augustine. Lactantius's position on this topic does not seem to be mentioned by any author before the 15th century, and during that century only a handful of authors discussed his views on the Earth and the antipodes to reject them.

Cardinal Fillastre's early reference to and dismissal of Lactantius

The first reference to Lactantius's dual denial of the sphericity of the Earth and the existence of antipodes and antipodeans, that I am aware of, appears in the introduction by Cardinal Fillastre (1348–1428) to *De chorographia* of Pomponius Mela, composed at the Council of Constance (1414–1418). While Mela speculated about the existence of antichthones south to the Equator (see Chapter 2, endnote 25), Fillastre noted the opposition of Lactantius who denied the existence of the antipodeans and the sphericity of the Earth, although Fillastre pointed out that "yet he proves nothing to the contrary, although he asserts that he can prove it" (Fillastre, *Introductio* 7, in Gautier Dalché, 1992, 357; Appendix, no. 8, available at www.routledge.com/9781032893068).[6]

Subsequently, Fillastre mentioned Augustine's denial of the antipodeans, as they "could not have been populated through Adam nor through Noah" (Gautier Dalché, 1992, 357). Fillastre is the first author who likely not only mentioned both ancient Christian authors in discussing the antipodes, but also linked them: "Augustine, *City of God* 16.9, who seems to have drawn on him [Lactantius], also denies it…" (Gautier Dalché, 1992, 357). The Cardinal affirmed, regarding the existence of those antipodeans disconnected with Adam/Noah: "that claim seems to lead to heresy, since it is asserted contrary to holy scripture" (Gautier Dalché, 1992, 357). To avoid the charge of heresy, the revised Reims version of Fillastre's text[7] assured the readers that what he said on this topic was "drawing both on the authority of holy scripture and the writings of learned men, as well as on my own slender intellectual powers" (Gautier Dalché, 1992, 357). Fillastre's next lines will be discussed in pp. 83–84.

Decembrio and Rho against Lactantius

Following Jerome's criticisms, some humanists pointed to the mistakes in Lactantius, as in Traversari's warnings in correspondence to Santio Ballo about Lactantius's theological errors (Rutherford, 2015, 477). Others defended Lactantius based on his rhetorical methodology, as Erasmus in his dedicatory letter to *Divine*

Institutes (Antwerp, 1555, 518; see Panizza, 1978, 83, note 29). In contrast to these scattered discussions, an early systematic and long-lasting influential criticism of Lactantius came from a couple of Milanese humanists, namely: Pier Candido Decembrio (1399–1477) and Antonio da Rho (c. 1395–c. 1447).

Candido Decembrio's early attack on Lactantius

Candido, as his father Uberto Decembrio, cultivated an interest in Plato's political work, the *Republic*, a preferable model for the oligarchic government of Milan than Aristotle's *Politics* (on the Decembrio family, Hankins, 1990, vol. 1, 117–154; vol. 2, 412–426). Candido wished to defend Plato against Aristotle's accusations, in book 2 of the *Politics*, of promoting a male shared possession of women, children and property. The situation was complicated because Lactantius and Jerome had also criticised Plato's political views on the same grounds. In addition, new studies of Plato's original Greek text in the 15th century revealed passages describing homosexual behaviour, abortion and infanticide.

Therefore, those defending Plato had to challenge the authority of Greek authors such as Aristotle and early Christian authors such as Lactantius and Jerome – not a minor task! However, Decembrio took an offensive posture, and in 1437–1438 affirmed to have written a work *In defence of Plato against Lactantius*. In 1464, Decembrio mentioned his *In defence of illustrious men against Lactantius*; regrettably, both have been lost, although the first may have been incorporated in Rho's criticism of Lactantius (see below).[8]

Antonio da Rho's Three Dialogues against Lactantius

The final assault of the Milanese Platonic intellectuals on Lactantius came from Decembrio's friend Antonio da Rho, a Franciscan, with knowledge of traditional scholastic theology and the new humanistic approaches. In 1444 Rho dedicated his *Dialogi tres in Firmianum Lactantium* (*Three Dialogues against Lactantius*) to Pope Eugenius IV.[9] The work was never printed but survives in at least seven manuscripts (Hankins, 1990, vol. 1, 148, note 83). It was a dialogue where Niccolò Arcimboldi (1404–1459), a jurist and diplomat friend of Decembrio and Rho, defended Lactantius against the attacks by two other characters: Decembrio and Rho himself!

Rutherford notes that "[m]any of the errors in theology were exactly those cautionary marginalia seen first in the thirteenth century that had become an optional apparatus that scribes might add after they had completed the primary text" (Rutherford, 2015, 478; see p. 51). Rho did not only denounce Lactantius's errors, he tried to identify their source which he found in Lactantius's assertions that "philosophy must be discarded, because what is needed is not the pursuit of wisdom, [...] but wisdom itself [Lactantius, *Divine Institutes* 3.16.7]" (Rho, *Dialogi* 3.163)[10] and that "wisdom resides in heaven and originates in heaven" (Rho, *Dialogi*, 3.164).[11] Indeed, for Rho, "[t]he innumerable errors and contradictions of the rhetorician Lactantius in theological matters were proof of the need for technical questions to be handled by professional theologians trained in philosophy and logic" (Hankins, 1993[2003], 489). Rho's verdict was that Lactantius should be "booed and expelled from the band of philosophers with true reasonings" (Rho, *Dialogi*, 1.141)[12] and "ought to return to rhetorical school where he taught boys" (Rho, *Dialogi*, 1.142).[13] Rho did, however, include Lactantius among the authorities on the *studia humanitatis*, like Cicero or Livy, and considered him to be a good literary model for his students (Rho, *Dialogi*, 1.14).[14]

Among the many errors discussed by Rho in *Dialogi tres in Firmianum Lactantium*, we find criticism of Lactantius's denial of the antipodes. The topic of the antipodes is covered in *Dialogi* 1.130–142 (this was error number 14 in the first of three dialogues that form Rho's book, see below p. 73).[15] It is essential to keep in mind that, at that time, Portuguese sailors had not yet reached the Equator. Consequently, the antipodes had to wait until the 1470s to see the Europeans's arrival. On what grounds, then, did Rho level his criticism of Lactantius over the topic of the antipodes?

Niccolò introduced the topic by reading Lactantius's attack on the antipodes, quoted at length (*Divine Institutes* 3.24.1–9). Candido replied, inquiring as to whether or not Niccolò accepted those ideas. In turn, Niccolò answered by quoting Augustine (*City of God* 16.9) in support of Lactantius. Candido agreed that no antipodes could be found in the ocean between Spain and the Indies; however, he emphasised that Augustine did not deny the antipodes altogether. After an excursus discussing the two opposite positions regarding the earth being higher or lower than the ocean (Candido sided with the latter position, referring to God's supernatural activity for the benefit of humankind), he claimed that the inhabitants of Spain were, in fact, west-east antipodeans in relation with those of the Indies. In contrast, there was no evidence of antipodeans in the north-south direction. In this way, Candido 'saved' Augustine while showering Lactantius with insults. This section ended with a curious twist as Niccolò – apparently forgetting Lactantius – mentioned that some contemporary philosophers held the view that there was more land than ocean. In so doing, Niccolò pointed out that this would open up the possibility that antipodes – in addition to the Spaniards and Indians – could exist on several dry places. In all those positions, we discover the influence of previously studied ideas in Chapter 2. The strategy of maligning Lactantius and saving Augustine becomes obvious (on this discussion, see Vogel, 1995, 297–302).

The aftermath of Rho's attack of Lactantius

Although not printed, the *Three Dialogues against Lactantius* was known, and generated a certain degree of polemic. One of the surviving manuscripts, dated in 1445, belonged to Niccolò Arcimboldi himself and contains a prefatory letter to Rho and his annotations.[16] Niccolò challenged the way he was portrayed in the book and recognised that Lactantius made mistakes. However, Niccolò justified his theological deficiencies due to the antiquity of Lactantius, before the great 4th–5th century Latin theologians (Jerome, Augustine, Ambrose)[17], and considered that his friends were exaggerated in their criticisms of Lactantius. Niccolò, moreover, challenged Decembrio's interpretation of the *Republic* (Hankins, 1990, vol. 1, 149, 152–153; vol. 2, 614–618).

The party of Decembrio and Rho had to confront the opposition of several people even inside the Milanese court, particularly Francesco Filelfo (1398–1481), who became an enemy of Decembrio after 1439 (Hankins, 1990, 129). Filelfo requested Rho's retraction from his unjust attacks against Lactantius in a 1443 letter that was widely distributed in a popular collection of Filelfo's letters from October 1427 to July 1473.[18] Filelfo, in 1443, also wrote a satirical text against Rho that was published as part of a collection of one hundred satires of one hundred hexameters each, divided into ten books (*Satyrarum* 3.6).[19]

The list of Lactantius's errors in the Renaissance

Some of the criticisms and debates outlined above about Lactantius were left unprinted and were not widely circulated. After the invention of the printing press in the mid-15th century, the praises by Renaissance authors of Lactantius became well known to the broader public. This led to numerous editions of his works. However, on the other hand, what had come of the rich array of criticisms of Lactantius?

The autograph manuscript of *Three Dialogues against Lactantius* has at the beginning a *list of Lactantius's errors* (53 *Errata* in total), on six pages and divided into three blocks (a sort of table of contents for the book), following the *Three Dialogues's* structure (Rutherford, 2015, 478, note 122; Rutherford and Schulten, 2017, 605). Although the *Three Dialogues* was never printed, this list of errors was incorporated in the first printed edition of Lactantius's *Opera*, which included the *Divine Institutes* (Subiaco, 1465) and reappeared in some early printed editions of Lactantius, becoming the most influential consequence of Rho's criticisms against Lactantius: "All but a very few of the fifteenth- and sixteenth-century editions, and many thereafter contained some version of the *Errata*" (Rutherford, 2015, 485). The interest in this list was so strong that it was copied "into earlier manuscripts and into the few printed editions that had omitted them" (Rutherford, 2015, 485). In the first block of errors, error number 14 reads as follows:

> The fourteenth error in philosophy is when he claims that – if antipodes existed – then people would have their heads downwards and their feet upwards. And where the antipodes truly are is taught there.
>
> (Lactantius, 1465, fol. 11^{r})[20]

That was not the end of this debate. A Latin epigram c. 1460, by theologian friar Adam Montaldus (1440s–1494), against Rho's *errata,* was included in the second edition of Lactantius's opera (1468) after the *errata*. Both appeared in many later editions of Lactantius.

Galeotto Marzio against Lactantius

Another example of the growing criticism towards Lactantius appears in a book by the Italian humanist, physician and astrologer Galeotto Marzio di Narnia

(c. 1423/7–c. 1494/7; for a biography, see Miggiano, 2008, 478–484): *De incognitis vulgo* (*DIV*, 1477). This book was never printed in full (for an introduction to Marzio and *DIV*, see Vasoli, 1977[1980]). According to Marzio's researcher Gabriella Miggiano, the oldest manuscript of *DIV* is probably the E.IV.11 of the Biblioteca Nazionale di Torino: *Galeotti Martii Narniensis Tractatus de incognitis vulgo*, saec. XV, cart., 145 cc. (effective 125). The *DIV* was partially printed for the first time (bilingual Latin/Italian) by Mario Frezza in Marzio, 1948 (Miggiano, 2013, 35, note 19).

The religious debate about the eternal fate of non-Christians

In *DIV* Marzio discussed controversial topics, questioning the traditional interpretation of the Bible, the authority of several prominent early Christian authors and medieval theologians, as well as some teachings of the Catholic Church. Among those questioned teachings were the sacraments, reduced by Marzio to adult baptism and the eucharist, as the Hussites that were not so far from Hungary, where Marzio worked at the service of King Matthias Corvinus (Miggiano, 1992, 133–136). Another topic of contention was the eternal fate of pagans who had not become Christians because they lived before Christ or afterwards in unreached parts of the Earth. This was a topic with a long history of debate (Marenbon, 2015).

As we saw in Beatus's maps, the medieval conviction was that, by their own time, the Gospel had been proclaimed throughout the entire oikoumene (in spite of Augustine's doubts discussed in p. 45), and only those who had rejected the Gospel (i.e. heretics, Jews and Muslims) were still outside the church and condemned to hell. All the church needed to do, drawing on a peculiar interpretation of Matthew 24:14, was to wait until the end of times. This conviction was expressed by early church writers, medieval theologians, popes and councils, and summarised with the words: *Extra Ecclesiam nulla salus* ("outside the Church there is no salvation"). Such formulation went back to official declarations of the Catholic Church, such as the IV Council of Letran (1215), the Council of Vienne (1311–1312) and the Council of Florence (1431–1449).

Marzio introduced into this context a discussion about those who lived at the periphery of the oikoumene, including antipodeans, which enlarged the question of the salvation of contemporary pagans still not reached by the Gospel. The existence of antipodeans added weight to Marzio's view that the orthodox position that left all those unaware of the Gospel apart from salvation was untenable. Instead, Marzio defended that non-Christians who had not rejected the Gospel could be saved.

This topic appears in Marzio's chapter 5, devoted to the nature of faith: "What faith is and what it consists of, and that everyone is saved by their faith" (Marzio, *DIV*, title of chapter 5, 1948, 70).[21] After discussing those "who lived before the preaching of Christ," he affirmed that a "different reasoning" is appropriate for those living after Christ (Marzio, *DIV* 5, 1948, 81).[22] Marzio differentiated: (i) those reached by the Gospel and either converted (i.a), or those who remained in their original religions due to some negligence or cultural misunderstanding (i.b),

(ii) those not reached by the Gospel that have always lived according to their own religion, (iii) those who accept the Gospel in part (like the Turks who accept Christ as a prophet, but not as God) and finally (iv) those who reject the Gospel. While he considered the last ones condemned by "faith," "philosophy" and "nature," Marzio tried to save groups (i.b), (ii) and (iii). Marzio introduced group (ii) as "those born at the ends of the world…" (Marzio, *DIV* 5, 1948, 81–82).[23]

Marzio criticised the philosophers in chapter 13 on theological grounds:

> What is the difference between Christian theologians and philosophers on the acquisition of happiness, in which it is proved that according to our religion no one can gain happiness unless they are predestined by God, while the philosophers attribute it to our will.
>
> (Marzio, *DIV*, title of chapter 13, 1948, 83)[24]

Marzio defended that salvation is not due to works or sanctity of life, "but only from the election and predestination of the creator" (Marzio, *DIV* 13, 1948, 83).[25] He claimed that "God could predestine and thereby save someone from any faith or from any place" (Marzio, *DIV* 13, 1948, 86).[26] God could save those outside of the community of the believers, whether the community of faith be the Jewish people in the Old Testament times (Marzio gave the example of the non-Jew Job[27]) or the Church after Christ (giving the legendary example of the pagan Roman emperor Trajan[28]). However, Marzio considered Jesus's statement a caveat: "no one can enter the kingdom of God unless they are born of water and the Spirit" (John 3:5, New International Version (NIV), quoted in Marzio, *DIV* 13, 1948, 87).

How could Marzio solve this contradiction? Having accepted the validity of Jesus's statement, "These things must be true because they are said by the truth" (Marzio, *DIV* 13, 1948, 87),[29] he criticised the literalistic approach to these words, using Jerome: "[in disputes about Holy Scripture] words are not as necessary as the sense" (Marzio, *DIV* 13, 1948, 87)[30] and Paul: "the letter kills, but the Spirit gives life" (2 Corinthians 3:6, NIV, quoted in Marzio, *DIV* 13, 1948, 87). Marzio then drew on biblical texts where "everything" and "all" cannot be taken literally (1 Corinthians 9:22; Philippians 2:21 and Jeremiah 6:13), ending with Jesus own words: "All who have come before me are thieves and robbers" (John 10:8, NIV, quoted in Marzio, *DIV* 13, 1948, 87). Were words such as "all" taken literally, the wicked would include David, Moses, and the prophets up to John the Baptist. Therefore, Marzio concluded: "The Saviour's words are to be interpreted with some caution" (Marzio, *DIV* 13, 1948, 87).[31] For Marzio the baptism was only needed for those who had heard the Gospel, and he added:

> No Christian has ever instructed those who live in the antipodes, has ever gone to the most deserted lands, has ever spoken about Christ with headless monsters [*acephali*], with imbeciles [*blenni*], with many races of humans beings, [...].
>
> (Marzio, *DIV* 13, 1948, 87)[32]

Here both the antipodes and the traditional monstrous peoples like the *acephali* or the *blenni*[33] – living at the edges of the oikoumene – are linked together (as in medieval maps by Beatus and others). Marzio then defended, as in chapter 5, that those not reached by the Gospel had a chance to be saved provided they were of virtuous character, expressed love for their parents, home and children, and abided by the golden rule, as those were human duties "even without Christ saying it" (Marzio, *DIV* 13, 1948, 88).[34] Such a point took the discussion further than the salvation of those not knowing Christ to the very heart of Christian theology: what was the real soteriological value of Christ's sacrifice? (Miggiano, 2004, 191).

The expansion of geographical knowledge

Marzio was well versed in geographical matters, as he had worked together with Archbishop Johannes Vitéz to correct a manuscript of Manilius's *Astronomica,* which was completed in 1469[35] and published by Johann Müller (Regiomontanus) in 1473. Later, he studied the *Geography* of Ptolemy with Regiomontanus and participated in preparing its second edition (1477).[36]

Ptolemy defended that Africa extended to both sides of the Equator, and in the first half of the 1470s, this was confirmed when the Portuguese reached the Equator and entered what technically were the lands of the *antoeci* 'antipodes'.[37] Marzio was familiar with ancient geography and was also well informed of the new explorations. The Portuguese voyages to the south were known in Italy, as well as Venetian voyages in the Arctic regions (Miggiano, 1992, 106). In December of 1488, the Portuguese expedition led by Bartolomeu Dias returned after reaching the Indian Ocean, and writing the following year, 1489, Marzio already referred to *navigare ad Indos* (Marzio, *De doctrina promiscua* 3, 1548, 37; see Miggiano, 1998, 217).

Marzio's attack on Lactantius over the antipodeans

The combination of theoretical and practical evidence (ancient texts and contemporary accounts by sailors) probably explains why Marzio was so aggressive in his defence of the antipodeans. He devoted the entire chapter 14 of *DIV*[38] to dispel the objection that the antipodeans did not exist,[39] and addressed directly to Lactantius: "That it is not true that Christ's word has reached the whole Earth: where the arguments of Lactantius saying that the antipodes cannot be found are ridiculed and confuted" (Marzio, *DIV*, title of chapter 14, 1948, 89).[40]

Marzio started by recalling the "antipodes" – he noted that they are also called antichthones (*antictonas*) – among those that seem unreachable by the Gospel. He added that this name reminded him of the "sharpest and most eloquent philosopher and orator Lactantius Firmianus" (Marzio, *DIV* 14, 1948, 89),[41] whom Jerome – as Marzio remembered – considered a great confuter of the adversaries of Christianity (see p. 51). He then attacked with irony:

> [...]. That famous confuter, in his dispute against mathematicians [astronomers] on the topic of the antipodes and the roundness of the Earth, alleges

> some arguments that, if I had found them written in another book without a title, I would never have believed that they were from Lactantius. [...].
>
> (Marzio, *DIV* 14, 1948, 89)[42]

Marzio indicated that although the form of his argumentation was good, it did not make any sense. Consequentially, he considered Lactantius unworthy of the denomination of "orator." He contextualised this judgement referring to the request by Cicero and Quintilian that the orator should know mathematics (astronomy), something that, as Marzio pointed out, was not the case for Lactantius. The reference to Jerome previously and to Cicero here seems designed to lash out at the "Christian Cicero."

At this point, Marzio copied *Divine Institutes* 3.24.1, stressing that in the same chapter, Lactantius denied the very sphericity of the Earth "due to the high altitude of the mountains" (Marzio, *DIV* 14, 1948, 89).[43] Marzio then displayed a comprehensive defence of the Earth's sphericity. He argued that the altitude of the mountains is nothing compared with the size of the Earth, concluding that the highness of the mountains is not able to "deform the spherical figure" of the Earth (Marzio, *DIV* 14, 1948, 90).[44] Surprisingly, Lactantius had *never* used this argumentation to deny the sphericity of the Earth. Probably Marzio was referring to a passage in Lactantius arguing for the symmetry of the Earth's surface in case the Earth was spherical: "But if the earth also were round, it must necessarily happen that it should present the same appearance to all parts of the heaven; that is, that it should raise aloft mountains, extend plains, and have level seas" (Lactantius, *Divine Institutes* 3.24.8a, see Appendix, no. 12).

Marzio returned to the antipodes retorting that the words from Lactantius were "delirious ideas" (*deliramenta*) and linked his views with those of the populace (*Lactantio et plebeis*). He then explained the antipodes according to "all geographers and mathematicians (astronomers)" (Marzio, *DIV* 14, 1948, 90).[45] He argued that inhabitants of the poles would be antipodes among themselves as well as the inhabitants of the Equator around the Earth.

Nevertheless, Marzio did not defend the existence of isolated antipodes in uncommunicated areas of the Earth, as in the Cratesian model rejected by Augustine. Instead, he went further, emphatically stating that the Earth is habitable in its entirety (more in line with a Ptolemaic model): "For the Earth, being very extensive and gently sloping away, so that it seems to be more or less flat, accommodates comfortable dwelling-places everywhere" (Marzio, *DIV* 14, 1948, 91).[46] That is hardly surprising, not only because of his familiarity with Ptolemy, but also strategically. In effect, his theological argumentation would be stronger if there were antipodeans that had not been actually reached with the Gospel, but that were potentially reachable. This moral dilemma was not merely theoretical, but an excruciating contemporary problem. Therefore, after rejecting the objection of Lactantius that the concept of antipodes was absurd, Marzio denied the objection that the poles and the Equator could not be habitable. He did so not only with astronomical, astrological and climatic reasons against the claim of temperatures incompatible with life in these places. Moreover, Marzio opposed the empirical evidence from "reality itself," relying on "those who went there" and found "agreeable dwelling-places both beneath the equator and beneath the poles" (Marzio, *DIV* 14, in Gautier Dalché, 2013, 213).[47] He aligned himself with researchers (*historicos*) who "said the truth and were neither vain nor liars" (Gautier Dalché, 2013, 213)[48] in defending that the whole globe was habitable. Here we have a blend of ancient authors that supported that view, and contemporary geographical discoveries in the equatorial and arctic regions.

Marzio then resumed his attack, explaining Lactantius's denial of the antipodes by his "ignorance of mathematics [i.e. astronomy]" (Gautier Dalché, 2013, 214).[49] At this point, Marzio moved on to the theological arena and stated that the Gospel had not reached the antipodes nor "many islands that experience discovered much later" (Gautier Dalché, 2013, 214).[50] That led Marzio to tackle, as in chapter 13, biblical texts that could be used against his position, in particular, a text from Psalms quoted by Paul: "their voice goes out into all the earth, their words to the ends of the world" (Psalm 18/19:4/4-5 quoted in Romans 10:18, NIV). As in chapter 13, Marzio explained those texts as superlative, hyperbolic statements that did not need to be taken literally, and that doing so did not entail a disregard for Paul's authority but rather, an understanding of "his way of speaking" (Marzio, *DIV* 14, in Gautier Dalché, 2013, 214).[51] Marzio referred here to other texts where Paul expressed himself in a similar hyperbolic style: 1 Corinthians 9:22 and Philippians 2:21, already mentioned in chapter 13 (see above). Marzio stressed that these ways of superlative, hyperbolic, expression are common in the Bible,[52] also mentioning Luke 2:1 (Gautier Dalché, 2013, 214). That can be understood in the context of Marzio's view of the Bible as using "ordinary language," "general terms" (Marzio, *DIV* 13, 1948, 83, 87)[53] and "many things that closely resemble poetic tales" (Marzio, *DIV* 23, in Miggiano, 2013, 40, note 33).[54] This is a view with a long tradition in Christian theology – including Augustine –, which is generally known as the *principle of accommodation* of the Holy Spirit to popular views (scientific and others).[55] Marzio concluded that Paul's reference in Romans (mentioned above) to the Word of God reaching all humanity has to be understood as referring to "a large part" of humanity which was characteristic of the apostle's use of language (Marzio, *DIV* 14, in Gautier Dalché, 2013, 214).[56]

Marzio's sparing of Augustine

Marzio's long and complex discussion focused on Lactantius and never mentioned Augustine on this topic (although Marzio quoted from *City of God*, for example, in chapter 13, in discussing predestination). This could be viewed in the light of the already mentioned strategy in Rho of 'saving' Augustine (see above).

Although the existence of humans in the Southern Hemisphere was demonstrated at that time by the Portuguese, the Cratesian model could be discarded as these antipodes were not living on an isolated landmass. Augustine's attack against antipodeans was linked to the idea of an independent human population in an isolated continent not reachable on foot from the Mediterranean oikoumene.

In these circumstances, attacking the flat-earther and anti-antipodean Lactantius, while sparing Augustine, made sense. After Columbus's arrival to America, things changed dramatically.

The aftermath of Marzio's polemical criticisms

Due to the scandalous content of his book, Marzio was imprisoned in late 1477 by the Venetian Inquisition, tortured, judged and forced to retract and publicly recant his views. Eventually, he escaped being burned at the stake as an heretic by the Pope's intervention under the pressure of Marzio's influential royal friends. Not surprisingly, his book was never printed, but it survived in manuscript form. Some portions were printed only in the 20th century (see p. 74 and endnote 38). Interestingly, although the Turin copy bears the marginal notes of a censor (calling Marzio "stupid heretic"), chapter 14 is not commented, and its contents do not appear in a list of 12 errors of this book composed at the time.[57]

Once free, in 1478, Marzio dedicated a new revised copy to his protector Matthias Corvinus, King of Hungary (as was the original manuscript), now preserved at Vienna. In this copy, he also described his sufferings at the hands of the inquisition. Marzio composed other works, but was still haunted by the topics of *DIV* that he continued re-elaborating. He dedicated his last work, *De excellentibus*, finished around 1491–1492, to the French King Charles VIII, and it survives in manuscript form in the National Library of France (Paris).[58] One-third of its chapters derived from *DIV*, with chapter 9 of *De excellentibus* linking, right from the title, Lactantius's double denial of the sphericity of the Earth and the antipodeans.[59]

Although Marzio's criticisms of Lactantius were never printed,[60] manuscripts of *DIV* were preserved in Turin and Vienna.[61] They were read and copied by a select readership of librarians and erudite academics, who referred to and quoted them in the following centuries (Miggiano, 2004; 2005). In fact, the manuscripts in St Gallen and Paris, and a second copy in Vienna, were later copies from the 16th century.[62] Moreover, there was the other treatise, *De excellentibus*, available in Paris, which also criticised Lactantius. In Chapter 4, we will see examples of Marzio's influence on later criticisms of Lactantius.

The sustained influence of Augustine on the antipodeans debate in the Early Renaissance

The City of God *during the Renaissance*

In addition to theological developments, the rise of interest in ancient Greece and Rome since the 14th century led to an "Augustine renaissance in the later Middle Ages"[63] and to the first printed edition of *City of God* (*CG*) on 12 June 1467; it was printed in the same press as the first printed edition of the *Divine Institutes* (Lactantius, 1465; see endnote 5). The second one (1468) added the most relevant medieval commentary, by the English Dominicans Nicholas Trevet and Thomas Waleys, who completed it by 1337. Interestingly comments to book 16, actually by

Trevet, jumped from chapter 6 to 17, ignoring chapter 9.[64] Other editions followed until Johannes Amerbach published individual works by Augustine and, finally, the *Opera Omnia Augustini* (Augustine, 1505–1506). Although Erasmus promoted the figure of Jerome instead of Augustine as the model theologian since 1515, he edited Augustine's works in 1528–1529 (Oberman, 1981, 71–72).

A pictorial reception of Augustine's **City of God** *16.9*

By the Late Middle Ages, there were illustrated editions of *CG* (Smith, 1974). An example of the interest that the topic of the antipodes generated can be shown by the remarkable illustration of an early copy of a famous French translation of *CG* by Raoul de Praelles, commissioned by the French King Charles V in 1371 and finished in 1375 (Smith, 1974, 34–35). In the late 14th–15th centuries, this translation became very popular, in both manuscript and printed form (the last edition dates from 1531).

One manuscript copy of this book (c. 1478) displays a graphic illustration of *CG* 16.9 (Nantes, Bibliothèque Municipale, BM 181, MS fr. 8, fol. 163^{v}),[65] depicting Augustine saying to a group of listeners (on top of a spherical Earth): "If the earth is round, nevertheless it is not to be believed that the antipodes exist because not all the earth is inhabited" (Randles, 1994[2000], 76).[66] These words are not a literal quotation. One could speculate that this was a method of densifying Augustine's message and of adapting it to the limited space of graphic material. Nevertheless, in the image itself, there is not an empty space on the underside of the Earth; instead, five people appear there in shadows. Was their presence intended as a satirical depiction of the impossibility of the upside-down antipodeans, or a veiled affirmation of their existence after all? Could they be echoing the famous lines in Vergil, *Georgics* 1.247–248 referring to the darkness after dusk, when the Sun goes to the antipodean underside of the oikoumene? On this famed text by Vergil, see Hiatt (2008, 23–24).

Depicting antipodean people had a long tradition in other contexts. Cosmas, for instance, in his 6th-century *Christian Topography* satirically depicted antipodeans as upside-down people in order to ridicule their very existence, by drawing four human figures around the spherical Earth facing the reader (Kominko, 2013, 77–83, 258).[67] On the other hand, their potential existence is affirmed with a similar illustration in a 1245 French encyclopaedia, *L'Image du Monde*, by the priest Gautier de Metz. Two back-to-back travellers leaving the Northern Hemisphere in opposite directions found themselves facing each other in the antipodes.[68] According to historian Klaus Anselm Vogel, Metz's image of a journey around the Earth illustrated a didactic thought experiment. A real journey was generally believed to be impossible due to the geographical barriers, e.g. oceans, heat, huge distances, etc. (Vogel, 1995, 228–239). However, a century later, Mandeville considered it possible to travel around the Earth (see p. 58, endnote 110 in this chapter and Chapter 4, endnote 10). Images with four human figures in the four cardinal points around the Earth, nearly identical to that of Cosmas but without his satiric intention, appear in some medieval texts.[69] Finally, a more complex picture appears in an early 13th-century manuscript of Hildegard of Bingen's *Liber Divinorum Operum* (1163–1173/1174), where the seasons appear over the spherical surface of the Earth with seven human figures around it (Bingen, *Liber Divinorum Operum* 1.4, fol. 38^{r}).[70]

Augustine's influence on Early Renaissance authors on the topic of the antipodeans

Here we will analyse a variety of representative authors of the Early Renaissance, from different backgrounds and with different interests that discussed the problem of the antipodeans in light of Augustine's criticisms. They struggled to incorporate and at the same time limit such criticisms. The cosmography of Dante (c. 1265–1321) in his 'scientific' *Convivio* (*The Banquet*, 1304–1307, unfinished) and later in his long literary work *Divina Commedia* (*Divine Comedy*, c. 1308–1320), presents the so called mount of Purgatory in the Southern Hemisphere, otherwise covered by water, exactly in the antipodes of Jerusalem. Purgatory is located on the slopes of this mountain, with the Garden of Eden at its summit. This mountain in the Southern Hemisphere is mentioned in *Inferno* 26 and described in *Purgatorio* (Moretti, 1994, 270–271). This configuration solved a long debate about the location of these two inaccessible places of medieval geography: the antipodes and the Paradise, the former of pagan origin and the latter part of the Christian tradition (Hiatt, 2008, 126, 129; Scafi, 2006).

Dante, in his *Inferno* (the oldest part of the *Divine Comedy*, c. 1304–1309, published in 1314), hints at the absence of antipodean inhabitants: "... of what lies behind the sun, of a world without people" (Dante, *Inferno* 26.117, from Hiatt, 2008, 128). According to historian Alfred Hiatt: "This line was interpreted by Pietro Alighieri in the second version of his commentary on the *Commedia* as Dante's affirmation of Augustine's rejection of antipodal inhabitation" (Hiatt, 2008, note 145, 143). Pietro's commentary dates from the mid-14th century. From the second half of the 14th century is a commentary to the *Divine Comedy* by Benvenuto da Imola, also mentioned Augustine's denial of the antipodeans in *City of God* (Vogel, 1995, 200).

Dante explains the existence of an emerged landmass due to a shift of the Earth, which rushed to the north upon the fall of Lucifer, who emerged in the Southern Hemisphere through the centre of the Earth (Dante, *Inferno* 34.118–26). The idea here is that the pre-lapsarian world with the Paradise was situated in the Southern Hemisphere. After the fall, humankind moved to live on the opposite Hemisphere, cut off from their former residence, thereby matching Augustine's ideas as expressed in *De genesi ad litteram* 8.7.14, a passage noted by Pietro's commentary (Hiatt, 2008, 143–144, note 149).

Critical of Augustine was the Dominican missionary and Archbishop William Adam (?-c. 1341). After years of traveling through Persia, India and Ethiopia, he returned in 1323 to Europe and is considered the author of the *Directorium ad Passagium Faciendum*, a letter to Philip VI of France dating from 26 July 1332 to propose a crusade. At the beginning of the text, after affirming that he had travelled to 24º south, Adam extracted four conclusions, the third defending the antipodes with a language clearly opposed to Augustine: "that there is neither frivolous nor false to accept the existence of the Antipodes" (Adam, *Directorium*, first book, first part, second motive for the crusade, in Beazley, 1907, 822).[71]

A complex case of reception of Augustine's views on the antipodeans is Petrarch (1304–1374), a key figure in the emerging Renaissance. In a marginal note in his manuscript of Pliny's *Naturalis Historia* (Paris, Bibliothèque Nationale MS Lat. 6802), where Pliny mentioned the debate on the existence of antipodeans (see quotation in pp. 31–32), Petrarch wrote: "About these things there is no dispute in the works of Cicero and many others; Augustine argues against this robustly" (quoted in Hiatt, 2005, 13; 2008, 129, slightly adapted). A similar idea reappears in a note in his manuscript containing Servius's commentary of Vergil's *Aeneid*. In this text Vergil mentions the underworld and Servius seized the opportunity to discuss the antipodes (Servius's *Commentary* 6.532). At this point, Petrarch noted in the margin a long fragment from the *Commentary on the Dream of Scipio* by Macrobius (2.5.22–28), who follows Cicero's *Dream of Scipio* in defending the inhabited antipodes. Petrarch added at the end of this marginal note: "Augustine in his *City of God* argued against what they discuss" (*Codex Ambrosianus*, Biblioteca Ambrosiana, Milan, S.P. 10/27).[72] As Hiatt points out, here we can see a confrontation "between *Romanitas* and Christianity" (Hiatt, 2005, 11, 13; 2008, 130), which is also a confrontation between "antiquity and modernity" in the form of a controversy between Cicero and Augustine (Hiatt, 2005, 13; 2008, 129; On Cicero in Petrarch, see McLaughlin, 2015).

In addition to marginal notes, in six of Petrarch's letters the topic of the unreachable antipodes was deployed for diverse literary purposes (Hiatt, 2005, 4–5; 2008, 130–134). The zonal theory was explained in his epic poem *Africa*, in a replay of the story of the dream of Scipio to diminish the importance of human glory.[73] Similarly, it also appears in his philosophical dialogue *Secretum*, in a conversation between Augustine and Franciscus, i.e. Petrarch himself. In book 3 of the *Secretum* (started in 1347 and completed in 1353; Bouloux, 2002, 42), 'Augustine' criticises Franciscus's thirst for human fame, pointing to human limitations in line with the dream of Scipio. Franciscus rejects what he considers "trite old fables of the philosophers." 'Augustine', in turn, reproaches him for saying so with a series of rhetorical questions. Upset, he asks Franciscus whether the idea of the limits of the habitable portion of the Earth's surface is "merely an *old fable*," reminding him of the zonal theory (Petrarch, *Secretum* 3.16.3, 2015[2002], 117). Then, 'Augustine' mentions the inaccessibility of the antipodes, literally the place "under your feet" (*sub pedibus*, Petrarca, 1992, 268). He also refers to his own *City of God*: "my conclusion is expressed in *The City of God*, which you must have read" (Petrarch, *Secretum* 3.16.3, 2015[2002], 118). Finally 'Augustine' points to Franciscus's/Petrarch's own description of the world (*orbis*) as a "tiny island" (*insula parva*) in his *Africa* (Petrarch, *Secretum* 3.16.3, 2015[2002], 118).

Moving from Italy to France, Nicole Oresme (1320/5–1382), an influential scholar and later Bishop of Lisieux (1377), showed the traditional theological concern derived from Augustine over the antipodeans in his *Traité de la sphère*, written before 1377 (Oresme, *Traité de la sphère* 30; the relevant text was transcribed in Vogel, 1995, 223, notes 46–48). More influential was the *Imago Mundi* (c. 1410, printed in 1480–1483; read and annotated by Columbus), composed by Pierre d'Ailly (1351–1420), who became Archbishop of Cambrai in 1411. In chapter 7, in a text derived from Oresme's *Traité de la sphère*, Ailly mentioned the problem raised by Augustine regarding the lack of communication with the antipodes:

> According to this opinion the population of this area [the antipodes] would not have known the preaching of Christ and the Apostles contrary to the sacred affirmation "and their word was heard throughout the universe" [This is a reference to Romans 10:18b, which quotes Psalm 18/19:4/4-5] But in his chapter 16 of the *City of God*, St Augustine condemns this theory.
> (Ailly, *Imago Mundi* 7; Latin-French edition by Buron, 1930, 198–201; my translation)

Ailly described this region as covered by waters and said that some authors considered the existence of antipodes a "fable" (*fabula* – in fact, a word used by Augustine!). Ailly supported such criticisms and contrasted fables with real "histories," a word used in the label of the figure seven at the beginning of this book (see p. 85). In so doing, Ailly affirmed that he preferred "to rely on experience and probable histories rather than the imagination" (Hiatt, 2008, 145; original text in Buron, 1930, 201–202), and engaged in a defence of the habitability of the Torrid Zone by the hand of Greek, Roman, late ancient and medieval (Latin and Muslim) authors. The emphasis on "experience" and "history" can be read as a response to the requirements demanded by Augustine thereby providing a means of removing the crucial motivations for his condemnation of antipodes – the torrid/oceanic barriers – (on Ailly, see Hiatt, 2008, 145–147).

Cardinal Fillastre's early 'solution' to Augustine's dilemma regarding the antipodeans

Under the influence of the recently 'rediscovered' *Geography* of Ptolemy,[74] Fillastre disagreed with the uninhabitability of the poles and the Torrid Zone. He also accepted the Ptolemaic view of the Earth and water as a single spherical planet, for which he also quoted Genesis 1:9–10. In this way, Fillastre interpreted this text differently from the supporters of the eccentric model. He, then, suggested the possibility of sailing west "through the part opposite to us, and in the opposite direction" (Fillastre, *Introductio* 15; translated by Hiatt, 2008, 158). This elimination of the equatorial and oceanic barriers (in line with Albert the Great, Bacon, Oresme and Ailly) could remove the objection of Augustine without discrediting him.

However, Fillastre would choose a simpler way to vindicate the antipodean terminology with knowledge already at hand. In p. 70, we saw that in his *Introductio* 7, Fillastre referred to Lactantius and Augustine on the problem of the antipodes and antipodeans. Then, in *Introductio* 16, he exploited Augustine's acceptance of the sphericity of the Earth and argued that true antipodes are found when comparing inhabitants of the extreme east in Asia and Western Europe of the known oikoumene (Fillastre, *Introductio* 16, Appendix, no. 8).

Fillastre affirmed that a spherical Earth inside spherical skies, according to Plato's *Phaedo* (probably 108e–109a), indicates a central direction. He considered this argument sufficient to respond to Lactantius's belief that if there were antipodes, they would have their feet upwards (Fillastre, *Introductio* 16, Appendix, no. 8). This poor counterargument, the sort of reasoning that Lactantius rejected, likely shows that Fillastre had more interest in Augustine than Lactantius.

The existence of west-east antipodeans in the known oikoumene had already been suggested by Hermann of Carinthia, and Ptolemy's *Geography* reinforced this position supporting an oikoumene over at least 180º of longitude. As after the return of Marco Polo (1295) and Niccolò de' Conti (1444) from their travels the knowledge of the Far East was enlarged in comparison to ancient times, it was easy to conclude that the longitude could be even larger than previously thought. Although Fillastre did not mention these authors, he went on in following sections to stress the continuity of lands, including in *Introductio* 17–18 a detailed description of the populations at both ends of the single continuous oikoumene, thereby resolving Augustine's question regarding humankind connection with Adam and Noah with migrations by land (see Appendix, no. 8).

The antipodes in Pulci's Morgante

Luigi Pulci (1432–1484) composed the long chivalric epic poem *Morgante*, commissioned by Lucrezia de Medici in the early 1460s. He completed its first part in 1479 (23 cantos) and its final part (cantos 24 to 28) in 1483 (on the complex history of its first editions, see Wilkins, 1951). The poem revolved around the knights of Charlemagne, with the antipodes appearing in the second part (on this poem and its references to the antipodes, see Moretti, 1994, 271–275).

One of the characters, the demon Astarotte, rejected the lack of communication between hemispheres, declaring the story of Hercules's Pillars to be a "long, feeble tale," "wrong" and the fruit of "those days so ignorant" (Pulci, *Morgante* 25.228–229, 2000, 1037–1038).[75] In contrast, the existence of antipodeans – who worship the Sun, Jupiter and Mars – was affirmed by Astarotte (Pulci, *Morgante* 25.228–231, 2000, 1037–1039). The knight Rinaldo immediately posed two pertinent questions: the pagan antipodeans relationship with Adam and whether or not such idolaters could be saved. The demon answered him that pagans, in contrast to Jews and Muslims – who rejected Christianity –, could find a way of salvation by practising their own religion. Although all salvation was by Christ, the door was always open for those that were genuinely ignorant of Christ (Pulci, *Morgante*, 25.232–238, 2000, 1039–1041).

Finally, there are again some geographical considerations that defended not only the existence of land in the antipodes, but also the assumption that these lands were populated: "therefore He [God] did not without people make / that world where downward go those starlit rings, / planets and signs and many lovely things" (Pulci, *Morgante*, 25.238, 2000, 1041).[76] Interestingly, this plenitude principle was used at the very same time by Columbus in proposing an oceanic crossing in Portugal (see p. 91 and endnotes 101 and 102).

It is very interesting to compare this theological argumentation with Marzio's conception less than a decade before (see pp. 74–76). On the one hand, it expresses a similar defence of religious universalism as in Marzio (and Mandeville, see p. 75 and endnote 27). At the same time, however, the idea of pagans unaware of Christ's message could trigger a missionary zeal. In fact, Rinaldo finally departs to the other hemisphere as an adventurer and missionary (Pulci, *Morgante*, 28.33–34, 2000, 1231–1232), something that a few years later characterised the motives of Columbus. In a dismissive interpretation of Pulci's words as a prophetic intimation of contemporary explorations, Hiatt points out that it is the Roman Empire and its Christianisation what are envisioned here, transposed to the antipodes (Hiatt, 2008, 172–174).

The antipodes debate in Early Renaissance mapping

As in previous centuries, the problems surrounding the antipodes were reflected in the mapping of this period. An example is found in the *Epilogus Mappe Mundi*, a collection of eight figures accompanying Ailly's *Imago Mundi* (c. 1410). The figure seven from a 15th-century manuscript (Brussels, Bibliothèque Royale, MS 21198–204, fol. 4^{r}), is described by Hiatt in these terms: "In this image Ailly combined elements of three genres of medieval map: the T-O map, the zonal map and the *climata* map" (Hiatt, 2008, 145). Regarding the antipodes, while the Southern Hemisphere was left blank, a note crossed vertically from the first *climata*, across the Equator and extending near the Antarctic circle: "Before the *climata* towards the equator and beyond it are many habitations. As it is learnt from authoritative histories" (Hiatt, 2008, 147).[77] On the other hand, an Augustine-driven denial of antipodes was added in the Renaissance to a zonal map from a 12th-century *Commentary* on Macrobius. The original map contains a description placed in the Southern Hemisphere: "temperate zone of the antoikoi." A reader, apparently of the 15th-century, added: "that there may be people living here Augustine rejects, Book 16, *De civitate dei*" (Paris, BNF lat. 6622, fol. 59^{v}).[78]

However, mapping during the 14th and 15th centuries was a mixture of the old and the new, the traditional T-O maps and *mappaemundi* as well as the innovative maps derived from the 're-discovered' Ptolemy's *Geography* and the new portolan charts (Woodward, 2013, 597–598).[79] As the 15th century progressed, the new geographical discoveries led mapmakers to expand Ptolemaic maps and discard the Ptolemaic closed Indian Ocean, as in Fra Mauro's map (1448–1450).[80] So too, with Portuguese hard geographical data, in maps by Henricus Martellus Germanus c. 1489 and c. 1491 (Duzer, 2019). In this way, the new information covered the map, displacing speculations about imaginary monsters and antipodeans, in a process that consequently ended up drawing the real South African 'antipodes'.

An exception – or rather a transitional situation – is the map in *La salade* of Antoine de la Sale (1386–1462). This text, composed between 1437 and 1444, was printed in 1521 and 1527, with a circular map where Africa extends more to the south of the Equator than to the north, with the extreme south labelled as *terra Incognita et deserta*. In parallel, Asia extends its Far East side with a huge peninsula that almost reach the south Pole. The subequatorial sea in between is labelled as *Mare Antipodes et Incognitum* (La Sale, 1521, between fols 28^v and 29^r, mistakenly marked as fol. 39).[81]

In addition to maps, we have the globe which Martin Behaim (1459–1507) made around the pivotal year of 1492 (1491–1493/94). This globe exemplifies geographical knowledge on the eve of the first journey of Columbus across the Atlantic. Behaim referred in the legends to Ptolemy, Marco Polo, Mandeville, etc., and recent discoveries, including Portuguese explorations as he himself had travelled under Diego Cão (1485–1486). His sphere showed only three continents on one side of the globe spread on both sides of the Equator, and an open Indian Ocean. The remaining surface was water (Vogel, 1995, 384–392; on 15th-century globes, Vogel, 1995, 363–367).

Although no other 15th-century globes have survived, we have pictures of globes giving a similar view, but leaving unclear the south end of Africa and not depicting their backside. Examples are found in Piero del Pollaiolo's *Justice* (c. 1470),[82] Donato Bramante's *Heraclitus and Democritus* (c. 1486),[83] Marco d'Oggiono's *Christ Blessing* (c. 1500)[84] and Rafael Sanzio's *School of Athens* (1509–1512).[85]

As Hiatt notes, the southern antipodes became "an extension of the ecumene. [...] areas that invite, rather than repel, the passage" (Hiatt, 2008, 172). This solved Augustine's objection opening the way south. This solution allowed accepting antipodes without contradicting Augustine, as his caveat was converted into reality, and the lands of the Southern Hemisphere in Africa were shown to be connected by land with the Mediterranean oikoumene. In Chapter 4, we will see how America would challenge that solution in the 16th century.

The scholastic eccentric model of the Earth

In the previous chapter (pp. 58–60), we noted the debates produced by Aristotle's incomplete explanation regarding the configuration of our planet and the distribution of elements within it. The poet Dante, in a technical text, *Quaestio de aqua et terra* (1320), also known as *De situ et forma aque et terre* (Boffito, 1902 and 1903), rejected both the idea that the water and earth spheres were eccentric and also the possibility that, being concentric, the water could be higher than the earth in one particular place (in the mode of a water 'protuberance'). For Dante, there was instead a protrusion of the earth in the Northern Hemisphere over the water, explained by an 'attraction' of the stars[86] due to God's command (and there, in his chapter 21, he conveniently quoted Genesis 1:9). Dante even provided a figure in chapter 19 very similar to our Figure 2.8C. This text was virtually unknown until it appeared in print in Venice in 1508 (Benavent in Alighieri and Boccaccio, 2016, 18). By that time, Dante's astrological explanation was out of favour and, in addition, there was already sufficient evidence for more than one emerged landmass; therefore, Dante's geographical hunches had minimal impact (Iommi, 2014).

Despite active opposition,[87] the popularisation of the eccentric model was largely accomplished during the 14th–15th century. For all authors, the emerged oikoumene was considered quite large, at least 180° (following Ptolemy) and sufficient to conclude that the Iberian Peninsula and China were on opposite sides of the Northern Hemisphere. In addition, all agreed that the remaining surface of the planet (typically three quarters or at least the two-quarters of the Southern Hemisphere) was covered by water making the existence of the antipodes/antipodeans impossible. The critical difference of the eccentric model – in relation to the terraqueous globe of Crates, Ptolemy and several medieval authors – was that water and earth formed two different spheres with separate centres.

In *Quaestiones super libris quattuor de caelo et mundo (Questions on the Four Books of the* De caelo et mundo, c. 1330),[88] Jean Buridan (c. 1300–1358) considered the Torrid Zone as uninhabitable. This was surprising, as Buridan quoted Avicenna (c. 980–1037), who defended the opposite view, which he did not endorse. Buridan also rejected Crates's views on the antipodes, defending a single emerged oikoumene, as no other had been discovered. Nevertheless, in an eternal universe (such as Aristotle's), any emerged land would have been already eroded. He discussed the providentialist view that the dry land would have emerged on God's command. However, there was no natural explanation for the eccentricity. So, too, the emerged area would be eroded over time (the same objection directed at Aristotle). Therefore, Buridan defended a third possibility, that the centre of gravity and the centre of magnitude of the Earth do not coincide because the earth is not homogeneous (as is the sphere of water). The emerged portions are exposed to the sun's heat and the mixture with air, becoming lighter than the parts underwater. As erosion would reduce the size of the dry portion of the earth, it would become lighter and shift the centre of gravity. This counterbalanced erosion so that emerged lands would remain afloat. In this way, the emerged portion was cyclically maintained by a natural balancing mechanism.[89]

Albert of Saxony (c. 1316–1390) followed these ideas in *Quaestiones in libros de caelo et mundo*, equating the emerged Earth with the tripartite oikoumene (Randles, 1994[2000], 33–34). In his *octo libros physicorum*, Albert treated the earth and water as an "aggregate" with a single centre of gravity; however, this did not mean that he accepted a terraqueous model (Grant, 1984, 25). In spite of that, Buridan's Parisian school was very influential in the 15th century (Vogel, 1995, 281–296).

Oresme recognised in *Le livre du ciel et du monde* (1370–1377) the problem of explaining the existence of dry earth above the water (Oresme, *Le Livre du ciel et du monde* 2.11, fol. 101ʳ–104ʳ, 1968, 394–405), which he justified using the eccentric model aided by a diagram of a sphere of water where a small top section is labelled as *terre* (Oresme, *Le Livre du ciel et du monde*, 2.31, fol. 153ᵛ, 1968, 569 [Figure 40]).[90] He also included the providentialist explanation: "God and nature have ordained that the earth should be thus exposed so that people and animals can live there" (Oresme, *Le Livre du ciel et du monde* 2.31, fol. 153ᵛ, 1968, 568–569, with slight modification). At the same time, Oresme mentioned with approval "some who think" that Spain and India "are close together and that the sea adjoining the two is one and the same" (Oresme, *Le Livre du ciel et du monde* 2.31, fol. 152ᵛ, 1968, 566–567). Oresme's oikoumene was, then, larger than 180°. These ideas reappeared in his *Traité de la sphère* (before 1377; Vogel, 1995, 219–225).

Ailly, in *Questions on the 'Sphere' of Sacrobosco* (first published as part of Ciruelo, 1498), adopted Albert of Saxony's 'aggregate' of earth and water. However, both spheres were still considered separate (Grant, 1984, 26). The eccentric model was also explained in his influential *Imago Mundi* (c. 1410, printed in 1480–1483), which included the discussion in chapter 8 on the quantity of the inhabitable earth. After collecting different opinions on the topic,[91] in chapter 49 he indicated that the "common opinion" was that three-quarters of the earth were underwater even though "some philosophers" held to the opposite opinion (Ailly, 1930, vol. 2, 423–424 [Latin-French]). In line with the minority opinion, Ailly seems to have thought that the west-east size of the oikoumene was larger than the Ptolemaic 180°. This was a position he took from Oresme's *Treatise on the sphere* and, to a larger extent, from Bacon's *Opus Maius*. Nevertheless, Ailly recognised that there were no data to know the real size of the oikoumene (Vogel, 1995, 255–263; see below Chapter 4).

Figure 4 of Ailly's *Epilogus Mappe Mundi* of the *Imago Mundi* showed an eccentric distribution of water and earth,[92] although in the text of the *Epilogus* he vigorously defended the idea of a greater oikoumene against the majority of philosophers. Consequently, he rejected that the waters covered the half of the earth opposite to the oikoumene, based on the authority of Aristotle, Averroes, Seneca, Pliny, 4 Esdras and Jerome, following Bacon (Vogel, 1995, 258–260). Although Ailly has often been accused of plagiarism conflating opposite cosmographical views without realising their incompatibility,[93] Vogel considers that Ailly could not solve the contradiction with the data available at his time (Vogel, 1995, 256–263).

The most widely used biblical commentary in the Renaissance: *Postillae perpetuae in universam S. Scripturam* (c. 1322–1331) was written by Nicholas of Lyra in c. 1270–1349. It was first printed in Rome in 1471–1472, and had over a hundred later editions. In discussing the third day of creation, he affirmed that God created cavities in the earth where part of the water that initially covered the earth retreated to uncover dry land (on this idea, see Chapter 2, endnote 46). For him, the amount of water did not exceed much more than that of the earth (Vogel, 1995, 264–265). A century later, Paul of Burgos (c. 1350–1435) wrote his *Additiones* (1429) to the *Postillae* (often, wrongly, attributed to Lyra). He defended the eccentric model in opposition to Lyra and mentioned that astronomers already accepted eccentric planetary orbits. Burgos marshalled biblical texts to substantiate this view: Genesis 1:9, Proverbs 8:27 and also texts such as Psalm 68/69:2/3, 103/104:6–9 and 23/24:2 which state that the sea is very deep and higher than the earth. This was understood as the large depth of the ocean on the opposite side of the emerged oikoumene, and that earth had been founded over the sea (imagining in this model the sphere of earth 'floating' over that of water). Burgos criticised naturalistic attempts to explain the disruption of earth and water, such as Buridan's, which could only explain the current eccentric distribution, but not its origin that could have only been the divine command. A drawing illustrated the movement of the waters in Genesis to uncover the dry land (Figure 3.1).[94]

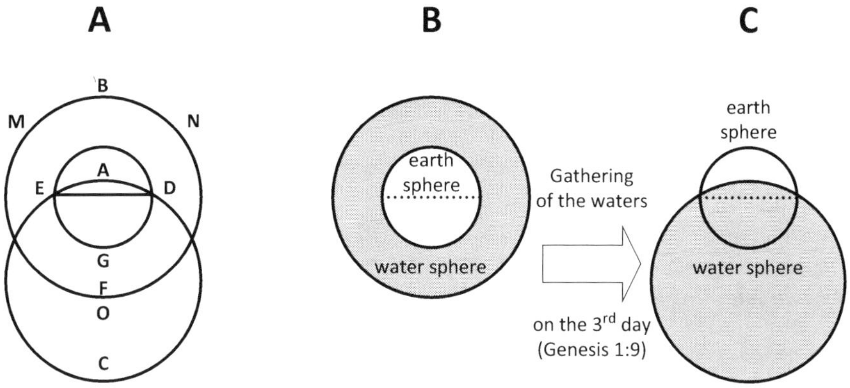

Figure 3.1 (A) The gathering of the waters on the third day of creation in the *Additions* of Paul of Burgos to the *Postillae* of Lyra.[95] To understand the dynamic process embedded in this image, the diagram has been divided into two, in line with the proposal of Grant (1984, 24, note 74). In (B) we can see the concentric arrangement of earth and water that is rearranged under God's command to uncover a portion of the earth sphere (oikoumene) (C) (B and C from Felipe (2012, 298, Figure 4), used with permission).

In 1458 Aeneas Sylvius Piccolomini (1405–1464) became Pope Pius II. He was a scholar who displayed an encyclopaedic approach in *Historia rerum ubique gestarum*, published in 1477 and read by Columbus. He mentioned the ideas of Crates, the scholastic eccentric Earth, and Ptolemy, along with the legend of the Pillars of Hercules and references to the encircling ocean of Homer. Piccolomini does not seem to favour any one view in particular, although he tends to approve the scholastic eccentric Earth in connection with the teleological explanation for the emerged oikoumene: "the Divine Will has decided this for the sake of people, an opinion worthy of approval by a Christian" (Piccolomini, 1477, fol. a.i^{v}).[96]

The Servite monk and teacher of philosophy and theology, Gasparino Borro (c. 1430–1498), discussed the heavens in the chapter *De cœli rotunditate* of his *Commentum electum et pretiosum super tractatum sphaerae mundi* (1494, printed c. 1500). First, he quoted Lactantius, *Divine Institutes* 3.24.4–8, "who denies the sphericity of the heaven" (*qui negat cælum eſſe rotundum*). Then, he mentioned the "other opinion of the philosophers, astrologers and theologians, who all agree that the heaven is spherical" (this and the previous quotation are from Borro, c. 1500, fol. c.iii^{r}),[97] with which Borro sided. He then moved to the topic of the antipodes, where he again discussed two opinions: first Macrobius's acceptance of the antipodes in the *Dream of Scipio* (Borro, c. 1500, fol. c.iii^{r}), and then Lactantius, "who denies the antipodes" (*qui negat antipodas*), quoting *Divine Institutes* 3.24.1 (Borro, c. 1500, fol. c.iii^{v}). In the ensuing discussion, Borro agreed "philosophically and theologically" (*approbo tam philoſophyce q̈ theologice*) with Lactantius's denial of the antipodes (Borro, c. 1500, fol. c.iii^{v}). From a philosophical point of view, Borro used Buridan's model of eccentric spheres; from a theological perspective, he discussed the Augustinian theological problem of the antipodeans's Adamic descent in detail, surprisingly without mentioning Augustine (Borro, c. 1500, fol. c.iiii^r; translated by Randles, 1994[2000], 40). After these words, in the same page, Borro begins directly the next chapter, "That the Earth is round" (*Quod terra fit rotunda*). Disconcertingly, historian William Graham Lister Randles considered Borro siding here "with the strict anti-Aristotelian Biblical orthodoxy of Lactantius" (Randles, 1994[2000], 40). As can be seen in Borro's text, this is rather Augustine's theological argumentation.

Visual representation also contributed to the popularity of the eccentric model. In particular, the image of a the small sphere of earth – with the structure of a zonal map – placed eccentrically inside a large sphere of water, both spheres touching on one side, and both surrounded by the spheres of air and fire. This representation was printed in a good number of books from the late 15th century onwards, such as *De Sphaera* of Sacrobosco (Venice, 1485, 1488, 1490, 1499 and others, see Figure 3.2).[98] Other models, rather more similar to Paul of Burgos's drawing where the eccentric spheres of earth and water cross each other in what seems a lateral view (Figure 3.1C), can be found in an edition of *De Sphaera* of Sacrobosco from Leipzig (1487 and later editions) and slightly different, with a larger earth, in another edition from Leipzig (1495 and later editions).

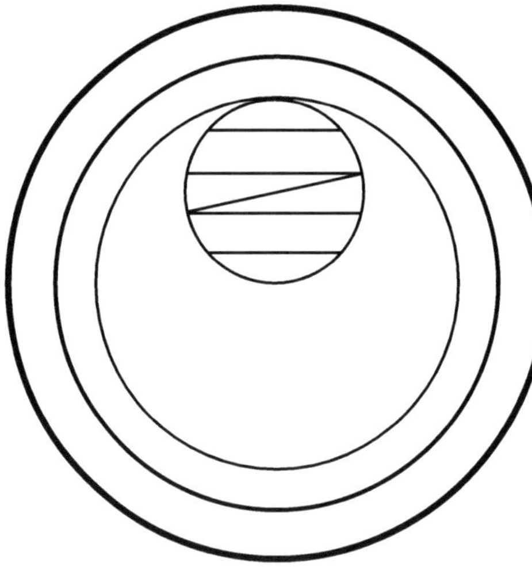

Figure 3.2 The eccentric model of the earth and water spheres in the *De Sphaera* of Sacrobosco (1485).[99] The original image is filled with flames in the external circle (representing fire), with clouds in the intermedium circle (representing air), and with horizontal waves in the central circle (representing water). The eccentric sphere is the element earth with the schematic lines of the zonal map and the ecliptic.

The antipodes, Augustine and Columbus at the meeting of Salamanca (1486–1487)

The meeting of the experts of the Catholic Monarchs with Christopher Columbus (c. 1451–1506) in Salamanca to discuss his plan to arrive in India (Far East Asia), crossing the Atlantic to the west, is probably the most famous inspiration for the flat error. That is due to Irving's portrayal of that meeting as a clerical 'council' that seemed to have been inspired in Galileo's trial – a trial which has also been portrayed with its own deformed mythology at the service of the science and Christianity conflict.[100]

We can understand Columbus's proposal to reach Asia, crossing the Atlantic, from the perspective of the debate between the terraqueous and the eccentric models of the Earth, Columbus favouring the terraqueous view. This allows us to understand the opposition he had already faced in Portugal (1483–1484), prior to moving to Castile (in modern Spain). According to João de Barros, the experts of King D. João II heard Columbus affirming that:

> [...] the West could harbour other islands and lands, because nature could not be so disordered in the composition of the Universal Orb as to give more space to the water element than to the uncovered land for the livelihood and breeding of animals.
>
> (Barros, *Da Asia* 1.3.11, 1552, vol. 1, fol. 37^{r})[101]

These last words seem to be an attempt by Columbus to deploy a modified version of the teleological scholastic argumentation (an approach traditionally used in the context of the eccentric model), combined with a plenitude principle.[102] Barros also considered reasons for Columbus's ideas, and for his hope of finding Marco Polo's Island of Cipango (Japan), the recent discoveries of the Azores and the explorations of Africa: "For as in the lifetime of the Infante Dom Henrique the Islands Terceiras [Azores] were discovered, and so much of the land of Africa [...]" (Barros, *Da Asia* 1.3.11, 1552, vol. 1, fol. 37^{r}).[103] However, Portuguese experts scoffed at what they considered "imaginations" ("jmaginações" (Barros, *Da Asia* 1.3.11, 1552, vol. 1, fol. 37^{r})).[104] As with Zaccaria Lilio, ten years later, they did not consider the existence of islands far off the Old World, such as the Azores, as a challenge to their scholastic model (Randles, 1993b[2000], 20; on Lilio, see pp. 119–121). To complicate matters, it should be remembered that there were also many islands believed to exist in the Atlantic. Columbus could expect to find these legendary geographical locations, even though their existence was regarded by others with scepticism (Brendan's island, Antilla/Isle of Seven Cities, Island of Brasil, etc.). Real and imaginary islands in the Atlantic and Indian Oceans were mixed up in maps and books at the time (Flint, 1992b). In fact, Cormack noted that: "we can see Columbus as the last of the great Medieval travellers, following d'Ailly who relied heavily on Roger Bacon, rather than as the first of the great renaissance explorers" (Cormack, 1994, 369).

Once he moved to Castile, we have little information about Columbus's meeting with the royal experts (*Junta* in Spanish). The *Junta* studied his project in

Salamanca in the winter of 1486–1487, but it was not meant to judge Columbus.[105] Though many of the academics and court advisors were clerics, the *Junta* had nothing to do with the Inquisition,[106] and it was not concerned with the sphericity of the Earth. Regarding the early Christians quoted against Columbus in Irving, only Augustine was mentioned. The first reference I found to a long list of early Christians being used against Columbus is Castellanos (1589) (see p. 131).

The report by Columbus's son, Ferdinand (*Vida del Almirante*, composed c. 1528, preserved in an Italian translation of 1571), lists six grounds for rejecting the project (see my translation in Appendix, no. 6, from where I quote below).[107] The *first* reason was just a claim of authority. The *second* had to do with ancient geographical ideas about the impossibility to cross a perilous – might be even infinite – ocean, mentioned by Seneca the Elder (c. 54 BC–c. 39 AD) in *Suasoria* 1. The *third* reason concerned the medieval eccentric model which linked the distance from Spain to India across the Atlantic to the west with the distance by land to the east.[108] The *fourth* point that led to the rejection of the project is surprising given that the experts apparently referred to Ptolemy (who held a radically different view, the terraqueous model)[109] to defend a strange incorrect view of gravity away from what was well known by that time as being a central force.[110] The *sixth* ground for rejection drew on another ancient/medieval model, the five zones (already mentioned in the third point), to defend the impossibility of the project, probably because Columbus's proposal was seen as coming dangerously close to the Torrid Zone or aiming to cross it into unknown southern antipodes.[111] This last idea was, in fact, the *fifth* ground for rejection, in this case not due to geographical but theological reasons: the 'doubts' of Augustine in *City of God*.

Columbus's Far East Asian target was hardly a new unknown antipodean continent, it was just "merely long unvisited" (Fernández-Armesto, 1991, 96). Therefore, it is interesting that Ferdinand states that the *Junta* mentioned that Augustine "considers it impossible that there should be Antipodes *and* that one can pass from one Hemisphere to the other" [my italics]. This last statement indicates that Augustine was used to reinforce the previous argumentation that pointed to the impossibility of crossing from Spain to Japan by sea. Indeed, Augustine had not just denied the existence of the antipodes, but considered the entire Western Hemisphere covered by waters (see p. 46). Of course, by that time, the Torrid Zone had already been crossed by the Portuguese. However, these trips were probably too recent and still too close to the old oikoumene to be taken into account by these experts.[112] As we will see in the next chapter, this was a vital point of the next *Junta* in 1491 at Santa Fe (Manzano, 1989, 351–360; see p. 107).

The varied objections to Columbus were precisely what were to be expected during times marked by confusion regarding geography, without the need to imagine flat-earthers. However, what the reports of these *Juntas* – written many decades later by enthusiast supporters of Columbus who already despised the expertise of the 15th century – hide carefully is that the keystone of Columbus's project was a profound mistake or bias in reducing the oceanic distance from Europe to Japan. A point which any expert in cosmography knew was wrong, whatever old-fashioned views they may have held regarding the structure of the planet.[113] Without the existence of the American antipodean (perioecian) continent, Columbus would have never returned. Precisely, as soon as he did return, a debate ensued to interpret what these lands were, as we will see in Chapter 4: Asia/India, new islands or even the fabled antipodes.

The confusion can be seen in a geographical book by Spanish grammarian Antonio de Nebrija (1444–1522): *In Cosmographiae libros introductorium* (Salamanca, c. 1503). An initial version – *Isagogicum Cosmographiae* – might be from late 1480s (Flórez et al., 1990, 231). The book by Flórez *et al.* contains a Latin-Spanish edition of a 1533 Paris reedition of *In Cosmographiae libros introductorium* (Flórez et al., 1990, 235–281; see quotations in Randles, 1994[2000], 41). The book opened its first chapter with the Ptolemaic definition of the terraqueous sphere. However, in the same paragraph, Nebrija stated: "the *orbis terrarum* known to us, emerges above the water like an island" (Nebrija, c. 1503, fol. ai^{v}–aii^{r}; Flórez et al., 1990, 240–241).[114] Although Randles considers this as another example of not noticing the clash of two contradictory theories, Nebrija did not clearly refer here to the scholastic view of the two eccentric spheres, which was not mentioned. Therefore, it could be understood as an attempt to reconcile the Ptolemaic view with the idea of a single emerged oikoumene, although Nebrija also mentioned the recently discovered "antichthones" islands in the Caribbean:

> Of the remaining part of our hemisphere and of the region opposite inhabited by the antichthones, nothing certain has been reported to our ancestors. But such is the boldness of the men of our time, that it will soon occur that they will bring us a description of this land, either that it is made up of islands, or that it is a mainland (*continentis*), of which a large part of its coastline has been revealed by our navigators, principally the part situated opposite the recently discovered islands (Hispaniola and Isabella) [=Santo Domingo/Haiti and Cuba] and the other neighbouring ones.
>
> (Nebrija, c. 1503, fol. aii^{v}; Flórez et al., 1990, 242–243; translated by Randles, 1994[2000], 56)

A similar passing reference to newly discovered lands in the extreme west and east is found in Yerva (1498, fol. aai^{v}; Flórez et al., 1990, 128–129). Indeed, Nebrija, who in general followed Ptolemy, was very open about the clash among scholars on other topics. One such example was the closed/open nature of the Indian Ocean, indicating that Ptolemy was wrong, as shown by the Portuguese (Nebrija, c. 1503, fol. aii^{r}; Flórez et al., 1990, 242–243). However, while Nebrija was writing, the discoveries in America were growing from a few islands to a whole new continent. In 1509, in the *Exhortatio ad lectorem* of a book which was published much later, Nebrija realised that the exploration of the whole globe was about to be completed, although still considering the new islands close to India (Nebrija, 1550, fol. Biii^{r}; quoted by Gil, 1989[2017], vol. 1, 285). It is to this topic that we turn our attention in Chapter 4.

Conclusions

The 14th and 15th centuries marked a turning point in the reception of the passages dealing with the antipodes in Lactantius and Augustine that would set the pattern for the following centuries. Lactantius, initially admired in the Renaissance, started to be seen in a less favourable light during the 15th century, with his rejection of the sphericity of the earth and the antipodeans becoming part of these criticisms. Augustine retained his overwhelming authority to the end of the 15th century. Even if the existence of the antipodes and antipodeans became increasingly accepted from the 14th century onwards, his position stood as a warning for good Christians of a line that was better not to cross.

However, new views finally emerged, derived from fresh news brought by late medieval travellers to the Far East, such as Marco Polo (1254–1324) and Niccolò de' Conti (c. 1395–1469), or from travellers to the south east, below the Equator, such as William Adam (?–c. 1341). Additional reasons for the expansion of geographical views included the study of the *Geography* of Ptolemy and other ancient texts, as well as the Portuguese exploration of the African west coast. That meant that the oikoumene-island had to accommodate an increasingly larger west-east oikoumene stretching across 180° of longitude (or even more). The expanding views of the oikoumene prompted people to speak of 'internal' antipodeans (Spain and China, a concept already present in Hermann of Carinthia in the 12th century). This idea allowed accepting the existence of antipodes, but only on the Northern Hemisphere [*perioeci*], in a single landmass, and without the need to cross either immense oceans or the fiery Equator. Moreover, such proposals were not in conflict with Augustine. According to the historian of geography Patrick Gautier Dalché, this was a common position in most universities in the 15th century (Gautier Dalché, 2013, 217). As the Portuguese continued sailing further south, the distance between the north of Europe[115] and the south of Africa increased and allowed the introduction of other internal antipodes [*antoeci*] by the end of the 15th century, a topic we will find in Lilio (see pp. 119–121). This very line of reasoning was anticipated, to a certain degree, a century and a half before, by William Adam.

Lactantius's clear, true or false argumentation permitted either only an outright rejection of his view or to ignore it. However, the more nuanced views of Augustine gave his ideas the flexibility to accommodate them with the new eccentric cosmographical model and with novel travel information. In fact, the eccentric cosmography could still be adapted to a large emerged oikoumene-island as well as to Augustine's concerns about the antipodeans, by placing water on the Western Hemisphere.

As mentioned in Chapter 1, the theory of a flat Earth was not supported by any serious author in the Middle Ages and the Renaissance (with the possible exception of Aethicus Ister, second quarter of the 8th century, unless he was writing a parody, see Aethicus, 2011). Furthermore, the flat Earth theory was hardly attacked until well into the 15th century, William of Conches could be considered one of the few exceptions (see p. 56). In this context, it is interesting to see that a staunch critic of Lactantius, such as Rho, did not even raise a concern regarding the flat Earth. Rather Rho centred his attacks exclusively on the debate over the antipodes. It is almost as if the anti-sphericity references in Lactantius were only seen as part of his overall argumentation against the antipodes. This perception changed during the 15th century, when Lactantius's anti-sphericity stance became increasingly embarrassing, which can be observed even before Rho. For instance, in Fillastre, and then more fully in the late 15th century with Marzio, who did attack Lactantius for his denial of the antipodeans *and* the earth's sphericity.

However, up until c. 1500, the authority of the key ancient authors was still intact. In particular the authority of Aristotle in philosophy, Ptolemy in astronomy and geography, and Augustine in theology. Lactantius, alone, seemed to remain the sole object of severe attacks. It is surprising that, even in light of new explorations and discoveries, the figure of Augustine was excused from all blame, although it was already possible to recognise the existence of antipodes against his opinion. The strategy followed was twofold: *First*, the existence of west-east antipodeans was recognised, but their absence in Augustine was explained due to the recent discovery of the extreme Far East; furthermore, they were in the same Euro-Asian landmass, and thus, the situation was entirely different from what Augustine wished to deny (uncommunicated antipodeans in isolated landmasses); and *second*, ignoring or rejecting any real antipodeans in the north-south direction, as the distance from the north of Europe to Africa was smaller than from the west of Europe to the east of Asia. Nevertheless, as the Portuguese crossed the Equator and extended the knowledge of the southern tip of Africa, the *antoeci* were discovered. Even these, however, were once again attached to the same landmass of the known oikoumene, and did not threaten Augustine's ideas.

The above allowed for a reasonable defence of Augustine, while it put Lactantius's reputation at stake. However, as we will see in Chapter 4, greater changes were to come. The revolutionary force of the new geographical explorations of the last quarter of the 15th century was to be quickly revealed after the Portuguese arrival in 1500 in Brazil, a true inhabited antipodean landmass separated from the old oikoumene. The undermining of the authority of the Fathers, a rare attitude at that time, would have significant consequences when it became a commonplace in the 16th century.

Renaissance historian James Hankins referred to that process when describing the earlier criticisms of Decembrio to the Fathers in a letter exchange with his patron, the Archbishop of Milan Francesco Pizolpasso, in 1438:

> […]. Where Decembrio really broke fresh ground, however, was not in his attack on the authority of Aristotle, but in his even more violent assault on the Fathers of the Church. [...]. Scholastics doctors regarded *auctoritates* in the light of laws or principles from which might be deduced a body of scientific doctrines; to subject these laws to historical criticism was therefore an extremely perilous undertaking. […].
>
> (Hankins, 1990, vol. 1, 145)

A similar attitude, distinguishing the authority of the early Christian Fathers in matters of faith from other topics on which "they spoke merely as human beings," was found in George of Trebizond's *Comparatio philosophorum Platonis et Aristotelis* (1458; Hankins, 1990, 244). However, Archbishop Pizolpasso "could not accept Decembrio's contention that Jerome and Lactantius had gotten their facts wrong," because that "would mean undermining the authority of the Fathers in other matters" (Hankins, 1990, 145). As we will see in Chapter 4, the 16th century had to face the fact that they were, indeed, wrong. Archbishop Pizolpasso's fears, repeated two centuries later by Cardinal Bellarmino in the face of Foscarini's and Galileo's criticisms of the authority of the Fathers in relation to the movement of the Earth, were prescient. We will chronicle precisely that in Chapter 5.

Notes

1 Petrarch knew Lactantius and the *Divine Institutes*, which he quoted in *De otio religioso* and *Invective contra medicum* (Lee, 2012, 141, 312; Nolhac, 1892[1907], 211–212).

2 "Boccaccio cited him [Lactantius] frequently in his works, especially the *De genealogia deorum* and he owned a copy of Lactantius's most important work, the *Divinae Institutiones*" (Stinger, 1977, 117).

3 This was a collection of Greek manuscripts brought to Italy in 1463 and printed in Latin translation in 1471, which were believed to be authored by an ancient Egyptian priest called Hermes Trismegistus. English translation in Copenhaver (1993; the key passage by Ficino is in p. 163). See also Mintz (1980, 138–169; 151 in particular) and Garnsey (2002[2004], 161–162). See also Ficcino's *De christiana religione* (1474; Copenhaver, 1993, 182).

4 However, it appears there indirectly, and it has been argued that the actual origin has to be attributed to Rudolf Agricola (1444–1485). For a discussion, see Rutherford and Schulten (2017, 618) and Schmitt (1972, 25). In any case, similar laudatory references were common among early humanists. Kendeffy has argued that Lactantius "deliberately prepared the label for himself" by his intensive use of Cicero (Kendeffy, 2015, 57).

5 The first edition of *Divine Institutes*, as part of Lactantius's *Opera*, was in Italy (Lactantius, 1465). This was the first dated book printed in Italy (29 October 1465). It was printed at the Benedictine monastery of Subiaco, near Rome, by two German printers (Sweynheym and Pannartz). They also printed Augustine's *City of God* (Augustine, 1467). On the history of the *editio princeps* of Lactantius, see Rutherford and Schulten (2017, 617; and particularly the bibliography there in note 102).

6 I thank Prof. Harry Hine (University of St Andrews, UK) for this translation.

7 This alternative manuscript version contains at several points additions, apparently by Fillastre himself, to vindicate the orthodoxy of his views to the canons of the Reims cathedral, where he had been dean since 1392 (Hiatt, 2008, 155).

8 Zaccaria (1956, 56), Rutherford (2015, 479) and Rutherford and Schulten (2017, 610).

9 The autograph by Rho is preserved at the Vatican as BAV, MS Otto. lat. 1903. Another Vatican manuscript contains the dedication with an image of Rho kneeling to present the book to the Pope: BAV, MS Vat. lat. 227; digitalised at: https://digi.vatlib.it/mss/detail/Vat.lat.227. This has a note in Decembrio's hand indicating that the Pope "had read most of the first dialogue in 1443" (Rutherford, 2015, 484). The *Dialogi* was first printed and translated in the Latin and English edition by David Rutherford and Paul Schulten (Rho, 2023).

10 Rho (2023, 706 (Latin), 707 (English)).

11 Rho (2023, 706, 707). See Rutherford (1998, 181; 2015, 482). On that and the polemic against Valla's preference for rhetoric against philosophy, see Rutherford (2015, 480–481; 2017, 158–159, note 9).

12 Rho (2023, 240, 241).

13 Rho (2023, 242, 243).

14 Rho (2023, 84, 85). See quotations and references from an earlier work, *Oratio ad scholares* (1430/1431) in Rutherford (1998, 185; 2015, 482–483), Rutherford and Schulten (2017, 604, 617).

15 Rho (2023, 226–243). Before this section, we find other criticisms to Lactantius's cosmological views, such as the belief in a single heaven where the Sun, Moon and the stars were all put together by God. These are considered errors 10th and 11th, as Rho defended the existence of multiple heavens where each heavenly body was placed separately, following the Aristotelian-Ptolemaic ideas. These two authors are hailed as two geniuses against Lactantius, who is recommended to go back to school to read Euclid and other mathematicians in order to understand Ptolemy (Rho, *Dialogi*, 1.82–97; Rho, 2023, 168–189).

16 Milan, Biblioteca Ambrosiana, MS D 105 sup. On that letter, see Rutherford (2015, 484). The letter and annotations are translated in Rho (2023, Appendix 1).

17 A point that was explicitly rejected in the book by Rho, pointing that the separation between Lactantius and them was no more than 60 years. Rutherford (1998, 183), Rutherford and Schulten (2017, 618–621).

18 A first partial edition of these letters, including this particular letter, was printed in *Francisci Philelfi Epistolarum liber primus* (Venice, 1473). Then, in 1502, the definitive, complete collection was also published in Venice (*Franciscus Philelphus. Epistolarum familiarium libri XXXVII*). The letter is translated in Rho (2023, Appendix 2).

19 The first edition was *Francisci Philelfi Satyrarum Hecatostichon decades* (Milan, 1476). The second was in Venice in 1502.

20 *Quartuſ decimus error in philoſophia quãdo dicit ſi antipodes eẽnt capita deorſum haberẽt pedes uero ſurſũ. Et ubi uere ſint antipodes ibi docetur* (I thank Prof. Harry Hine (University of St Andrews, UK) for this translation). In the second edition by Sweynheym and Pannartz (1468), they were reordered according to their appearance in the seven books of the *Divine Institutes*. This particular *error* appears there as the first of book 3 (Rutherford, 2015, 485).

21 *Quid sit fides et in quibus consistat et quod unusquisque in fide sua salvatur*. Here, and in the quotations of Marzio below, I thank Prof. Harry Hine (University of St Andrews, UK) for his translations.

22 *Hoc igitur videretur iis convenire qui ante Christi doctrinam vixerunt, sed eorum qui post redemptoris nostri salutarem adventum fuere, alia est ratio.*

23 *qui sunt in extremo orbi nati...*

24 *Quae differentia sit inter theologos christianos et philosophos de acquirenda beatitudine, ubi probatur quia nemo secundum relligionem nostram potest consequi beatitudinem nisi fuerit a deo praedestinatus cum a philosophis tribuatur arbitrio nostro.*

25 *sed ex electione et praedestinatione creatoris.*

26 *Unde ex quacumque fide, ex quacumque loco deus potuit praedestinare et hoc modo salvare.*

27 This biblical character is also used by Mandeville (*Travels* 32; on this book, see p. 58) to open salvation to non-Christians (even Muslims and pagans! A remarkable view in the light of the religious conflict at that time). See Mandeville (1983[2005], 180).

28 The salvation of Trajan appears back at least to the 13th century. See Launay (2018, 30).

29 *Quae cum a veritate dicantur vera esse necesse est.*

30 *Non tam necessaria sunt verba quam sensus* (Jerome, *Ep.* 29.1, *ad Marcellam*, as quoted by Marzio).

31 *Cum cautione quadam verba salvatoris intelligenda sunt.* This usage of the text of John 10:8 can be traced back to Origen, who used it to support his allegoric method (*Commentary on the Gospel According to John* 1.274).

32 *Numquam enim quisquam christianus antipodes erudivit, numquam desertissimas adiit, numquam cum acephalis, cum blennis, cum multis generibus hominum sermonem de Christo habuit,...*

33 Although in classical Latin the rare word *blennus* means "stupid" and it is not used for any of the monstrous peoples, it could be possible that this word was used by mistake instead of the name *Blem(m)yes* or *Blem(m)yi*, used for an imaginary headless Ethiopian tribe (I thank Prof. Harry Hine (University of St Andrews, UK) for a clarifying

discussion on this topic). This is an attractive conjecture, as the author was clearly referring before to headless beings and afterwards to the diversity of human peoples, all that after mentioning the antipodes. It makes sense to think that this entire sentence is related to beings living in the periphery of the oikoumene, where creatures unreached by the Gospel could still be expected to live. Monsters and antipodeans, also appeared next to each other in Augustine' *City of God* (see above Chapter 2).

34 *Etiam tacente Christo* (see Miggiano, 1998, 208–226; 1992, 107–109 in particular).

35 Biblioteca Apostolica Vaticana, Cod. Pal. lat. 1711, fol. 88[v]. This text has in book 1 a traditional exposition of the sphericity of the Earth and the existence of the antipodes.

36 The full-text was translated into Latin by Jacopo d'Angelo de Scarperia in 1406–1410 (*Cosmographia*). The first printed edition (Ptolemy, 1475) was without maps, the second one (Ptolemy, 1477) was the first with maps (26). Both these editions were entitled *Cosmographia*.

37 In 1474, Toscanelli, a friend of Regiomontanus, suggested in a letter to Alfonso V of Portugal an Atlantic crossing that inspired Columbus (Randles, 1990a[2000]; 1990b[2000]).

38 In Turin's E.IV.11 manuscript, the anti-Lactantius chapter 14 is found in fol. 73[r]–78[r]. In the edition by M. Frezza (Marzio, 1948), the Italian translation of fragments of chapter 14 appears in pp. 48–52 and the original Latin texts in pp. 89–91. While the portion published by Frezza corresponds to the beginning of the chapter, the last part is published in partial form (with French translation) by Gautier Dalché (2013, 213–218; although the reference on p. 214 indicates that the Latin text is taken from Turin ms. E.IV.11, however, the pagination indicated, fol. 54[v]–57[v], corresponds with a 1478 copy of the manuscript, also in Turin: Bibl. Univ. I.I.89, where chapter 14 is found in fol. 53[r]–58[r], this is precisely the copy mentioned in Gautier Dalché, 2011). To check the complete text of this key chapter, I have been able to obtain reproductions of the 15th-century copy in Vienna: *Marcius Galeotti De rebus incognitis* ad Mathiam regem Hungariae, Vienna, Österreichische Nationalbibliothek, Lat. 3166, fol. 296[r]–392[v] (chapter 14 is in fol. 334[r]–338[r]; available at: http://data.onb.ac.at/rec/AC13978351) and the derived copy of the 16th century also in Vienna: Lat. 12509, fol. 1[r]–119[v] (chapter 14 is in fol. 51[r]–56[r]; available at: http://data.onb.ac.at/rec/AC13970012). I thank Dr Miguel Marañón (Instituto Cervantes, Madrid, Spain) for transcribing the Latin texts of chapter 14 from these two manuscripts.

39 Briggs has a mention to the *Invectivae in Franciscum Philelphum*, an earlier exchange of polemical letters between Marzio and Filelfo (1464–1466), and indicates that there Marzio "affirma se croyance aux Antipodes" (Briggs, 1974, 75–84; quoted p. 77). These letters were published in Marzio (1932), and Filelfo and Sforza (2015), but I have not had the chance to verify this information. We saw in p. 73 Filelfo's polemic with Rho over Lactantius.

40 *Quia non est verum quod verbum Christi pervenerit in omnem terram, ubi irridentur et confutantur rationes Lactantii dicentis non posse antipodes reperiri.*

41 *acutissimo et eloquentissimo et philosopho et oratore Lactantio Firmiano.*

42 *Is enim confutator egregius adversus mathematicos de antipodibus et terrae rotunditate adducit quaedam quae, si alio in codice scripta sine titulo comperissem, numquam credidissem esse Lactantii.*

43 *propter celsos altosque montes.*

44 *figuram sphericam deformare.* On this topic, see Chapter 4, endnote 97.

45 *omnes geographos et mathematicos.*

46 *Terra enim amplissima et clementer devexa ita ut plana quodammodo videatur ubique commodas recipit habitationes.*

47 *Ad rationes tales res ipsa respondet, quoniam qui illic fuerunt, refferunt et sub equinoctiali et sub polis comodas habitationes.*

48 *historicos vera dixisse, et non esse vanos neque mendaces.*

49 *ex matheseos ignorantia.* The word *mathematicus*, up to at least the 17th century was used for astronomy (and astrology).

50 *sicuti ad plurimas insulas quas longo tempore post cognovit usus.* That could be a reference to islands in the Atlantic, found in the 14th and 15th centuries.

51 *Quia in omnem terram exivit sonus eorum apostolo par est credere, sed secundum suum dicendi modum.*

52 *et tales locutiones superlative, id est hiperbolice, in scriptura sunt frequentes.*

53 *tota Sacra Scriptura ad communem loquendi usum formata [...] Scriptura hanc habet dicendi consuetudinem ut generaliter loquatur.*

54 *in Sacra Scriptura multa simillima poeticis fabulamentis reperiuntur.*

55 On the long history of the principle of accommodation among Jews and Christians, see Benin (1993), who devotes chapter 4 to Augustine and the influence of this principle in his *City of God*. Benin describes this principle in this way:

"Divine accommodation/condescension alleges, most simply, that divine revelation is adjusted to the disparate intellectual and spiritual level of humanity at different times in history. [...]. The Lord accommodates or condescends, freely and benevolently, to the human level lest his salvific message go unheard and unheeded. The Lord, as it were, had to 'come down' to earth in order to effect the proper unfolding of his universe" (Benin, 1993, xiv).

Although the principle of accommodation can be seen as opposed to concordism, there is a "weak" principle of accommodation that bridges both approaches (see Felipe, 2018).

56 *Apostolus igitur [...] cum ad omnes verbum dei delatum ait more suo, ad magnam retulit partem.*

57 Miggiano (2004, 203, note 3), and Gautier Dalché (2011, § 25; 2013, 217). The list of errors (*Errores ex libro Galeoti Narniensis inter alios principales sunt isti*) is found at the Biblioteca Apostolica Vaticana, Vat. Lat. 8865, sec. XV, cart. misc. fol. 48^{v}–49^{r} and in Cod. Lat. 170 (182), fol. 135^{r} preserved at Biblioteca Universitaria di Bologna. The list was published by Kardos (1933, 3–4). As the documents of the trial of Marzio disappeared in 1479, during a fire in Venice, these are the only related surviving documents.

58 *Galeotti Martij Narniensis Liber excellentium ad Serenissimun Carolum Iuniorem regem Galliarum*. Paris, Bibliothèque Nationale de France, Nouv. Acq. Lat. 731, sec. XV, membr., fol. 297; digitalised at: https://archivesetmanuscrits.bnf.fr/ark:/12148/cc714331.

59 *De errorib⁹ Lactãtij ꝗ negauit rotũditatẽ terre et antipodas.* Chapter IX spans pp. 73–84. A note on *De excellentibus*, including the titles of all its chapters, was published by Rive (1785, 12–16).

60 Others of his works did go into print with several editions up to the 17th century, and he became a well-known author. There was an attempt to print *DIV* in the mid-16th century by Vadianus (an author on whom Marzio's criticism of Lactantius had an influence, as we will see in Chapter 4, endnote 115). Vadianus brought a copy of the *De incognitis vulgo* from Vienna to St Gallen (where it still is, see https://aleph.unibas.ch/F/?local_base=DSV05&con_lng=GER&func=find-b&find_code=SYS&request=000170529) and recommended the publication to Lorenzo Torrentino. However, the book was never printed. For more details, Miggiano (1992, 141–151; 2004, 194–196).

61 References in p. 74 and endnote 38.

62 It seems that the Paris copy has been somehow altered to avoid some of the heterodox content (Miggiano, 2004, 196, note 2).

63 Oberman (1981, 64–110). For the concept of "Augustinian Renaissance" and other related terms, see Saak (2013, 63–67).

64 The section of *CG* 16.9 on the antipodes was included at the very end of the entry "homo" in the florilegium *Milleloquium sancti Augustini* by Bartholomew of Urbino in 1347, printed in Urbino (1555, columns 984D–985A).

65 Digitalised at https://catalogue-bibliotheque.nantes.fr/ark:/73533/Nantes_246/v0163.

66 *Et si terra sit rotunda non tamen axdcndum (= accedendum) est antipodas et quare tota terra non est habitabilis.*

67 See a reproduction of the image at: https://ru.wikipedia.org/wiki/%D0%A4%D0%B0%D0%B9%D0%BB:Cosmas_-_antipodes.jpg. Kominko discussed other similar representations in early medieval Byzantine Psalters (Kominko, 2013, 80–82).

68 For a review of the antipodes images in this popular medieval book, translated to different languages, including an English translation printed in 1480–1481, 1490 and 1527, see Goldie (2010, 66–68). See, for example, the 14th-century copy at BNF Fr. 574 fol. 42r; available at: https://gallica.bnf.fr/ark:/12148/btv1b84526412/f97.itemr=BNF%20Fr%20574. The explicit affirmation of the Earth's sphericity in this text was not incompatible with depicting T-O maps in f. 49r and 49v (for a quotation from Gautier's text, where travelling around the Earth is compared with a fly walking a round apple, see Relaño (2002[2019], 25 and note 19 in p. 42).

69 For example a 9th-century copy of Isidore's *De natura rerum* (manuscript Ms. 422 of the Bibliothèque municipale de Laon, see figures in fol. 5v and 6v; available at: https://gallica.bnf.fr/ark:/12148/btv1b8492138z/f22 and https://gallica.bnf.fr/ark:/12148/btv1b8492138z/f24). Another example is a 13th-century manuscript of Sacrobosco's *Tractatus de sphaera* (British Museum, Egerton MSS. 843, fol. 23r; available at: https://web.archive.org/web/20211017085600/https://www.bl.uk/catalogues/illuminatedmanuscripts/ILLUMIN.ASP?Size=mid&IllID=10143). On that last image, see Taylor (1935, 65–67).

70 In Biblioteca Statale di Lucca, MS 1942; available at: https://2.bp.blogspot.com/-h4z4hq0SStw/VzkOabwy9KI/AAAAAAAAAvI/zbewr4i6HmUlMQCKRHX5tR4nRc2D6bJpACKgB/s400/Hildegard-of-Bingen_Liber-Divinorum-Operum_I-4_Lucca_MS_1942_fol_38r.jpg.

71 *Tertia, quod non est friuolum neque falsum Antipodes assignare* (translated in Kish, 1978, 250).

72 *Adversus quod disputant Augustinus de Civitate dei* (quoted in Hiatt, 2005, 16; see also Bouloux, 2002, 41–42; my translation). See the manuscript at https://ambrosiana.comperio.it/opac/detail/view/ambro:catalog:109131.

73 Hiatt (2005, 22–29; 2008, 130). This poem was composed in the late 1330s and early 1340s and abandoned around 1343, being posthumously circulated in 1396 (Everson, 2001, 98–101).

74 See endnote 36 for the traditional history of that rediscovery in the Renaissance. However, the knowledge of Ptolemy was not entirely lost during the Middle Ages, among Christians, such as Boethius, Cassiodorus or Bacon, and Arabic authors (Gautier Dalché, 2009). It is interesting to note that the astronomical work of Ptolemy, *Almagest*, waited for printing until 1515 in Venice.

75 "Un error lungo e fioco." "Sappi che questa oppinïone è vana." "Era più grossa allor la gente umana." (Translation in Pulci, 1998, 621).

76 "sì che il mondo non fe' sanza persone / dove tu vedi andar laggiù le stelle, / pianeti e segni e tante cose belle" (Translation in Pulci, 1998, 623; the description of the antipodes continues in 25.239).

77 *Ante climata uersus equinoctialem et ultra multas habitationes continent. Ut ex historiis autenticis compertum est* (transcribed and translated by Hiatt, 2008, 147; image in Figure 19, p. 146). In a 1483 printed edition in Leuven, available at: https://gallica.bnf.fr/ark:/12148/bpt6k52802c/f9.image, the note is inserted horizontally on the left side of the Torrid Zone below the Equator.

78 *quod hic sint homines habitantes reprobat Augustinus 16 libro de ciuitate dei* (Hiatt, 2008, 73, translation slightly modified).

79 Different maps appear in the Italian textbook *Sfera*, by the Florentine merchant Gregorio Dati (1362–1435), which has also been attributed to his brother, the Dominican Friar Leonardo Dati (1365–1425). He depicted a zonal map, a T-O map and a tripartite mappaemundi plus regional maps and fragments of portolan charts (Clemens, 2008). The text also describes the T-O maps (Relaño, 2002[2019], 11 and note 23 in p. 19).

80 The open Indian Ocean appears in maps of the 15th century by Andrea Bianco (1436), Andreas Walsperger (1448), Giovanni Leardo (1452–1453), the Catalan world map (c. 1450), the Genoese world map (1457), etc. From texts accompanying Fra Mauro's map, we know that he supported the eccentric model and its supernatural explanation (Vogel, 1995, 312–318). The Walsperger and Catalan world maps are drawn with an eccentricity to the south and east, respectively, which Vogel interpreted as a depiction of the eccentric model (Vogel, 1995, 318–320).

81 This map is discussed in Hiatt (2008, 170–172; 2011, 33). For a reproduction of the map, see: https://gallica.bnf.fr/ark:/12148/bpt6k1527259d/f75.

82 Gallerie degli Uffizi, Florence (no. 1890 n. 496), digitalised at: https://www.uffizi.it/en/artworks/justice-pollaiolo.

83 Pinacoteca di Brera, Milan (no. 1240), digitalised at: https://pinacotecabrera.org/wp-content/uploads/2015/06/Bramante-Eraclito-e-Democrito-1.jpg.

84 Galleria Borghese, Roma (no. 435), digitalised at: https://www.collezionegalleriaborghese.it/en/opere/salvator-mundi-2.

85 Room of the Segnatura at the Vatican Museums, digitalised at: http://www.museivaticani.va/content/museivaticani/en/collezioni/musei/stanze-di-raffaello/stanza-della-segnatura/scuola-di-atene.html.

86 The involvement of stars into the emergence of dry landmass over the water was already mentioned at least by Robertus Anglicus in *Commentary* to Sacrobosco (1271) and several 13th to 16th centuries authors, as I found this explanation in Pedro de Medina's manuscript *Libro de cosmographia* 70 (1538), fol. 56v (Medina, 1972, 136 [Spanish] and 202 [English translation]).

87 Apart from Dante, among those who rejected the eccentric model in the 14th century and defended that water and earth formed a single sphere, we can mention Andalò di Negro (1260–1334; Thorndike, 1949, 35; Wootton, 2015, 117, 135–136). Wootton also mentions other authors which I have not studied, such as Giles of Rome (1243–1316; Wootton, 2015, 114, 133), Themo Judaei (c. 1325–1371; Wootton, 2015, 117, 135) and Marsilius of Inghen (1340–1396; Wootton, 2015, 117).

88 Some quotations and a summary of Buridan argumentation in book 2.7 appear in Randles (1994[2000], 31–33). Most of Question seven is translated in Grant (1974, 621–624) and the whole Latin treatise was published by Moody (Buridan, 1942 [1970]). On Buridan's views, see Moody (1941), Hooykaas (1980, 309) and Grant (1984, 24–25).

89 Grant reminds us that this model is not entirely original to Buridan, but has precedents in Avicenna and even Xenophanes (6th century BC). Grant (1974, 621 note 1; also pp. 615–620).

90 See at: https://gallica.bnf.fr/ark:/12148/btv1b8447191h/f314.

91 The pseudo-Ptolemaic *On the division of the spheres* (one-sixth uncovered), the Ptolemaic *Almagest* (a quarter), Aristotle and Averroes (more than one quarter), Seneca and Pliny the Elder (a short sail to cross the Atlantic) and 4 Esdras (only a seventh underwater).

92 Image at: https://gallica.bnf.fr/ark:/12148/bpt6k52802c/f6.image.

93 Grant (1974, 633), Randles (1994[2000], 39), Relaño (2002[2019], 27).

94 On the polemic, see Vogel (1995, 266–275), Randles (1994[2000], 44–45, 76), Wey Gómez (2008, 124–130).

95 Redrawn from the edition in Nuremberg (1481, fol. 28v), see at: https://digital.onb.ac.at/RepViewer/viewer.faces?doc=DTL_423959.

96 *Idq3 diuinam mentem hominum caufa ftatuiffe digna fentẽtia quam chriftianus approbet.* (Translated by Randles (1994[2000], 38), slightly modified). See also Vogel (1995, 321–327).

97 *Alia ẽ opi. philofophorũ atq3 aftrologorum theologorum q3: qui omnes concedunt cœlum effe rotũdum.* For Borro's text here (and below), I have examined the original Latin text from scanned copies generously provided by the New York Public Library,

US: KB 1494 (Sacro Bosco, J. de. *Commentvm Electvm Et Prae / ciosvm Nvperrime*). Rare Book Division. The New York Public Library. Astor, Lenox, and Tilden Foundations). I thank María Jesús Prieto (Secondary School *IES Malala Yousafzai*, Madrid, Spain) for the translations of Borro.

98 Oronce Fine changed this diagram in his 1524 edition of Sacrobosco with a representation of the terraqueous model with the image of a landmass (apparently Africa) spanning both the tropics and the Equator (Sacrobosco, 1524, fol. a.ii^{v}; for discussion see Axworthy, 2020, which depicts Fine's image in p. 214). Image available at: https://media.springernature.com/original/springer-static/image/chp%3A10.1007%2F978-3-030-30833-9_8/MediaObjects/471249_1_En_8_Fig6_HTML.png. Curiously, Fine retained a zonal diagram depicting the Southern Temperate Zone with the label *zona incognita et habitabilis* (Sacrobosco, 1524, fol. b.iii^{v}) and a further diagram showing the seven climates in the Northern Hemisphere, while in the Southern there was, below the Torrid Zone, again a temperate *incognita zona* (Sacrobosco, 1524, fol. c.iii^{v}).

99 Redrawn from the edition in Venice (1485, fol. 13^{r}), see at: https://daten.digitale-sammlungen.de/0003/bsb00036841/images/index.html?id=00036841&seite=9&fip=193.174.98.30&nativeno=&groesser=300%25. For a coloured version of this kind of image see: https://detalesanewton.files.wordpress.com/2018/02/sacrobosco.png.

100 Major studies of the history of the interpretations over the last four centuries of Galileo's problems with the Vatican include Finocchiaro (2005), the collective work edited by McMullin (2005), and Heilbron (2010).

101 "[…] aſſy poderia mais ao ponente auer outras jlhas 7 tȩrras, poꝛque a natureʒa nam auia de ſer tão deſoꝛdenáda na cõpoſiçam do óꝛbe vniuerſal, que quiſeſſe darlhe mais párte do elemẽto da águoa que da tȩrra deſcubȩ́rta, pera vida 7 criaçam dos animáes" (my translation).

102 Historian John M. Headley (1997a) discussed the geographical plenitude principle (of Platonic roots) in the 16th century; however, it is already here in Columbus (Wey Gómez, 2008, 138–139). Rubiés (2017, 58), mentions how Pietro Bembo put in the mouth of Columbus a speech based on it (Bembo, *Historia Veneta* 6, Venice, 1551, fol. 82^{v} and Paris, 1551, fol. 125^{r-v}; composed in the late 1530s). Rubiés considers it anachronistic, but, as we saw in pp. 84–85, this view was not alien to the late 15th century. In any case, it is also true that Barros's language reflects more the 15th-century context, with its reference to the eccentric Earth model, than Bembo's.

103 "Poꝛque como em o tempo do jnfante dom Anrique ſe deſcobꝛiram as jlhas terceiras, 7 tanta párte de tȩrra de Africa […]" (my translation). Wey Gómez, who considers that the target of Columbus was in the Torrid Zone and west-southern Atlantic rather than more directly in the west-northern Atlantic, renders "ilhas Terceiras" as the tropical "Cape Verde" (Wey Gómez, 2008, 137), while other historians consider that a reference to the Azores further in the North Atlantic (Randles, 1990b[2000], 5; 1994[2000], 47; 1993b[2000], 20). In fact, one of the Azores islands is still called "ilha Terceira" (literally "Third island," a shorthand reference from its former full name "Ilha de Jesus Cristo das Terceiras"). The name of "terceiras," that is, "the third(s)," comes from them being the third archipelago discovered in the Atlantic, after Canaries and Madeira. I thank Prof. Onésimo Almeida (University of Brown, US) for a helpful discussion on this issue.

104 Ironically, King D. João II received a similar proposal from Münzer in 1493 (Münzer, 1493a), unaware that Columbus had just completed his first voyage for the rival Catholic Monarchs in Spain.

105 See Manzano (1989, 83–104, 139–147) and Wey Gómez (2008, 109–112) for information surrounding this meeting.

106 Sadly, Irving confusingly referred to this meeting not only as "junto," but also as a "council" or "council of clerical sages," and introduced references to the Inquisition in

his fictional biography of Columbus (Irving, 1828, vol. 1, bk. 2, ch. 4), which at times seems inspired by the trial of Galileo, as mentioned in p. 91.

107 It is also discussed in Felipe (2012), where I divided the argumentation into five, fusing what I consider here second and third points.

108 The overland distance between Spain and Far East Asia was a heatedly debated topic at the time. This distance measured the size of the oikoumene and, in the eccentric model, it was related to the proportion of water to earth and the eccentricity of both spheres. For Ptolemy, the size of the oikoumene was 180º (half the circumference, in contrast with Marinus of Tyre 225º); however, Marco Polo described the Far East Chinese lands and the existence of a large island further east (Japan). Therefore, several late 15th-century scholars and mapmakers extended the oikoumene up to 270º: Toscanelli (in his 1474 letter to Alfonso V of Portugal), Martellus (in his c. 1491 Yale map), Behaim (in his 1492 globe) and others (Duzer, 2019; Randles, 1990a[2000]; 1990b[2000]; 1992[2000]; Vogel, 1995, 339–352). Columbus embraced this idea, expecting to find Japan where he found the Caribbean islands. Such a large emerged oikoumene was difficult to fit in the eccentric model. In addition, Columbus accepted a measure of the Earth radius – ultimately from Ptolemy – smaller than the famous measure by Eratosthenes (which is closer to our modern measure). The expectation from both views was that of a small ocean from Spain to Japan (which could become even shorter taking into account real and imaginary Atlantic islands, see p. 91 and endnotes 101 and 103).

109 Ferdinand said at the beginning that "the response and information they [the members of the *Junta*] gave their Highness were as diverse as were the variety of their wits and opinions." That explains why they did not necessarily concord.

110 Randles (1984, 440), wants to see there an influence of Lactantius's theories, but as in the case of Allegro (2017; refuted by Nothaft, 2017), there is no evidence for that across the Middle Ages (indeed, in this text there is a reference to Augustine for the denial of the antipodes, while Lactantius is not mentioned). More likely there has been a confluence on this 'common sense' view which apparently was compatible with the belief in the sphericity of the Earth, not only seen in the mention there by Ferdinand of the worries of 15th-century Portuguese in their initial voyages to Guinee, but also seen in Mandeville, who considered travelling from the west of Europe to Jerusalem a way uphill and then going further east, downhill, implying that the Western Hemisphere was at the bottom (Mandeville, *Travels* 20); despite that, Mandeville considered possible to sail around the Earth (on Mandeville, see p. 58).

111 Columbus interest in the Torrid Zone and the south-west Atlantic has been highlighted by Wey Gómez (2008, 136–138, 142, 157–158). He mentions Columbus's 1483–1484 proposal in Portugal, which suggests a target west of the Portuguese's Africa (according to Barros, but see here endnote 103), and the later proposal to England in 1488, which included a map which stressed the importance of the Torrid Zone and its recent exploration (according to Ferdinand Columbus and Bartolomé de Las Casas).

112 It is important to keep in mind that the Portuguese government imposed a policy of secrecy on their geographical discoveries, and that the Spanish navigations at that time did not reach beyond the Canary Islands, as the rest of the African coast was controlled by Portugal. However, Randles (1980, 35), has shown that Lorenzo Bonincontri – who taught at the University of Rome – knew in the early 1480s that the Portuguese had proved Ptolemy's *Almagest* 1.3 wrong by crossing the Equator and finding inhabitants on the other side of it (Bonincontri, 1484; see above Chapter 2, endnotes 20 and 21, for Ptolemy's views on the habitability of the Equator, denied in *Almagest* but supported in *Geography*). Columbus had first-hand information, as he had lived many years in Portugal and travelled to the Guinea.

113 The other main source of information on that meeting is Las Casas's *Historia de las Indias* 29 (1527–c. 1560, unpublished until 1875). It seems likely that Las Casas copied

that from Ferdinand Columbus, as the same six arguments against Columbus are there with similar wording (Las Casas mentioned at the end of his chapter 5 that this was copied from Ferdinand). Other reports, such as that by one of the experts attending the Junta, the former professor of law in Salamanca, Rodrigo Maldonado, are much shorter. In 1515 he just recollected that all agreed in rejecting Columbus's proposal as impossible (Fernández-Armesto, 1991, 54; for the original testimony of Maldonado in English, see Phillips, 2000, 131). This was the common opinion at that time, both in Portugal and Castile (see beginning of this section and the various reports collected in the mid-1515s, e.g., in Manzano, 1989, 128–129).

114 *orbis terrarum nobis cognitus, insulae instar, aquis eminet.* (Translated in Randles, 1994[2000], 41).

115 The incorporation of Scandinavia to Christendom produced an expansion of the oikoumene towards the North Pole by the Late Middle Age. Recently, it has also been discovered a rare reference to the first Scandinavian explorations of North America (the land of *Marckalada*) in the *Cronica universalis*, written c. 1340 by the Milanese friar Galvaneus Flamma (Chiesa, 2021). This appears in a defence of the habitability of the equatorial and polar regions, apparently following Pietro d'Abano's early 1300s *Conciliator* (Chiesa, 2021, 92, note 13).

4 Old texts and the geographical challenge of the voyages (1434–1522)

From ancient texts to new worlds: the 'geographical revolution'

Before the famed astronomical 'Copernican revolution' started in 1543 with the heliocentric proposal of Copernicus (1473–1543), there was an earlier, less-known 'geographical revolution', which claimed the Earth to be a single terraqueous globe, as we discussed in Chapter 1. With roots c. 1500 and even before, it became the consensus during the first decades of the 16th century, explaining the mixed configuration of land and water over the surface of our planet and its solid massive internal structure until today.

The rediscovery of the complexity of ancient geography

There was no unified medieval image of the world. Several ideas from different periods of the classical past lingered with a difficult co-existence, and many medieval authors just transmitted everything. Ideas already discarded in Antiquity were preserved in encyclopaedias or in antiquated primary sources.[1] The two main views were: first, a *single oikoumene* in the Northern Hemisphere around the Mediterranean with an eastern land projection to India, China, and even Japan; and second, the *Cratesian view* of four isolated landmasses of which the Old World oikoumene was just but one. By the late medieval period, the single oikoumene view was fused with the scholastic eccentric model and became dominant.

The availability in the early 15th century of the full-text of the *Geography* of Ptolemy (c. 100–c. 170) transformed it into a new strong influence (see Chapter 3, endnote 36). Ptolemy supported several key points. First, a terraqueous sphere, "... the continuous surface of land and water is (as regards its broad features) spherical..." (Ptolemy, *Geography* 1.2, 2000, 60). Then, a single 180º wide oikoumene, surrounded by water in the north-west and by unknown lands at the other edges. Finally, Ptolemy considered that the African landmass extended into the Southern Hemisphere across a habitable Torrid Zone, which southern tip was connected to the eastern Asia (enclosing the Indian Ocean).

The introduction in 1439 of Strabo's *Geography* in Italy – by Gemistos Plethon (c. 1355/1360–1452/1454) – challenged two key ideas of Ptolemy: the habitability of the Torrid Zone and the enclosed Indian Ocean (Anastos, 1951[1979];

DOI: 10.4324/9781003545682-5

1952[1979]). However, Plethon noted that "Strabo is unjustified in holding that there is a torrid zone devoid of human habitation" (quoted in Anastos, 1951[1979], 7). In addition, the open nature of the Indian Ocean could be supported by the testimonies of Marco Polo and Niccolò de' Conti that implied an oceanic connection between the Indian Ocean and the coast of China, Japan and the Indonesian Spice Islands (Goldstein, 1980, 35–39).

The challenge of the new geographical discoveries by Portuguese voyagers

This Renaissance project of recovery of ancient *geographical texts* during the 15th century ran in parallel with the Portuguese project of African exploration, which yielded fresh *geographical data* after they rounded Cape Bojador in 1434. The information conveyed by ancient authors was a double-edged sword. For the Portuguese project of sailing around Africa to reach India, two points were of interest: the habitability of the Torrid Zone and the communication of the Indian Ocean. Ancient reports of African circumnavigations and the *Geography* of Ptolemy supported the possibility of rounding the equatorial coast of West Africa. However, Ptolemy's bridge of land between the south of Africa and the east of Asia was a barrier for reaching India (Randles, 1986[2000], 2–3), which could be countered with the open Indian Ocean of Strabo, who, on the other hand, denied the habitability of the Torrid Zone. While theoretical geographical ideas hindered or stimulated the navigations,[2] data from the boats arriving at the Iberian Peninsula influenced the mapmakers[3] and changed geographical texts.

By the mid-15th century, the Portuguese having approached the coastal African Torrid Zone without dying and without encountering Crates's Equatorial Ocean, the views of Ptolemy (and others, including Plethon's criticisms of Strabo) seemed vindicated.[4] After crossing the Equator in the early 1470s and finding that Africa extended below it, the denial of the *antoeci* antipodeans was shown wrong. When in the late 1480s it became clear that the Atlantic and the Indian Oceans were connected, Strabo was proved right on this point,[5] and a modified Cratesian model regained interest.

Finally, after Spanish expeditions arrived to America in 1492 and circumnavigated the globe in 1522, the sailors triumphed over the ancient texts, as none accurately portrayed what they encountered. The ancients's legacy of contradictions, both among themselves and with respect to the new navigations, was hinted at by Guicciardini in the late 1530s (published in 1561): "These voyages have also revealed what some of the ancients believed and others denied, namely that there are other inhabitants under our feet whom they called the Antipodes" (Guicciardini, *L'Historia d'Italia* 6, 1561, 451; translated in Guicciardini, 1969[1984], 182).

Columbus's challenge of the eccentric model and Augustine's denial of the antipodes

According to historian William Graham Lister Randles, the "first outright recognition" of the clash between the pseudo-Aristotelian scholastic eccentric model and the Ptolemaic is found in the *Commentum in Psalmos* (1484, fol. ii.iii^r) by Bishop Jacob Perez de Valencia (c. 1408–1490). Being aware of the "experience" of the Portuguese sailors (Randles, 1980, 25; 1984, 440–44), he rejected the eccentric model, praising Ptolemy as:

> [...] much more rational, more in conformity with Holy Scripture and closer to truth, as is obvious to the senses and as has been proven by the experience of a good number of people who have travelled and sailed across different parts of the world.
>
> (translation from Randles, 1994[2000], 41–42; 1980, 22–25)

As we mentioned in pp. 91–82, this clash of geographical views was behind the debates surrounding Columbus's Atlantic crossing project in Portugal (1483–1484) and later in Spain, in Salamanca (1486–1487) and Santa Fe (1491), as the experts of the Spanish Catholic Monarchs followed the scholastic eccentric model (from Burgos's additions to Lyra's Bible) and resorted to the authority of Augustine to deny the antipodes. However, by 1491, the Portuguese had reached the Cape of Good Hope, demonstrating the existence of the antoecian 'antipodes', and Columbus was well aware of this, as he was present when they arrived at Lisbon (1488).[6] That was also known by Alessandro Geraldini (1455–1525), who intervened in favour of Columbus, bringing out the discoveries of the Portuguese at that last *Junta*.[7] It has been speculated that this was one of the reasons why there was finally a change in favour of Columbus.[8] Small wonder then that Columbus retorted in the margin of chapter 4 in Ailly's *Imago Mundi*, where the eccentric model was explained (copied from Oresme's *Traitié de la sphère*): "Water and earth together make round body" (translated by Grant, 1974, 633).[9] Ferdinand Columbus affirmed that his father: "believed that since all the water and land in the world form a sphere, it would be possible to go around it..." (Columbus, *Historie* 6, 1571, fol. 12^v; translated in Columbus, 1959[1992], 15). And Geraldini affirmed that Columbus did not believe that the largest part of the world was underwater (Geraldini, *Itinerarium* 14, 1631, 203). A friend of Columbus, Andrés Bernaldez, described in 1513 Columbus's views regarding the Earth as formed by "only earth and water embraced in sphericity" (Bernaldez, *Historia de los Reyes Católicos* 118).[10] The terraqueous model was then at the centre of Columbus's project, as it would be 50 years later at the heart of Copernicus's project (see pp. 159–161).

Upon the return of Columbus, Peter Martyr d'Anghiera, an Italian historian and advisor to the Catholic Monarchs, hailed the Caribbean islands as a "new world" of "antipodes" (see pp. 117). On 25 September 1493, the papal Bull *Dudum siquidem* divided the world with a line from pole to pole across the Atlantic, giving the Catholic Monarchs the territories "found or yet to be found towards the west and south" (Symcox et al., 2001, 42). As historian Alfred Hiatt noted, this included the antipodes (Hiatt, 2008, 175).

However, scholastics still defended the two eccentric spheres model, as Zaccaria Lilio (1496) and Pedro Ciruelo (1498). Also in 1498 Paul of Venice's early 15th-century work was printed defending the scholastic cosmography (Randles,

1993b[2000], 22, 24, note 35). At that time, Columbus had discovered only islands; therefore, they argued that no real continent had been found in the far west. Regarding the far south extreme of Africa, they argued that it was contiguous with the old oikoumene and in the same hemisphere (Randles, 1993b[2000], 20). Others, however, had no problem discussing hypothetical antipodeans as Rodrigo Basurto (or Vassurto), in a 1494 book, a year before taking on the chair of astrology/astronomy at Salamanca.[11]

However, in the summer of 1498, Columbus was on his third voyage (1498-1500), exploring the Caribbean coast of South America (which he began to conceive as a new unknown continent separated from Asia).[12] On the arrival of the first news of a vast landmass, north of the Equator towards the south, in late 1498, the Florentine merchant Simon dal Verde wrote enthusiastically from Cadiz to Mateo Cini in Venice (2 January 1499). He remarked "that the Earth has a spherical shape," as was already known by "the philosophers," and that something had been discovered dismantling the single oikoumene/eccentric models: "another world opposite us" (i.e. antipodes), where he had believed none "could live, because I thought there was only water and not land" (Berchet, 1893, 82).[13]

In 1500 the Portuguese arrived in Brazil, and in 1500–1501 to Newfoundland, while in 1502, Columbus explored a large portion of the Central American coast in his fourth voyage (1502–1504). With the diffusion of the news of the discovery of Brazil by the letters of Vespucci from 1502–1503 onwards,[14] affirming the existence of antipodes *and* antipodeans where water was expected, Europeans were astonished that such a vast part of the globe had remained unknown to them. Volterra pointed that "the Christians have opened up our globe (*orbe*) in length and in breadth," referring to the west to the Spanish exploration "beyond the Fortunate Isles (Canary Islands)" reaching the "Perioecians," and to the south mentioning the Portuguese voyages "beyond the Tropic of Capricorn" reaching the "Antoecians" (Volterra, 1506, fol. v^{v}; translated in Randles, 1994[2000], 57). Immediately, Matthias Ringmann and Martin Waldseemüller's *Cosmographiae introductio* (1507, including Vespucci's *Four Voyages*) named the new antipodean landmass "America". The exploration of America completed the destruction of two old theories already in crisis after the Portuguese exploration of Africa: the uninhabitability of a fiery Torrid Zone and the non-existence of the antipodeans.

The rise of the terraqueous globe

The term 'terraqueous' belongs to the 17th century, with the first use I know by the French Jesuit Jean Leurechon as "terraqueous globus" (Leurechon, 1622, 19). However, this model of the Earth was consolidated in the first half of the 16th century. Following the explorations mentioned above, we find a transition in the *Libellus de insulis oceanis* (c. 1512–1516; published in 1954) by Palacios Rubios, a jurist and advisor to the Catholic Monarchs. He contrasted Augustine's denial of the antipodes and what "some think" (*ut aliqui opinantur*) considering as antipodeans the inhabitants of the new western islands discovered by Columbus (Palacios Rubios, 2013, 52–53 [Latin-Spanish]). He also mentioned that part of the new territories was no longer considered island (*insula*) but continent (*continens*) (Palacios Rubios, 2013, 50–51).

The existence of a large landmass at the western end of the Atlantic revived the Ptolemaic view of the oceans as enclosed lakes surrounded by land, as in *Esmeraldo de Situ Orbis* (c. 1508; published in 1892) by Duarte Pacheco Pereira, quoting Perez de Valencia, or in João de Castro's *Tratado da Esfera* (c. 1535, published in 1940 and 1968).[15] This could be reinforced by linking America and Asia towards the North Pole and by a southern Antarctic landmass enclosing the Atlantic. This last feature appears in the 1513 Piri Reis map fragment, which called the Atlantic Ocean a lake, and the 1519 Lopo Homem full map (Hiatt, 2008, 201).

Nevertheless, the parallel revival of the Cratesian theory[16] encouraged the search of a passage to the Pacific in the Equatorial Zone of the newly discovered America. This was sought by Columbus, but only discovered by Magellan's expedition (1520) in the south of America (mirroring the rounding of Africa by Vasco da Gama to arrive to India in 1498). Although the North Passages proved elusive, these discoveries in the south showed that it was possible to circumnavigate the whole planet, as the oceans enveloped landmasses all around the Earth. This new distribution of land and water over the surface of the Earth destroyed the scholastic eccentric model (Randles, 1994[2000], 58–61). Although not published at the time, the Magellan-Elcano's circumnavigation (1519–1522) – crossing several times the Equator – led Pietro Pomponazzi in Bologna (1523) to realise that this was also the end of Aristotle's authority on cosmography (see pp. 128, 159).

The last Aristotelian/scholastic resistance to the terraqueous globe in the 16th century

By 1500, the rejection of the 10:1 relation of water to earth[17] appeared in the encyclopaedias by Gregor Reisch (1503; see p. 121) and Volterra (1506, fol. 5v). However, staunch Aristotelians would continue to defend it and the eccentric model, in some cases without mentioning – but in others indeed being aware of – the new geographical discoveries. Examples like Ciruelo, associated with the University of Paris, were John Major (1519, composed in 1510), Juan de Celaya (1518) and Gaspar Lax (1527).[18]

These conservative Parisian scholastics were heavily criticised by Antonius de Ferrariis, known as Galateus (1558, written in 1501). He was very much aware of the Iberian navigations, but as a humanist, defended the terraqueous model in the name of classical Antiquity against the innovations of the 'moderns', terminology used for the medieval scholastics of the eccentric model who were unaware of Ptolemy and Strabo. This approach was in line with the humanist anti-medieval/pro-classical agenda and used the recent evidence of the navigators selectively, silencing its incompatibility with many aspects of ancient geography (Vogel, 1995, 426–444). Similarly, Johannes Glogoviensis (1506), quoted by Randles, 1980, 37, defended Ptolemy on the habitability of the Torrid Zone, referring to the Portuguese navigations (although he expanded his oikoumene to include recently described arctic regions).

Even in Paris, Major's student, Jean Dullaert (1514), already initiated the change, realising that the newly discovered landmass to the south and west demonstrated that a large part of land emerged towards both poles (Randles, 1993b[2000], 22). Another former student of the University of Paris, Pedro Margalho (1520), defended that the Earth was not a perfect sphere but contained protrusions projecting out of the seas (Margalho, *Phisices Compendium,* 1520, fol. 3va). Margalho copied Volterra's criticisms of the eccentric model, pointing out that earth and water elements were "mixed" and that the value of a degree was the same on land as at sea (Margalho, *Phisices Compendium*, 1520, fol. 7rb).[19] This last idea reappeared in 1537 with Pedro Nunes, *Tratado da Sphera* 1 (Randles, 1993b[2000], 22) and was popularised by Pedro Mejía, *Silua de varia lecion* 3.18 (1540).[20]

The triumph of the terraqueous model

Outside Paris, the opposition mounted. In *Esmeraldo* 1.2 (c. 1508; published in 1892), Pacheco defended the terraqueous model: "the Ocean is only a very large lake set in the concavity of the earth and that the earth and sea together form a sphere" (Pacheco, *Esmeraldo* 1.2, c. 1508).[21] However, one of the most influential early public defences of the terraqueous globe was in a letter to Rudolf Agricola the Younger by Joachim Vadianus (1515, with several 16th-century re-editions, see pp. 122–123). In a passing reference, Erasmus ridiculed the eccentric model, and the denial of the antipodes, in a letter to the two dukes, Frederick and George of Saxony (5 June 1517), published at least at early as in the preface of Erasmus, 1533, fol. 3^{v}.[22] The explorer Martín Fernández de Enciso supported the terraqueous model, "the water and the earth together make up one body" (Enciso, 1519, fol. a.iij^{v}),[23] as did Fernán Pérez de Oliva. A former student and teacher in Paris, he referred to the experts measuring the land and the sea (*Terremarisque mensores*) who consider "a single sphere consisting of water and earth" (1526–1527, manuscript published in Pérez de Oliva, 1985, 74).[24] The terraqueous model can be found in popular books, such as a didactic play by John Rastell (c. 1520)[25] or a science textbook by Peter Apian (1524).[26] Verrazzano, reporting in a letter to the French King Francis I his explorations of the North American coast (1524; published in the 19th and early 20th centuries), rejected that there was less earth than water (Verrazzano, 1841, 51–54; 1970, 141–143). The conviction that the Southern Hemisphere was not underwater was shared by Franciscan monk and cosmographer Franciscus Monachus (c. 1526–1529), "to me it seems very likely that that part of the earth is not covered and enveloped by ocean" (Monachus, c. 1526–1529, fol. B.6^{v}),[27] who also rejected the eccentric model (Monachus, c. 1526–1529, fol. A.8^{r}). Jean Fernel defended the terraqueous globe based on "the writings of learned men and the reliable evidence of navigators" (Fernel, 1528, fol. B.j^{v}).[28]

João de Castro, in *Tratado da sphaera por perguntas e respostas* (c. 1535; first published in 1940), rejected the eccentric model, resorting to modern navigations to defend that water and earth "together make one single sphere" (Castro, 1968a, vol. 1, 51),[29] words used by Pedro Nunes in 1537: "These two elements, water and earth together, both make one sphere" (Nunes, *Tratado da Sphera* 1, 1537a, fol. a.iiij^{r}, side note).[30] Francisco Faleiro used similar language: "water and earth lie together [...] they make a round body" (Faleiro, 1535, fol. a.v^{r}).[31] Faleiro also rejected that the Southern Hemisphere was under water (Faleiro, 1535, fol. c.iii^{r}). Also in 1535, Francesco Maurolico wrote a cosmography book, published later, where he explained the terraqueous model in the first pages (with references to the Iberian navigations) and illustrated it with a diagram showing the insignificance of the protrusions projecting from the surface of the Earth compared to its entirety (Maurolico, 1543, fol. 8^{r}). In a later treatise, Maurolico criticised the "ignorance of antiquity" (*antiquorum ingnorantia*), opposing the recent geographical discoveries to the eccentric model (Maurolico, 1558, 7). Alessandro Piccolomini affirms that "all agree that the Earth together with the Water make a spherical and round body" (Piccolomini, 1540, fol. 10^{v}).[32] The preface of Sebastian Münster's edition of the *Geographia* of Ptolemy opposed the recent geographical discoveries against the eccentric model (Münster, 1540, fol. aa2^{v}),[33] and Pedro de Medina, in his famed *Arte de nauegar*, entitled chapter 13 as "How both elements the earth and the water make a round body" and continued "Note that these two elements, earth and water, both together make a round body" (Medina, 1545, fol. ix^{r}),[34] as well as Pedro Mejía in his *Dialogo de la tierra* 5 (*Coloquios o dialogos*, 1547, fols cxli-clii) and Rembert Dodoens in *Cosmographica*, completed in 1546 and published in 1548 (fol. D.ij^{r}).

By the mid-16th century, when Copernicus published *De revolutionibus orbium coelestium* (1543), which had a moving Earth based upon the terraqueous model, this was already a well-established model of the Earth (see pp. 159–161). In fact, Copernicus's ideas go back to his unpublished *Commentariolus* (c. 1514). From the 16th century onwards, the denial of the antipodes was seen as a bizarre idea until the 19th-century flat-earthers (Garwood, 2001; 2007[2008]).

From admiration to criticism of ancient geographers and praise of modern discoverers

The geographical/cosmographical revolution that we saw in the previous section was the first outright challenge to the authority of the ancients. Their defeat would give rise to an essential change of attitude towards the ancients. The exaltation of the moderns and the rise of their concept of superiority in parallel with a condescending view of the ancients was a necessary step in the process that led to the rise of the flat error c. 1600. We will investigate this process in connection with geography, as it was in this field that the contrast between the received wisdom and the new discoveries was first exposed. Eventually this affected theology, and as in geography, strategies were deployed to resist first – and later to integrate – the new challenging findings, which ended in the erosion of the ancient authorities. Here we will see how the transition challenged 'secular' authorities in geography; while in the next section (pp. 119–135), we will explore how this change affected the traditional Christian authorities in theology.

The geographical errors of the ancients

The new geographical information led to four possible outcomes for the reception of ancient knowledge: confirmation, supplementation, correction or rejection. The last three implicated some kind of 'error'. As Europeans realised how wrong most of their inherited geographical traditions were, they increasingly turned their old veneration for the past into contempt and criticism. The Portuguese Chronister Eanes de Zurara or Azurara (1453; published in 1841) already indicated that at the time of the Portuguese Prince Henry the Navigator (1394–1460), nothing "was known with any certainty" beyond Cape Bojador, and mentioned the "delusion in which our forefathers ever lived" and "their former error" (Zurara, 1841, 44, 359–360; 1896, vol. 1, 27, 229–230). Another Portuguese, Diogo Gomes (c. 1484–c. 1502; published in 1847), in describing his voyages c. 1460, realised the error of Ptolemy in considering the Torrid Zone as uninhabited (Gomes, 1847, 23). Columbus himself used the Portuguese discoveries against some of what he read in Ailly's *Imago Mundi*,[35] as Geraldini did against Lyra/Burgos in 1491.[36]

Although the Portuguese navigations were challenging the inherited geography in the 15th century, the rounding of Africa had been already achieved before for those accepting ancient written testimonies. Therefore, the moderns would have just equalled the nautical achievements of the ancients (Hooykaas, 1980, 11; see pp. 113–115). Such a caveat was smashed after the discovery of America, never mentioned in ancient literature, a clear achievement where the moderns surpassed the ancients, in finally finding both the antipodes *and* antipodeans. All this did not go unnoticed, as can be seen from the success of Vespucci's *Mundus Novus* (1502–1503).

The historian Anthony Grafton used as an example of these criticisms the words of the Spanish Jesuit José de Acosta (Grafton, 1992[1995], 1–2). He quoted *Historia Natural y Moral de las Indias* 2.8, 1590, 102, where Acosta reported that in his first crossing of the equator, he felt cold and laughed at Aristotle's *Meteorology* and his Torrid Zone. We will see below in detail Acosta's criticisms of the ancients, however, by his time, there was more than a century of criticisms of ancient errors. Grafton is right in that such criticisms did not destroy the confidence in the ancients, but we cannot ignore the quantity and strength of those attacks. Failure to do so is largely responsible for the mistaken impression that the challenge of the ancients belongs to the 17th century.[37] To show that this is not a vain claim based on some anecdotal evidence, I have investigated criticisms of the ancient errors in geography from the late 15th century to the first half of the 16th century.

I found such criticisms in the following authors, from all over Western Europe:[38] the Portuguese Gomes Eanes de Zurara (1453),[39] the Portuguese Diogo Gomes (c. 1484–c. 1502),[40] the German Hieronymus Münzer (1493a,[41] 1493b[42]), the Italian Peter Martyr d'Anghiera (1493–1530),[43] the Alsatian Sebastian Brant (1494), the Spanish Antonio de Nebrija (c. 1503), the German Valentim Fernandes (1503),[44] the Italian Amerigo Vespucci (1502–1503, c. 1505), the Alsatian Mathias Ringmann (1505, and with the German Martin Waldseemüller, 1507), the Italian Raffaello Maffei da Volterra (1506), the Portuguese Duarte Pacheco Pereira

(c. 1508),[45] the Portuguese Tomé Pires (1512–1515),[46] the Swiss Joachim von Watt [Vadianus] (1515), the Flemish Peter Giles (1516), the Duth Erasmus of Rotteram (1517),[47] the Spanish Martín Fernández de Enciso (1519), the Portuguese Pedro Margalho (1520), the Transylvanian Maximilian von Sevenborgen (Maximilianus Transylvanus, 1523), the Italian Pietro Pomponazzi (1523),[48] the Italian Giovanni da Verrazzano (1524),[49] the Spanish Gonzalo Fernández de Oviedo (1526, 1535), the Spanish Fernán Pérez de Oliva (1526–27),[50] the Flemish Franciscus Monachus (c. 1526–29), the Spanish Bartolomé de Las Casas (1527–c. 1560),[51] the French Jean Fernel (1528), the Spanish Ferdinand Columbus (c. 1528),[52] the Portuguese João de Barros (1532), the Portuguese João de Castro (c. 1535,[53] 1541,[54] c. 1545[55]), the Germans Simon Grynaeus and Johann Huttich (1532), the Portuguese Francisco Faleiro (1535), the Portuguese Pedro Nunes (1537), the Italian Alessandro Piccolomini (1540), the German Sebastian Münster (1540), the German George Joachim Rheticus (1541b,[56] 1541c[57]), the Italian Francesco Maurolico (1543, 1558), the Spanish Pedro de Medina (1543, 1552, 1561), the French Jacques Cartier (1545), the French Peter Ramus (1546), the Spanish Pedro Mejía (1547), the Flemish Rembert Dodoens (1548), the Italian Giovanni Battista Ramusio (1550), the Italian Pietro Bembo (1551),[58] the Spanish Martín Cortés (1551), the Spanish Francisco López de Gómara (1552), the English Richard Eden (1553, 1555), the Welsh Robert Recorde (1556), the Italian Francesco Guicciardini (1561–1564), the Portuguese Garcia da Orta (1563), the French Jean Bodin (1566), etc.

When the Portuguese writer Luís de Camões, in 1572, wrote that "the rough sailors" believed, based on their "long experience," certain things that scholars "judge to be false or misunderstood," this was a well-trodden topic (Camões, *Os Lusiadas* 5.17, 1572, fol. 82ʳ).[59] Forty years earlier, Gonzalo Fernández de Oviedo, the Spanish soldier and chronicler of America, had stated with irony: "What I have said cannot be learnt in Salamanca, Bologna or Paris" (Oviedo, *La historia general...* 2.9, 1535, fol. xiiiiʳ).[60] Proud of his navigation skills, Oviedo affirmed that he would be able to sail better than the famed scholar Ciruelo, with all his theoretical knowledge, and that he could show Pliny his errors in considering the Torrid Zone without inhabitants (Oviedo, *La historia general...*, 21.5; although this portion of the book was only published in Oviedo, 1852, vol. 2, 126–127).

The defences of the ancient wisdom

Any analysis of the impact of the navigations should consider that the new geographical data (as well as the new science in astronomy, medicine, physics, chemistry, etc.) had to confront both scholastics and humanists for its independence from Antiquity. This confrontation was long and complex, as the ancient texts displayed remarkable resilience in the face of new knowledge and ideas, as historians John H. Elliott, Anthony Grafton, and others have shown (see pp. 135–140).

Indeed, some scholastics and humanists tried to ignore, minimise or dismiss the problem, remaining within the limits of the ancient and medieval worldviews. The Aristotelian philosopher, Alessandro Achillini (1505, fol. 149b), finished his discussion on the habitability of the Earth leaving the actual geographical and anthropological details – which he considered of low importance – to the *historiographies* such as that of Marco Polo (Nardi, 1965, 41; Matsen, 1974, 31). Cochlaeus, in the preface of his edition to Mela's *Cosmographia* (1512), affirmed on the news about America: "Whether it is true or a lie, it has nothing ... to do with Cosmography and the knowledge of History" (Elliott, 1976, 14). Even authors in contact with Portugal followed this view, supporting the ancients and downplaying recent achievements (Hooykaas, 1979, 16–21; 1980, 233–239). Some attacked and scoffed at the new achievements and knowledge developed in the wake of the discoveries, such as Sá de Miranda (Albuquerque, 1984, 428–430) and Diogo de Sá (see pp. 124–125).

For those who wanted to accommodate the new discoveries, one strategy was to choose among ancient authorities those who agreed with the new findings, and ignore the others, to maintain the superiority of the ancients. With so many ancient authorities, and such a diversity of views, it was possible to play this game endlessly.[61]

A sophisticated related strategy defended that even what seemed now completely new was somehow already known. Scholars examined ancient texts to find even tenuous 'anticipations' of any new development in explicit or symbolic and allegorical ways. A pertinent example was the circumnavigation of Africa, for which several reports – even if disputed – existed from Antiquity (see Chapter 2, endnote 22). A navigator such as Pacheco, in the dedication of his *Esmeraldo de situ orbis* to the King D. João II (c. 1508), accepted ancient circumnavigation reports, being content with the merit to have lived during "a new beginning," witnessing the recovery of a navigation achievement "that during 1500 years or more it was altogether forgotten and dead" (Pacheco, 1937, 2).[62]

This was also how the initial impulse of Portuguese navigation was framed by Damião de Góis (1567). By connecting Prince Henry with reading ancient circumnavigation reports, Góis could claim his voyages to be a *rediscovery* of maritime routes (Hooykaas, 1980, 240–241). The same idea appeared in a 1491 letter of the Italian Angelo Poliziano to King D. João II (Poliziano, 1498, fol. n.ii^{v}–iiii^{v}; see Gautier Dalché, 2011, § 32). However, the opposite was also possible, as Hieronymus Münzer affirmed of Prince Henry that: "The Infant [Henry, the Navigator] rejoiced to know by experience that there were inhabitants against the [opinion of the] cosmographers, who denied that there were inhabitants in the Torrid Zone."[63]

Amazingly, this approach was even applied to America, and by Columbus himself! Even the Bible was searched very soon for shreds of evidences of transoceanic trips to America (in particular references to navigations to the mysterious Tarshish and Ophir).[64] The most dramatic challenge to ancient learning, the existence of a whole unknown continent (America), was 'found' *a posteriori* in ancient texts.[65] In this way, the new discoveries amounted just to making the moderns equal, not superior, to the ancients.

In contrast with these 'ideological' defences, some authors recognised that some ancients had indeed defended correct ideas, but that they frequently had been a minority. This accurately reflected the contradictions and debates of the past, and was recognised at the time by some authors, such as Guicciardini (see p. 106).[66]

Another, fairer, approach was to abandon finding ancient precedents for specific modern discoveries and to emphasise the necessity of the background of ancient learning (geometry and astronomy in particular) for modern breakthroughs. The navigator João de Castro, in the prologue to Prince D. Luís of his *Roteiro* of the Red Sea (1541), was conscious that even if his contemporaries were more advanced than the ancients in navigation and geography, "in all other arts we are without doubt surpassed by them," praising Hipparchus, Archimedes, Ptolemy, Euclid, Aristotle and Pliny (Hooykaas, 1980, 245, see also 402).[67] Other authors with similarly balanced views were Barros, Camões, Damião de Góis, Münster, Nunes and Orta (Hooykaas, 1979, 16–17; 1980, 243–246; in Rheticus, 1984, 127).

Other authors simply recognised the limitations of the ancients who were, after all, humans. Some, as Anghiera (1516, fol. f.iiii^{v}) or Monachus (c. 1526–29, fol. A.3^{v}), even referred to the famous words by Pliny: "that one born a human being should not possess all human knowledge" (Pliny, *Natural History* 3.1.1, 1942[1961], vol. 2, 3).

The praise of the modern discoveries and their makers

For those on the side of the moderns, the recent discoveries not only cast doubts on the real wisdom of their predecessors, but promoted the modern discoverers, sailors and explorers as the new heroes of their time. After centuries of accepting the superiority of Antiquity, some Europeans felt – for the first time – that they were living in times equal if not superior to those venerable old ages, entering for the first time in genuinely new worlds. The expression "new world," "another world," or similar were not only used for a new landmass, separate from the old world oikoumene, as in the case of America. Much earlier similar terminology was used for radical new places, even if they were close or attached to the old world, as in the case of the Canary Islands, Cape Verde Islands or the new African continental south.[68] However, after the *Mundus Novus* of Vespucci (1502–1503), this term became identified specifically with America.

To the delight of some and the horror of others, moderns saw themselves in possession of wisdom unavailable to the best ancient philosophers and geographers. The Portuguese mathematician Pedro Nunes affirmed with pride that "There is no doubt that the navigations of this kingdom in the last hundred years are the greatest, most wonderful, most elevated and most important events performed by any people of the world" (Nunes, 1537d2, fol. B^{r}).[69] The physician Garcia da Orta affirmed: "I say that you can get more knowledge now from the Portuguese in one day than was known to the Romans after a hundred years" (Orta, *Colóquios* 15, 1563, fol. 60^{r}).[70] Similarly, in the manuscript *Livro da fábrica das naus* (*Book on the Construction of Ships*, c. 1580, published in 1898), the priest Fernando Oliveira affirmed that the Portuguese have done more for navigation in 80 years than the Greeks and the Romans in 2,000 years (quoted by Contente Domingues, 2007, 475). As historian Jorge Cañizares-Esguerra summarised: "The Portuguese and the Spanish confidently saw themselves as the first 'moderns,' superseding the ancients" (Cañizares-Esguerra, 2004, 86).

The French author Jean Bodin, who conceded great ingenuity to the ancients, could also conclude that "No one, looking closely into this matter, can doubt that our discoveries ought to be compared with the discoveries of our elders; many ought to be placed first." He then mentioned that while the ancients "lived entirely within the Mediterranean basin, our people, on the other hand, traverse the whole earth [...] into another world, [...] to open up the farthest recesses of India." For Bodin, this showed "how much advance has been made" "in geography" as "information about India which used to seem fabulous to many (for Lactantius and Augustine said that those who believed in the antipodes [*antichthonas*] were crazy) have been verified by us [...]" (Bodin, 1566, 360; 1945, 301, slightly modified). He also discussed recent achievements such as the magnet, the printing press, etc.

The contrast between ancients and moderns was echoed in the disparity between the old and the new world. Indeed the sudden (over)use of "new world" and "antipodes," terms traditionally associated with impossible fables, points to the amazement of the discovery and its perception as a discontinuity, an enormous leap forward, in human history. As the Spanish historian Francisco López de Gómara famously stated in the dedicatory to Emperor Charles V of his *Historia*: "The greatest event since the creation of the world (excluding the incarnation and death of Him who created it) is the discovery of the Indies" (Gómara, 1552, fol. ii^{v}).[71]

Indeed, in a letter from 1448 to Prince Henry, the Papal Secretary Poggio Bracciolini flattered him, asserting that "you [...] went where none of the ancients, neither emperor nor king, ever went, as far as we have heard or read" (Bracciolini and Varthema, 1963, vii). The Portuguese Vasco Fernandes de Lucena affirmed to Pope Innocent VIII, in the 1485 *Oration of Obedience* of King D. João II, that his father D. Alfonso V, brought to the Church "new provinces, new kingdoms, new islands, and, as it were, new and unknown worlds" (Lucena, c. 1485, fol. 5^{r}).[72] Although he sided with the idea of a rediscovery of lands long forgotten, the Italian Angelo Poliziano, in a 1491 letter to D. João II, praised him for "... having discovered other lands, another sea, other worlds, indeed other stars and for having made them emerge from the eternal shades of ancient chaos to the light of day" (Poliziano, 1498, fol. n.iii^{r}).[73]

Similar listing of Portuguese achievements appeared in Geraldini (1491[1631]), Luis Vives (1531), Pedro Nunes (1537), António Luis (1539), Camões (1572), and in Góis (1593). Curiously, references to these "new" worlds were also applied to Copernicus, as in a letter from Gemma Frisius to Bishop Dantiscus (20 July 1541), affirming that this new astronomy "will bring us a new earth, a new sun, new stars, in fact, a wholly different universe."[74]

The connection between this new world or these new worlds and the antipodes was already used by the Italian Geraldini in favour of Columbus in 1491 (Geraldini, *Itinerarium* 14, 1631, 204–205, 209; see p. 121) and by the Venetian ambassador Pietro Paschalici in praising the king of Portugal in 1501 for "having created an empire in the antipodes" uniting "two worlds" (Randles, 1961[2000], 12). The expression "new world" appeared in the text and title of Vespucci's *Mundus Novus* (1502–1503); indeed, in the original 1502 third letter to Lorenzo di Pierfrancesco de Medici, Vespucci affirmed, "I was in the region of the antipodes" (quoted by

Hiatt, 2008, 193). The German Valentim Fernandes described Brazil in 1503 as "another orb unknown to all authors" and as "antipodes" (Fernandes, 1958, 374, 375).[75] The Portuguese Castro, in *Tratado da sphaera por perguntas e respostas a modo de dialogo* 1 (c. 1535), pointed out that only the new navigations disproved the eccentric spheres model and solved the old doubt about the existence of antipodeans (Castro, 1968a, vol. 1, 50–51).

With Columbus's explorations, the same rhetoric entered Spain. Focusing specifically on early references to the antipodes, some 1493–1494 letters by Anghiera (known at the time, and later published in Anghiera, 1530a) mentioned the discovery of new "islands," "India," etc.[76] and praised Columbus as the "discoverer of a new world (*novi orbis*)," "antipodes," "Western antipodes," "Western and Antarctic antipodes," "Western antipodes of a new Hemisphere," "antipodes in the Western Hemisphere." Columbus wrote to the Catholic Monarchs, regarding his third voyage to America (1498), that they "have become the masters of another world" (Columbus, 1847[1870], 148); and in a letter to the nurse of Prince Juan of Castile (1500), he described this voyage as "to the new heavens and new earth, which had been hitherto concealed" (Columbus, 1847[1870], 154). In 1510 Anghiera praised Spain for revealing "ſoo many thowſandes of *Antipodes* which leye hyd befoꝛe and unknowen to owre foꝛefathers" (Anghiera, *Decades* 1.10, 1516, fol. d.i^{v}).[77] These modes of praise soon became influential:

> [...]. Peter Martyr's opinion seems to have been the dominant one among the Italian Humanists. In a sermon at Rome in 1497 one of them described how Columbus had taken the Name of Christ to the Antipodes "which previously we did not even think existed," and no long afterwards another in Florence designated Columbus's discovery as "the other world opposite to our own."
> (Fernández-Armesto, 1991, 97)

In the reply from the directors of the Bank of Saint George in Genoa (1502) to a previous letter from Columbus, they affirmed that "you have discovered such a considerable portion of this earth and sphere of the lower world, which during so many years past and centuries had remained unknown to its inhabitants."[78] Giacomo Filippo Foresti (1503) considered the land discovered by Columbus as *extra orbem* (Randles, 1980, 70), while Vespucci (1502–1503) and Waldseemüller (1507) called it, in Isidorian terminology, the fourth part of the world (Randles, 1980, 71–72). References to the new lands inhabited by antipodeans appeared soon in Nebrija's cosmography (c. 1503; see p. 93) and Fernando Téllez's *Oration of Obedience* to the Pope (1508).[79] A supposed 1519 portrait of Columbus has a legend indicating that he was "the first to enter by ship into the world of the antipodes."[80]

In line with the greatness of such deeds, kings and sailors were praised as finally surpassing ancient Graeco-Roman heroes, such as Alexander, Roman emperors or Jason and Hercules.[81] The Franciscan Monk and cosmographer Franciscus Monachus considered that the empire of Alexander was not even a tenth of that of Charles V, which extended into the antipodes (Monachus, c. 1526–29, fol. B.5^{v}). Cosmographer Pedro de Medina called his own times "a golden age that the ancients called of Saturn" (Medina, 1552, fol. 44v).[82] In the same book, a letter from Alonso de Chaves, chief pilot and cosmographer of the Spanish Monarchy, points out that the Spanish territories in America were larger than the territories of "Greeks, Romans and Carthaginians" (Medina, 1552, fol. 45r).[83] The reference to Saturn was already present in Martín Cortés de Albacar, who considered Charles V superior to this ancient legislator (Cortés, 1551, fol. iir v) and affirmed that "If the ancients had reached what the moderns achieve, the Indies would not be awaiting to be discovered" (Cortés, 1551, fol. iiijr).[84]

In parallel with a 'secular' interpretation, a theological providentialist framing was observed in descriptions of Portuguese explorations of Africa as well as the Spanish arrival in America and world circumnavigation. Kings and sailors were compared with biblical characters, such as Noah, Abraham, Moses, Joshua, David, Salomon and even the Queen of Saba,[85] and were considered elected by God for these great deeds. Monachus even applied to Charles V the words of Psalm 18/19:4/4-5 (Monachus, c. 1526–29, fol. B.5v).

In finding a reason for such dramatic developments, one line of argument connected the novelty of the recent discoveries with a missionary justification of these expeditions to expand the Christian faith. As early as Zurara's *Crónica do descobrimento e conquista da Guiné* (1453; published in 1841), Prince Henry's achievements were exalted for bringing to these distant lands "the divine light" (Zurara, 1896, vol. 1, 7).[86] This religious, missionary justification was repeated endless times in Portugal and Spain.

Of course, all of the above could be framed in a larger eschatological context, as in the case of Columbus in his later years. He saw himself as guided by the Holy Spirit and the discovery of America as a providential event, designed to complete the propagation of the Gospel to the whole Earth in preparation for an imminent Parousia.[87] Other events in Columbus's eschatological scheme leading up to the Parousia were the finding of the Garden of Eden (which he believed to have located in modern Venezuela during his third voyage) and a last Crusade to reconquer Jerusalem and rebuild the temple under the leadership of a universal Spanish ruler.[88] Indeed the language of new heavens and new earth was filled with biblical eschatological references that Columbus exploited.[89] These ideas seem to have been more overwhelming in the last years of Columbus's correspondence and his *Book of Prophecies* (c. 1501–1506, unfinished at the time of Columbus death; first published in full in Columbus, 1894).[90] Interestingly, the "geoeschatological"[91] ideas of Columbus were paralleled by a 'chronoeschatological' framework dominated by the conviction of an approaching second coming of Christ according to the millennial week scheme of 6,000 years of human history that he took – through Ailly – from none other than Augustine, *City of God* 22.30.[92] These ideas also resonated with the Franciscan Joachinist views.[93] It is important to note that Columbus had strong links with Franciscans in Spain – in his last days, he even wore their habit – and that this order had an important influence in the initial stages of the Catholic missions in America (Flint, 1992a; Segundo, 2012).

This gave the 'moderns' an edge of superiority over previous generations unable to carry out this mission. The superiority of the moderns did not downplay religion (as in later centuries), but rather, the opposite. What could be seen as a failure of Christianity (for ignoring or denying the antipodes) was transformed into missionary zeal. The result was the ideal of a providentially designed universal Christian empire that surpassed the limited Hellenistic and Roman empires (Rubiés, 2017, 57–61). Of course, different nations aspired to take this position, first the Catholic Iberian nations and later Protestant countries of northern Europe. A curious case is that of Jean Pierre de Mesmes, who in his *Les institutions astronomiques* (1557) indicates that in the last century and a half a series of great personalities have increased the astronomical knowledge and hopes for God to send even more enlightened researchers (see summary in Thorndike, 1941, vol. 5, 20–21).

By the end of the 16th century, the superiority of the moderns was taken for granted in most fields, even though the 17th century still saw the famed *querelle des anciens et modernes*, mostly centred on the arts, as Elliott noted:

> The Europe of 1600 was confident of itself-more confident than the Europe of a hundred years before. [...]. By 1600, having conquered America and brought it within the confines of his intellectual world, the European could survey the earth with pride, confident in his own spiritual and technical superiority, his military capacity and his economic power.
>
> (Elliott, 1970[1992], 53)[94]

This revolutionary change in the European mindset led eventually to transform the traditionally pejorative terms 'new' and 'modern' into positive adjectives.

From the defence to the criticism of ancient Christian authors on cosmography

Soon, the criticisms directed against ancient geographers and philosophers were also addressed to early Christian authors, for centuries considered authorities second only to the Bible. This change in attitude towards them enabled the rise of the flat error. The following will sketch how this change in attitude was produced.

Lilio and the 'last' rejection of the antipodes at the end of the 15th century

The canon Zaccaria Lilio (?–c. 1501–1506) launched what was probably the last defence of Lactantius's and Augustine's denial of the antipodes. His *Orbis breviarium* (1493) was a handbook of – by that time – already outdated geography from ancient sources. The book started with a description of the five zones, and Lilio remarked that, according to Macrobius, nothing was known of the Southern Temperate Zone (the word 'antipodes' was not even used), and nothing could possibly be known in the future, from where it was concluded that only the Northern Temperate Zone was inhabited. This was followed by traditional zonal and T-O maps. Close to the end of the book, he described the inhabited space as longer than wider, quoting Cicero to affirm: it "is a small island" (Lilio, 1493, fol. q.ii^{v}).[95]

Contra Antipodes is the second of five treatises, published in a single volume in Florence in 1496, where Lilio tackled what he left out in 1493: the antipodes. He denounced, with Lactantius and Augustine, that the antipodes/antipodeans were just a *logical inference* (a "conjectural reasoning" [*ratiocinãdo cõiectãt*]) and that there were no *historical reports (hisſtoriæ)* about their existence, as ancient explorations did not find them (Lilio, 1496, fol. eiiii^{r}).

Therefore he criticised the Earth model implied in the belief of the antipodes in a section entitled "That the Earth is not round" (*Qvod terra not sit rotvnda*, Lilio, 1496, fol. e.iii^{v}–eiiii^{r}). As historian Klaus Anselm Vogel points out (1995, 408), this was opposed to one of the traditional sections at the beginning of astronomy/cosmography textbooks, such as book 1 of Sacrobosco's *Sphaera*: *Qvod terra sit rotvnda*, which was followed by *Qvod aqva sit rotunda* (Thorndike, 1949, 81, 83). However, Lilio by no means endorsed Lactantius's flat Earth.[96] This is a crucial point to keep in mind to avoid a misleading view of the reception of Lactantius on the sphericity of the Earth in the late 15th century. Lilio changed the meaning of the term "Earth" to discuss not the planet (as the more astronomical/cosmographical text of Sacrobosco) but the oikoumene (as the geography texts). He first mentioned Ptolemy and Pliny defending that the oikoumene was longer than wider. In the second place, having only one oikoumene filled with mountains and plains, Lilio concluded that the planet could not be *perfectly* spherical – *abſoluti orbis forma* (Lilio, 1496, fol. eiiii^{r}). Therefore the denial of the sphericity of the Earth was only a denial of the *perfect* sphericity of the Earth, an objection already dispelled by Pliny![97]

Lilio mentioned the Iberian navigations to India, although denying that they were a new thing (Lilio, 1496, fol. f.ii^{v}). As he defended the eccentric model of the Earth, how would he then accommodate the recent discoveries in South Africa? Lilio enlarged the emerged portion of the Earth in this model. Making again good use of Macrobius and Cicero's *Dream of Scipio* (as in 1493), Lilio asserted that those in the south were not truly antipodes of the Europeans. They just lived in an inclined (*devexa*) land, home of *antoeci*, although Lilio used the term *obliquos*. However, the existence of true antipodes, in the lower – Western – Hemisphere, was for Lilio "absurd and impossible" as he considered it underwater (*quod abſurdum uidetur & ĩpoſſibile*, Lilio, 1496, fol. f.iii^{r}). By the date this book was published, Columbus was still returning from his second voyage to America. The news about a few islands in the Western Ocean found during his first voyage (see pp. 107–108), which prompted Anghiera to speak of western antipodes (see p. 117), were probably of insufficient weight for Lilio to change his cosmography.

There were also theological concerns. Lilio noted on the first pages that, according to the Bible, human history had two population bottlenecks, Adam and Noah, and that there was no reference to people going to the antipodes at that time (Lilio, 1496, fol. e.iii^{r}). At the end of the book, he recalled two critical events from the New Testament: Christ's birth and death in the oikoumene, and the apostles's proclamation of the Gospel everywhere without going to the antipodes (Lilio, 1496, fol. f.iiii^{v}). A list of all the apostles and their destinations in the oikoumene followed and the section ended with words from Psalm 18/19:4/4-5 used in Romans 10:18 (Lilio, 1496, fol. f.v[v]). He closed his text as he opened it, with a desperate denunciation of the popularity of the idea of the antipodes, showing that he wrote defensively (Lilio, 1496, fol. f.v[v]).

Columbus, Geraldini, Reisch, and the defence of Augustine at the end of the 15th century

Although Augustine's denial of the antipodes was used against Columbus, it is revealing of the complexity in the usage of authorities at the time that Columbus, far from rejecting the authority of the famed Bishop, fought back to enlist his support. In the marginal notes of some of his books[98] and in a famous letter to the Catholic Monarchs during his third voyage to America in 1498,[99] Columbus defended the conviction that most of the Earth's surface was above the water and wanted to use in his support 4 Esdras 6:42, 47 (mentioned by Ailly), which affirmed that water covered only one-seventh of the Earth surface.[100] However, as the canonicity of this pseudo-epigraphical book from the 1st to 2nd centuries AD was debated, Columbus enlisted to support his opinion Augustine's *City of God* 18.36 and Ambrose's *Hexaemeron* (Ailly, 1930, vol. 1, 210–215; Columbus, 1847[1870], 146).[101] Augustine – including his *City of God* – was also frequently quoted in some of Columbus's texts and in his *Book of Prophecies* ("Introduction" in Columbus, 1997; see p. 118 and endnote 90). Like Galileo, more than a hundred years later, Columbus did not shy away from theological discussion and used Augustine as supporting authority.

We do not know Columbus's defence from Augustine's objection to the antipodes. However, we know how Geraldini, at Santa Fe (1491), approached none less than the Cardinal Mendoza:[102] "[I] suggested to him that Nicolaus of Lyra was indeed a remarkable expounder of sacred theology and Aurelius Augustine a great man by virtue of his teaching and saintliness, but that they had lacked any knowledge of cosmography," (Geraldini, *Itinerarium* 14, 1631, 204).[103] This rhetorical device – they were great men, but did not know science – would be a key solution for the following centuries, setting the pattern for other science and Christianity disagreements, and was famously used by Copernicus (see pp. 164–165). Geraldini cannot be considered Copernicus's source, as his book was not published until the 17th century.[104]

Reisch's *Margarita Philosophica* (1496, published in 1503), as usual at that time before the American continental landmass was found, combined the eccentric model (7.1.42) with the sphericity of the Earth (7.1.44), the five zones (7.1.45) and a Ptolemaic map of the oikoumene (7.1.47).[105] In 7.1.48, the populations of Western Europe and East Asia were regarded as antipodeans.[106] To that, Reisch added this excursus: "However, the divine Augustine denies their existence in *Civitate Dei* 16.9, perhaps because such a distant habitable region had not yet become generally known" (Reisch, 1503, fol. 150[r]).[107] This is a similar justification to that in Geraldini.

Vadianus and the open criticism of the ancient Christian authors in the early 16th century

As the ancients became increasingly exposed as erring in light of modern discoveries, the tide of discontent also reached ancient Christian authors such as Lactantius and Augustine, well remembered for denying the antipodes. What to do now? This was the question set by the Humanist Rudolph Agricola the Younger (1490–1521) in 1514 to his friend, the poet and Humanist Joachim von Watt (Joachimus Vadianus, 1484–1551).[108] Like Petrarch, Agricola pondered the contradiction between Augustine and his beloved Lactantius with ancient secular texts regarding the antipodes, "Pliny, Pomponius, Strabo, Macrobius and many others" (Agricola in Vadianus, 1515, fol. A.iii^{r}; Appendix, no. 17, available at www.routledge.com/9781032893068)[109] as well as modern authors, such as his teacher Laurentius Corvinus. Vadianus was working on a commentary of Pomponius Mela and took the chance to write a detailed criticism of Lactantius and Augustine. Both letters, from 1514, were printed (Vadianus, 1515), and Vadianus's letter was later attached to his commentary on Mela (Mela, 1518, with several re-editions).

Vadianus discussed the existence of the antipodes with a detailed exercise of erudition, recognising the contradictory views of the ancient sources (Vadianus, 1515, fol. B.ii^{r}; C.i^{r}), the Portuguese discovery "that the Antoecians of our climate are in the furthest parts of Africa" (Vadianus, 1515, fol. C.i^{v}; Appendix, no. 17), and the recent news about America from Vespucci (Vadianus, 1515, fol. B.iii^{v}, C.ii^{v}, C.iii^{r}). The early use of *America* betrays the influence of Saint-Dié. Vadianus denounced the errors of the ancients, criticised the eccentric model and defended the terraqueous globe (Vadianus, 1515, fol. B.ii^{r}, B.iii^{r}; Appendix, no. 17).

As expected, his opinion was negative regarding Lactantius's and Augustine's denial of the antipodes.[110] However, he distinguished between both authors:

> [...] as much as I approve the opinion of Augustine about the Antipodes, I cannot at all defend or agree with the dreams of Lactantius. The former certainly had a reason for denying their existence; the latter did not know that which he was denying.
>
> (Vadianus, 1515, fol. C.ii^{r}; Appendix, no. 17)

As we have seen before, the justification for Augustine's error was his dependence on the Cratesian revival of Late Antiquity. Vadianus, who knew that, exemplified it with Macrobius, and pointed beyond him to the misleading influence of Cicero (Vadianus, 1515, fol. C.iir^{v}, C.iii^{r}). Vadianus did not doubt that if Augustine had known what was known in the 16th century, he "would have put his trust in the demonstration of the most reliable of sciences," mathematics, accepting the antipodes (Vadianus, 1515, fol. C.ii^{v}; Appendix, no. 17).

For Lactantius, although recognising his eloquence, Vadianus had nothing but scorn, labelling his rejection of the antipodes as "inept" and "ignorant," the result of inadequate overconfidence (Vadianus, 1515, fol. C.iii^{v}). Vadianus considered as Lactantius's first mistake his rejection of the sphericity of the Earth, with a point by point refutation of Lactantius and a long erudite discussion on the antipodes (Vadianus, 1515, fol. C.iii^{v}–D.i^{r}). In contrast, he noted that Augustine "did not want to deny that the earth was round" (Vadianus, 1515, fol. C.iv^{r}).[111] He mentioned the negative opinion of the Viennese canon Andreas Stöberl (Stiborius, c. 1464–1515) regarding Lactantius on this point, again, after labelling him "a writer who was otherwise so great," repeating the accusations of ignorance and arrogance (Vadianus, 1515, fol. C.iv^{v}–D.i^{r}; Appendix, no. 17). Vadianus ended his argumentation against Lactantius with the words of Jerome to Paulinus of Nola (Vadianus, 1515, fol. D.i^{r}; Appendix, no. 17; see above p. 51).

At the beginning of his commentary on Pomponius Mela, Vadianus referred to his letter to Agricola (added at the end of the book) and affirmed that there he "said many things in support of Augustine, and against Lactantius" (Mela, 1518, fol. 5^{r}).[112] In what, according to Vogel, "may represent the beginning of the *querelle des anciens et modernes* (argument of the Ancients and Moderns)" (Vogel, 2006a, 494), Vadianus strongly sided with the experience and voyages of the moderns in his letter to Agricola and his detailed commentary on Mela.[113] This led him to a debate with his former teacher Johannes Camers (1447–1546), who regarded Christian and secular ancient authors as the fundaments for Geography and was more interested in reconstructing the old texts than criticising them in the light of recent discoveries (as in Camers's commentary on Solinus's *Geography*, 1520).[114]

Abandoning Lactantius and saving Augustine during the 16th century

Vadianus's strong criticisms have been regarded by Classical reception scholar Susanna de Beer (2013, 177), as an "exception" in the early 16th century. This exceptionality could be traced back to the late 15th century, since Marzio influenced Vadianus.[115] Indeed, we will have to wait decades to see something similar, particularly in the last decade of the century, as in Acosta. However, criticisms were common in the 16th century, even if milder.

One of the most important commentaries on the *City of God* in the 16th century was commissioned by Erasmus to Juan Luis Vives. In *City of God* 16.9 (Augustine, 1522, 488–489), Vives mentioned first that Augustine denied the antipodes (in the 1531 revision it was noted that he did so following Lactantius, fol. 195^{r}). He blamed Cicero and other pagans who taught the impossibility to access other landmasses, leading Augustine to deny the antipodes, an explanation similar to that of Vadianus.

In *Cosmographicus Liber* (1524), Peter Apian[116] first praised Lactantius for his learning, only to point later to his mistake in attacking the astronomers and the sphericity of the Earth, while Augustine was characterised as following him in attacking the antipodeans. Apian finished assuring the readers that the apostles had indeed been in antipodean locations: James (Spain) and Thomas (India). He thus illustrated a traditional example of antipodeans in a theological frame (Apian, 1524, 54–55). This text reappeared in a new edition (Apian, 1584, 56),[117] which also had a reference to the more obvious antipodean character of some places in South America (Peru) in relation to Asia (India and Sri Lanka), some islands in

Indonesia (Moluccas Islands) in relation to West Africa (Guinea), and North America (Mexico) in relation to other places in Asia (Yemen) (Apian, 1584, 57).[118] Another addition on the antipodes criticised Augustine and Lactantius. This criticism concentrated on Lactantius, again initially praising him only to point later to his errors and ignorance (Apian, 1584, 59).[119]

In his *Tratado da Esfera* (c. 1535, published in 1940), João de Castro, after carefully explaining the concept of the antipodes and its implications, indicated "that this point was so difficult to understand until our times, that many of the ancients could not be persuaded that there were such antipodeans." He saw, as the key reason for such denial, the lack of experience and described the criticisms of Lactantius without particular anger. However, he was surprised that "Saint Augustine has the same opinion," although he "without doubt was one of the greatest philosophers and the greatest minds that existed," but still "such mind as of St Augustine considered the antipodes as fables." Castro then pointed then out that only the experience of modern times solved the doubt (Castro, *Tratado da esfera* 1, 1968a, vol. 1, 57–58; my translations). That was something with which Nunes (1537d2, fol. B.i$^{r\ v}$) and Acosta agreed (*Historia* 1.8, 1.10, 1590, 33–36, 41–43).

In his *Cosmographia*, Francesco Maurolico recalled Lactantius's and Augustine's denial of the antipodes (Maurolico, 1543, fols 3^{v}, 13^{r}, 16^{v}–17^{v}) and Pliny's reference to debates over the antipodes (Maurolico, 1543, fol. 13^{r}). He explained the denial of Augustine due to the ancients's geography and the biblical idea of a single descent (Maurolico, 1543, fols 16^{v}–17^{r}), while scoffing at Lactantius (Maurolico, 1543, fols 17^{v}).

Pedro Mejía, in the third Dialogue of his *Coloquios o dialogos* (1547), introduced the topic of the antipodes, which triggered the question of what to do with the denial of the antipodes by Augustine and Lactantius (no reference is made to the sphericity of the Earth). He correctly showed that Augustine did not negate the possibility of antipodes, but just noted the lack of evidence for them at his time. Mejía pointed out that a similar situation existed in his age regarding the Antarctic region, which at that time was not known to be inhabited, but that this could change if explored in the future. He indicated that in Augustine's age, the Torrid Zone was believed to be uncrossable and that this led to the theological problem of the unity of humankind. Therefore, he concluded that Augustine's rejection was of no importance. He then dispatched Lactantius with one sentence, recognising his eloquence and sanctity, but affirming that he misunderstood this aspect and erred, as in other more important topics, although with no evil intention (Mejía, *Coloquios* 3, 1547, fol. 106^{r}–107^{v}).

After a brilliant military career in the Far East Portuguese colonies, Diogo de Sá became associated with the universities of Coimbra and Paris. He followed the Aristotelian scholasticism in *De Navigatione Libri Tres* (1549), where he stressed the hierarchical view of knowledge that gave primacy to Philosophy and criticised the mathematical ideas of Nunes (1537). Although Sá praised experience (particularly nautical experience), it was always subordinated to the scholastic cosmographical views.

This book contains a significant amount of patristic material, including Lactantius and Augustine. Both authors and the discussion of the antipodes and the sphericity of the Earth appeared in the dedication to Portuguese King D. João III. As expected in a conservative author like Sá, there is no recrimination towards these two Christian authors, and although he recognised the errors of the ancients (even Ptolemy), he excused those errors based on their ignorance and lack of knowledge of distant lands. Discussing the antipodes, he affirmed of the Southern Hemisphere, that "they [the ancients] had no knowledge of this part" (Sá, 1549, fol. 3^{v}).[120] Mathematician and historian Luís de Albuquerque noticed (1987, 68) that this justification was similar in other Portuguese contemporary authors such as Pacheco[121] or Nunes[122] in order to explain the mistakes of secular and Christian authors that were finally superseded by new knowledge, justification which we have also found in the late 15th century in Geraldini and Reisch (see p. 121). That allowed Sá to 'save' not only Augustine, but also Lactantius, a unique feature among all authors here investigated.

Following Lactantius's logic, *De Navigationi* 2.1 launched a furious attack first against the arrogance of mathematics, but then also against secular philosophy from a Christian standpoint, as a preparatory step to expose the high value of Christian faith. He relied heavily on numerous quotations from Lactantius's *Divine Institutes*, particularly book 3 (even quoting in support of his own attacks word for word *DI* 3.24.2), carefully avoiding the portion of the antipodes and the Earth's sphericity (Sá, 1549, fol. 38^{r}).[123] The use of *DI* 3 at this point in his work means that Sá understood (and imitated) the role of this text in the apologetic strategy of Lactantius.

Francisco López de Gómara published *Historia General de las Indias* (1552), a hugely influential chronicle of America. In chapter 3, he criticised not only pagan, but also Christian authors for denying the habitability of the Torrid Zone and Scandinavia.[124] Gómara mentioned several ancient philosophers (Anaximander, Anaximenes and Democritus) and one Christian (Lactantius) who believed in a flat Earth, also affirming that they "deny the antipodes" (Gómara, *Historia* 2, 1552, fol. iv^{r})[125] and, more specifically, the antipodeans (Gómara, *Historia*, 4, fol. v^{r}; see p. 126 and endnote 129). Further, he also explained that Augustine and Isidore rejected the antipodeans, and tried to save Augustine's reputation before mentioning Clement of Rome as accepting the antipodeans (Gómara, *Historia*, 1, 1552, fol. iii$^{r–v}$; *Historia*, 4, 1552, fol. v$^{r–v}$).[126] The defence of Augustine (and Isidore) was initially based on the lack of evidence for antipodeans's existence; in particular, the lack of biblical records for Augustine and lack of profane history for Isidore (apparently not realising the connection of both authors). However, Gómara gave greater weight to the theological problem of Adam's descent and humanity's unity. Correctly, Gómara linked this issue to the geographical view of uncommunicated landmasses due to the large size of the oceans and the heat of the Torrid Zone. Gómara summarised his explanation saying: "Lactantius and Isidore had no cause to deny them [antipodeans]. St Augustine had the ones I said" (Gómara, *Historia*, 4, 1552, fol. v^{r}).[127] Therefore, the priest Gómara seemed to consider that the only reasonable argument to deny the antipodes was the theological one, and on that basis saved Augustine's reputation and did not hesitate to discard both Lactantius and his famous Spanish compatriot and Bishop, Isidore of Seville.

However, there is already in Gómara confusion on the debates of the antipodes and the shape of the earth. In his first reference to Lactantius, Gómara considers those who deny the antipodes automatically to be flat-earthers, which is wrong: "However, Anaximander and Anaximenes and Lactantius, and those who deny the antipodes, affirm to be flat this round body made up of water and earth" (Gómara, *Historia*, 2, 1552, fol. iv^{r}).[128] A few pages later, he makes the reverse statement, considering that those who were flat-earthers denied the antipodeans, which is correct: "Leaving aside pagan authors, I say that there are also Christians that deny the existence of the antipodeans. Those who considered the earth flat denied them. And Lactantius Firmianus speaks against the existence of the antipodeans…" (Gómara, *Historia*, 4, 1552, fol. v^{r}).[129] This double correspondence between denying the earth's sphericity and the antipodes is a crucial element of the flat error, as it could automatically transform the ancient and medieval authors denying the antipodes into flat-earthers. However, as we can see, Gómara does not use this argument against Augustine, whom he never accused of being a flat-earther. It does not seem that Gómara was aware of having produced a flat error. Probably, he was not trying to use the expression "and those who deny the antipodes" in order to include all the authors from the Middle Ages in this category (as his examples are all ancient authors who denied both the sphericity and the antipodes). In any case, this shows how the ambiguity and lack of precision were creeping in.

In the letter to the reader of *A treatyse of the newe India* (1553), Richard Eden described Lactantius as "well learned," but disregarded his "lyghte and opinionate argumentes" (Eden, 1553, fol. A.iiii^{r}). However, Eden saved Augustine, arguing that he also made mistakes in his main area of expertise, i.e. theology, which he had to correct in his *Retractationes*. Eden was also sure that had Augustine had more training in astronomy and more experience (as the modern travellers), he would not have denied the antipodeans. Augustine would not have had any problem retracting that error as well, and Eden attacked those "ſuperſticious" who did not dare to contradict Augustine "in any poynte" (Eden, 1553, fol. A.vr). He later went back to Lactantius, sympathising with his aim of undermining pagan philosophy, but calling attention to the limitations of his learning and attacking the blunder of his denial of the Earth's sphericity. Eden also criticised Lactantius and Augustine in his additions to the translation of Anghiera's *Decades* (Eden, *Decades* 3.1, 1555, fol. 94^{r}).[130]

In his cosmographical treatise, *Castle of Knowledge* (1556), Robert Recorde ignored Augustine and centred his attacks on Lactantius for denying the sphericity of the heavens and Earth, while his denial of the antipodes, although mentioned, received little attention (Recorde, 1556, 103–104, 107–109; see Williams, 2011, 132–142). In the same year, Johann Dryander (Eichmann) wrote an introduction to the very popular narrative by Hans Staden of his voyages in Brazil (*Warhaftige Historia*, 1557). Dryander attacked together Lactantius and Augustine who although were "two holy scholars well versed in both theology and in the liberal arts," they "doubt and refute the existence of the antipodes." First he responded by pointing to the existence of antipodeans in the "upper half of the globe," comparing Spain and India and linking these places with the resting places of two apostles (as in Apian). He then attacked the eccentric model of Lyra (in fact, Paul of Burgos), "otherwise regarded as an excellent man," by referring to the "many voyages of the Spaniards and Portuguese." Interestingly, Dryander pointed his readers to Marzio's book for more information on these topics (Staden, 1557, fol. B.i^{r}–B. ii^{r}; 2008, 12–14 [translation]). In a noted historiographical book, Jean Bodin also mentioned the denial of the antipodeans by both, Lactantius and Augustine, in 1566 (see p. 116).

Popular Spanish writer Antonio de Torquemada, in *Jardín de flores curiosas* (1570), criticised Lactantius and Augustine, accusing the latter correctly of doubt about the antipodes in connection with Adam's offspring, and the former for his denial of the sphericity of the Earth, with the traditional words of surprise about such an error in the "wise and prudent" Lactantius (Torquemada, 1570, fol. 206^{r}; 1600, 114^{v} [translation]). Famed Jesuit astronomer and mathematician, Christopher Clavius's *In sphaeram Ioanis de Sacro Bosco* (1570) did not mention ancient Christian authors except for Lactantius, whose views on gravity he rejected (Clavius, 1570, 250). Clavius also defended the terraqueous sphere and is considered to have "put an end to any further rearguard resistance from the Aristotelians" (Randles, 1994[2000], 74).[131]

Before ending this review of 16th authors, we need to mention how the Protestant historian Henri Lancelot Voisin de La Popelinière discussed this topic in the first pages of his *Les trois mondes* (1582). He knew the Christian discussion regarding to the antipodes, noting the connection in Augustine with the descent of Adam and Eve (La Popelinière, 1582, fol. ſti^{v}) and he even mentioned Marzio in fol. ũiij^{v}. On the topic of the shape of the Earth he attacked generically the theologians for stepping out of their field in this discussion, mentioning "Saint Augustine in particular, Lactantius and many others" (La Popelinière, 1582, fol. ũiv^{r}).[132] In fol. b^{r} reappeared a critical reference to the theologians for mocking the antipodes. However, in a more specific discussion he noted that because the philosophers did not give any demonstration regarding the antipodes, Augustine accepted the sphericity of the Earth, but denied the antipodes. It seems that La Popelinière tries to give some excuse for Augustine, as in contrast, he considered that what moved Lactantius's denial of the sphericity of the Earth was the drive "to better refute the opinion of the pagans" (La Popelinière, 1582, fol. ũiv^{v}).[133] From this observation La Popelinière reflected on the "fragility of human nature," which wishing to "annihilate the pagan doctrine in order to advance our own" rejects ideas not sufficiently studied (La Popelinière, 1582, fols ũiv^{v}–ſti^{r}).[134] Then La Popelinière continued with the well-known story of Virgilius, for which he mentioned Aventinus as a source (La Popelinière, 1582, fol. ſti^{r}; on Aventinus, see pp. 170, 172).

A more detailed research could probably uncover more authors discussing these topics, but it seems that most writers follow the pattern shown here.

The theological implications of the discovery of the new American landmass

From the beginning, theology was part of the discussion surrounding the new discoveries, primarily because of Augustine's theological dismissal of the antipodeans. The theological concerns used to counter the possible existence of antipodeans backfired once they were found to be no longer a fable that could easily be denied, but a hard fact never to disappear. An early example is a lecture at the University of Bologna by the famed philosopher Pomponazzi on Aristotle's *Meteora*, on 18 March 1523, where he confronted the uninhabitable Torrid Zone of Aristotle and Averroes with the latest news of the Earth's circumnavigation (see p. 159). The next day he turned to Augustine's *City of God* 16.9 and posed the difficult question: "And if these antipodes are children of Adam like us, why has Christ not thought about their salvation?" (My translation from the Italian by Nardi, 1965, 43, who does not provide an actual Latin quotation). Having been condemned by the Pope in 1518 for denying the soul's immortality, he avoided the debate with an elusive answer affirming his submission to the Church: "To this I reply by saying that the friars should solve this, since I believe in the holy church," but not without sneaking the old idea of an additional crucifixion of Christ in the antipodes, a nightmare for theologians: "I do not know what the answer is, unless perhaps that Christ had himself crucified in the other pole also!".[135]

Not everyone found a reason to worry there, Augustino Giustiniani, in his *Psalterium* (1516, fol. C.vii^{r}–D.i^{r}),[136] took the opportunity from Psalm 18/19:4/4–5 to gloss over the life of Columbus, portrayed as an instrument of God to bring Christ to this New World. This interpretation, which Giustiniani said went back to Columbus himself,[137] cast a blind eye over previous interpretations of this Psalm verse and other texts from the Bible, which for centuries were used to suggest that the whole world was already known and evangelised (see Hesychius and Augustine, p. 45, Oresme, p. 83, Abulensis, p. 133, Marzio, p. 78 and Lilio, p. 121).

Others recognised that there had been a change in geographical views with potential theological implications. In 1527 the leading theologian Cardinal Cajetan discussed the global evangelisation in his commentary of Matthew 24:14. He affirmed that in his own time new nations were found by Spanish ships under the Antarctic pole, unknown in previous centuries when the habitability under the Equator was hardly known (Cajetan, 1530, fol. 55^{r}).

In his *L'Historia d'Italia* (1537–1540, published in 1561–1564), Francesco Guicciardini devoted a chapter to describing the Portuguese and Spanish geographical discoveries and how they "have demonstrated the falsity of these opinions and presuppositions" held by the ancients (Guicciardini, 1561, 448; 1969[1984], 178 [translation]). Guicciardini was very aware of further implications, not only for ancient geography but also for Christianity. He emphasised that the traditional interpretation of Psalm 18/19:4/4-5 (quoted in Romans 10:18) was no longer tenable, and he drew challenging conclusions: "it is unworthy to be believed, either that faith in Christ had existed there before these times, or that so vast a part of the world had never before been discovered or found by people of our hemisphere"

(Guicciardini, 1775, vol. 2, 57; 1969[1984], 182 [translation], slightly modified). The last sentence echoed Augustine's problem of the ancestry of the indigenous peoples of Americans. Small wonder this passage was expurgated in all editions and translations until its republication in 1775–1776. In this way, the book of Guicciardini was published with only a general criticism of the "ancients." The two problems of the Adamic ancestry and the completion of the world's evangelisation were hotly debated during the 16th century. By its end, a solution for both was found by Acosta and Góis.

However, an additional problem was the crisis of authority. Even if, in the end, there were ways of solving the specific problems raised by the new discoveries, the traditional authorities –and in particular the theological authorities – had already been questioned. Despite the efforts to save Augustine, there was no turning back, as his authority was tarnished: everyone remembered his criticisms and doubts regarding the antipodeans. In this context, the terminology of doubt was recovered by Iberian authors. In 1537 Nunes wrote that regarding the antipodes, "even the Saints doubted" (Nunes, 1537d2, fol. B.i^{v}).[138]

A more explicit connection with Augustine was found in none other than the son of Columbus – Ferdinand – who, c. 1528, in describing the objections raised to his father's voyage, remembered that:

> Finally, all held to the Castilian proverb, which in what does not seem reasonable, is wont to say, "Doubts Saint Augustine," because that Saint in the IX chapter of the XXI book [*sic*, it should be XVI] of *Civitate Dei* reprobates and considers it impossible that there should be Antipodes, and that one can pass from one Hemisphere to the other.
>
> (Columbus, 1571, fol. 33^{v})[139]

Augustine's doubts were no longer considered a cautious scepticism, but a sign of narrow-mindedness, stubbornness, etc. Bartolomé de Las Casas, frequently relying on Ferdinand Columbus, also tells this anecdote in *Historia de las Indias*, composed in 1527–c. 1560, but only published in 1875–1876: "Others alleged Saint Augustine, who, as mentioned above, denied that there were antipodes, who are those that walk opposite to our feet, and thus brought as a saying: 'Saint Augustine doubts'" (Las Casas, 1986, vol. 1, 159).[140] Acosta, in 1590, also criticised Augustine for remaining "yet still in doubt" about the sphericity of the heavens (Acosta, *Historia* 1.1, 1590, 15).[141]

To complicate matters, the same rhetorical device we saw in pp. 123–127 used to protect Augustine (wise in theology, but ignorant in science) could be turned on its head, leading to the conclusion that the wise Bishop and saint should have known better! Monachus, a monk, criticised Augustine for, despite "his divine sharpness of intellect," having joined the "heresy" of rejecting the antipodes (Monachus, 1526–1529, fol. A.8^{r}).[142] An amazing reversal of the situation of Columbus! There were also further theological implications. Monachus considered the references to Jerusalem as the centre of the world, springing from a literal interpretation of

Psalm 73/74:12, to be an error, and pleaded that instead it should be understood as a prefiguration of Christ bringing salvation to all humanity (Monachus, 1526–1529, fol. A.8^{v}).

Senensis, Mazzoni, Castellanos, Acosta, Góis, and the production of lists of mistaken ancient Christian authors during the late 16th century

Until the first decades of the 16th century, only Lactantius and Augustine were mentioned in discussions regarding the shape and habitability of the Earth. However, in the second half of the 16th century, new authors started to make their way into the discussion of these topics. The availability of new ancient Christian texts (such as Procopius of Gaza), and fresh readings of the already familiar texts (such as John Chrysostom), illuminated the historical evolution of cosmological ideas, weakening what used to be a simpler consensus. Procopius was a Christian rhetorician (465/470–526/530) and writer of biblical commentaries, whose *Epitome* or *Commentary on the Octateuch* (published in Procopius, 1555) was analysed regarding cosmographical matters in 1566 by Sixtus Senensis, a renowned Italian Catholic theologian, and after him mentioned in 1588 by the Spanish Jesuit José de Acosta.[143] In contrast, the famous Chrysostom, Bishop of Antioch (349–407), was the author of *Homilies on Hebrews*, that denied the sphericity of the heavens and thus indirectly also of the earth, and which was well-known in the western medieval tradition.[144] However, it was only with Senensis that his denial of the sphericity of the heavens seems to be connected with that of Procopius and other authors to reconstruct the ancient cosmological beliefs of some Christians who denied the sphericity of the earth, the antipodes, etc. In fact, Senensis even realised the connection and dependence of Procopius's cosmology from Chrysostom: "Procopius seems to have got this view, virtually word for word, from Chrysostom's 14th and 27th *Homilies on the letter to the Hebrews*" (Senensis, 1566, 520B).[145] In this way, Senensis was becoming aware of the cosmological implications of the Antiochian School, a fact that apparently was not completely understood until the 18th century with the rediscovery of Cosmas and his connection with the Antiochian School (see p. 141). The lists of ancient authors articulating contradictory views on important cosmographical issues (mirroring other theological debates), compiled by 16th-century scholars (Table 4.1), have further been expanded by scholars from the 17th to the 21st century. Moreover, they are still reviewed in specialised Bible commentaries or in histories of biblical interpretation.[146]

Sixtus Senensis (1520–1569), in discussing the first verses of Genesis in his landmark *Bibliotheca sancta* (1566), quoted several ancient Christians on the different cosmographical concepts put forward by the Bible (Senensis, 1520–1569, 92A–B, 520B-D, 526B-D, see Table 4.1). Senensis contrasted the evidence based on the sailings of "our merchants" with the reasonings of those who denied the antipodes. However, he also attempted to 'excuse' their mistakes based on their ignorance of the new discoveries and because the science of their day considered the ocean uncrossable. Therefore, to maintain the monogenicity, they "wisely denied the Antipodes." That did not apply to Procopius's objections, which for Senensis were due to a misunderstanding of Psalms 23/24:2 and 135/136:6 (Senensis, 1566, 526C-D).[147]

The Italian philosopher Jacopo Mazzoni (1548–1598) engaged in a lengthy defence of Dante in *Della difesa della Comedia di Dante. Distinta in sette libri* (1587, vol. 1). Mazzoni had an historical view of the development of cosmographical topics, and in book 3, chapters 29 (illustrated with a figure of the eccentric model, p. 539) and 33, provided a long list (see here Table 4.1) of ancient and medieval authors against the antipodes. He also mentioned that Lactantius rejected the sphericity of the Earth to strengthen the denial of the antipodes (that was correct, as discussed in p. 39). He pointed to some authors favourable to the existence of the antipodes, but apart from Virgilius of Salzburg, all were pagans. After a detailed discussion of the list, he investigated why the antipodes provoked such rejection. Mazzoni blamed the ancient pagan authors who denied the crossability of the Equator.[148] From that, he concluded that Augustine and "many of the ancient Fathers" ("molti de'Padri antichi," Mazzoni, 1587, 555) denied the antipodes to avoid problems relating to monogenism and to avoid a double incarnation of Christ, explaining that those were also the motives for the condemnation of Virgilius (which he knew from Aventinus, see pp. 170, 172) and likely the reasons that prompted the rejection of the antipodes in Petrarch.

The Spanish priest Juan de Castellanos (1522–1607) depicts in his *Elegías de varones ilustres de Indias* 1.2 (1589, 19) a vigorous religious opposition to Columbus (refering to Augustine, Lactantius, Isidore, Scotus and Durandus's rejection of the antipodes, although without mentioning the sphericity of the Earth). Curiously, Castellanos placed the discussion not at the meetings with royal experts (*Juntas*), but during a mutiny in the first oceanic voyage. As far as I know, and besides the testimonies about the discussions at the *Juntas*, this is the first display of a theological rejection addressed at Columbus, which already includes references to Lactantius and Augustine.

José de Acosta (1540–1600), who lived in America (1572–1587, Peru, Bolivia, Chile and Mexico), offered a remarkable criticism of early Christian writers, as he opened his famous *Historia Natural y Moral de las Indias* (1588, 1590)[149] with an attack on those who denied the sphericity of the heavens and the Earth, as well as the existence of the antipodeans. After his first introductory lines to this problem (Acosta, *Historia* 1.1, 1590, 13–14), Acosta moved on to criticise several ancient Christians: Chrysostom, Theodoretus, Theophylactus, Lactantius (whom he related to Epicurus), Jerome (he was unsure here and mentions that "it ſeemeth that S. *Ierome* draweth neere to this opinion," Acosta, *Historia* 1.1, 1590, 14),[150] Procopius and Augustine (see here Table 4.1). He gave references for all of them except Procopius's text, which he had not seen, mentioning Senensis as his source. Acosta did not give much weight to these ideas, as these authors were not much concerned with the sciences. However, he was astonished by Augustine's doubts (Acosta, *Historia* 1.1, 1590, 14–16). He returned to the topic of the antipodeans in Augustine in Acosta, *Historia*, 1.6, 1590, 27. And he not only discussed the famous text of *City of God* 16.9 on the antipodes, but also other texts where Augustine seemed to consider the views of the flat-earthers seriously (Acosta, *Historia*, 1.1-1.2, 1590, 15–19).

This can be seen as the start of the tradition of considering the possibility of Augustine as a flat-earther – which was also discussed in Góis, 1593, 202 – that has been debated until today.[151] Acosta also discussed Augustine's doubts about the source of the moonlight (Acosta, *Historia*, 1.2, 1590, 17; see Marone, 2007) and the eccentric model of the Earth, which he attributed anachronistically to Augustine, explaining his reasoning in line with what we saw in pp. 39–47 (Acosta, *Historia*, 1.6, 1590, 27–28). Acosta then devoted the entire chapter 7 to attack Lactantius.

Despite his criticisms, Acosta also tried to save ancient Christian authors, considering them more concerned with theology than natural philosophy (Acosta, *Historia*, 1.1, 1590, 14–15). Discussing the eccentric model, Acosta justified Augustine based on a lack of experience by the ancients (Acosta, *Historia*, 1.6, 1590, 27–28). He then compared Lactantius's and Augustine's rejection of the antipodes and affirmed that although they were both agreed in their unfavourable opinion, they used very different argumentations. As usual, Acosta displayed little interest in saving Lactantius, of whom he said that he "followes the vulgar" opinion, instead of following "the force of reafon" (Acosta, *Historia*, 1.7, 1590, 31, 32).[152] In contrast, Acosta affirmed that Augustine's reason for denying the antipodes "was very different", and devoted chapter 8 to discuss Augustine's concern, which he "drew … from the bowels of divinitie" (Acosta, *Historia*, 1.8, 1590, 33–34).[153] Acosta correctly identified the root of the problem in the ancient insistence on the impossibility of crossing the ocean, mentioning Pliny the Elder, Cicero, Macrobius and Pomponius Mela and even Aristotle, "who although he were a great Philofopher, yet was hee decyved in this poynt" (Acosta, *Historia*, 1.9, 1590, 36).[154] In chapter 10, Acosta chose as a title: "That Plinie, and [most of][155] the ancients, held the fame opinion with Ariftotle." He explained that the reason for that belief was that as one moved towards the south there was more heat, and from that, they concluded that the tropical region would be too hot to be inhabitable (Acosta, *Historia*, 1.10, 1590, 41).[156]

Acosta later discussed the problem of the arrival of humans in America (Acosta, *Historia*, 1.16, 1590). As Augustine feared, the existence of a separate landmass populated by humans opened up the question of the unity of human origins, whose monogenism or polygenism was hotly debated during the 16th and 17th centuries (Gliozzi, 1977).[157] Acosta rejected the arrival of humans in America by navigation and tentatively defended the peopling of America by land (Acosta, *Historia* 1.6, 7.3, 1590).[158] Therefore, Augustine's real theological concerns were solved despite the existence of antipodeans. Monogenism was still tenable with the possibility of peopling the whole globe from a single human origin (still today the preferred explanation for science, with an African origin). Acosta also used that explanation for the arrival of animals from a single Old World source after the flood, a point discussed by Augustine in *City of God* 16.7 (Acosta, *Historia*, 1.20, 4.34, 1590).[159]

Nevertheless, if Acosta wished to defend Augustine, he had even more interest in saving the reputation of the Bible from the (ab)use of the flat-earthers and the deniers of the antipodeans. He devoted to this objective the third and fourth chapters of book 1, where he attempted a traditional concordism, even though at one point he used the principle of accommodation that "but the holy fcripture frames itfelfe to our manner of conceiving and ſpeaking" (Acosta, *Historia*, 1.3, 1590, 21–22).[160] In discussing later how pagans saw the Old World as the only emerged land, and set the columns of Hercules in Cadiz as a literal end of the world, he added: "And not onely prophane writers ſpeake in this ſort, but alſo the holy Scripture, to apply it ſelfe to our phraſe ſaith [more literal: to accommodate itself to our language]" (Acosta, *Historia*, 1.8, 1590).[161] He gave examples of hyperboles, like the census in *all* the world (Luke 2:1) and one text suggesting the evangelisation in *all* the world (Colossians 1:6). This discussion is similar to what we found in Marzio (see p. 78).

The Portuguese Manuel de Góis (1543–1597) was part of a group of Jesuits in Coimbra who produced late scholastic commentaries on Aristotle, well-known in the 17th century (on the Conimbricenses, see Marinheiro, 2012). These topics are discussed in two places in his influential 1593 commentary on Aristotle's *On the heavens*. First, in the commentary on book 2, chapter 4, question I, the article II is devoted to the sphericity of the heavens listing authors for and against it, with Lactantius criticised for his faulty understanding of gravitation and his denial of the spherical Earth. Augustine's doubts on this later topic are highlighted as in Acosta.[162]

In the commentary on book 2, chapter 14, question I, article IV, Góis tackled in detail the question of the antipodes, mentioning the negative opinion of Lactantius, Augustine, Bede and Lucretius. He then summarised the argumentation of Lactantius, Augustine and Abulensis, before responding to each of the three. The weight of the reply was based on the evidence of "the voyages of our century," as he was aware that behind Augustine's dismissal of the antipodes was the fact that in ancient times "the geographers decided that the Pillars of Hercules were the furthest boundary of the Earth" (Góis, 1593, 321)[163] On Augustine's theological problem, Góis went to none other than Augustine himself to find solutions. The first from *City of God* 16.7: animals could have been transferred by angels. The second could be a migration on foot from the Old World, a caveat considered in *City of God* 16.9 and discarded in the light of the geography of that time. A third possibility was arrival by navigation, referring to old reports of large distance navigations in classical and Biblical accounts. Góis sided with Acosta's solution: migration on foot. The article closed with a reflection on the concept of evangelisation by the apostles "into all the earth," as discussed by Abulensis (Alonso Fernández de Madrigal, also known as Tostatus) before the arrival in America, based in Romans 10:18, quoting Psalm 18/19:4/4-5,[164] as in other authors.[165] That was challenged and solved with an historical-contextual view of the Bible: "whether or not one of the apostles crossed the equinoctial circle, nevertheless, within the life-time of the apostles, the Gospel reached into all the earth, that is, into the principal regions of the earth that were then known," in this way finally abandoning the literalistic reading of these words (Góis, 1593, 322).[166] This solution was similar to that by Acosta, already proposed by Marzio!

Early 17th-century negative views of ancient Christians and science

The *Vita Sancti Macarii Romani servi Dei, qui inventus est iuxta Paradisum, auctoribus Theophilo, Sergio et Hygino* was part of the volume *Vitae Patrum*, published by the Jesuit Herbert Rosweide in Antwerp (1615) and Lyon (1617).[167] At the beginning of chapter 3 of the *Vita Sancti Macarii* there is a mention of the travel of three monks to reach "the place where the sky meets the Earth" (Rosweide, 1615, 225).[168] There we find note number 5:

> There was an opinion among many ancient thinkers, that the earth was flat and that there were no antipodes; and that the consequence of this was that heaven was somewhere joined to the earth. Indeed, many have even claimed that Paradise was located in a certain region of the earth, on a somewhat elevated plane, where heaven is joined to the earth.
>
> (Rosweide, 1615, 232, note 5)[169]

When in chapter 16 the three monks find Macarius and tell him of their intentions (Rosweide, 1615, 228),[170] he explains that it is impossible to reach Paradise, as a Cherub guards it with a fiery sword.

Is Rosweide's note a case of flat error? If we take into account that this is originally an eastern Greek text, possibly from the 5th century,[171] the reference to ancient thinkers is not related to the Middle Ages (as in the typical flat error) but to the ancient or late ancient period when the text was composed. At that time, there were patristic Greek authors (connected with the theological school of Antioch, see p. 35) who, in fact, believed "that the Earth was flat and that there were no antipodes," as Rosweide noted. This reference illustrates the growing knowledge in c. 1600 relating to the cosmographical views of the early Christians, as can be seen in Table 4.1. In fact, Rosweide completed his note 5 by offering an obscure evidence from the poem of a 5th-6th centuries Bishop of Vienne (France): "*Alcimus Avitus lib. I. de Genefi*", who connected the place where heaven joins the earth with India at the edge of the world (*Ergo vbi tranfmiffis mundi caput incipit Indis, / Quò perhibent terram confinia iungere cælo,*).

The problem is that he seems to have overstated how many ancient Christians believed in a flat Earth.[172] In fact, it is questionable whether Rosweide was correct in identifying the *Vita Sancti Macarii* as an example of flat-earthism in the first place. This text never mentions the flat Earth, and the expression "the place where the sky meets the Earth" does not need to be read cosmologically, but rather theologically, in line with Genesis 2-3. In this context, *cælum* was not the sky of the planets and stars, but the abode of God meeting humans in a particular place, the earthly Paradise, as can be concluded from the context. This legend is only one among many describing all sorts of adventurous characters searching for the ancient Garden of Eden, supposedly guarded by angels in some inaccessible place in the Far East, frequently on a mountain, a traditional place to meet the divinity as its height makes it closer to the heavens (Scafi, 2006, 51–52; Caudano, 2016, 187). Indeed, Rosweide explains correctly in his note that Paradise was supposed to be located in a high place, closer to heaven,[173] but he unnecessarily mixes this with the unrelated idea of reaching a physical juncture between the descending edge of the skydome and the edge of a flat Earth, which seems a modern fancy.[174]

The famed English populariser of science, Francis Bacon (1561–1626), in his book entitled *Of the proficience and advancement of Learning, divine and humane*

(1605) displayed a standard praise of the moderns, who discovered the antipodes during their navigating expeditions (Bacon, *Of the proficience*... 2, 1605, fol.15^{v}). However, immediately after he composed two similar texts, which were published much later: *Filum Labyrinthi* 7, 1734, 457 (composed in 1606; Appendix, no. 3) and *Cogitata et Visa*, 1653, 11–12 (composed in 1607; Appendix, no. 4).[175] What appears in these two texts of 1606–1607 is a more critical evaluation of early Christian authors, crucially mixing the issue of the sphericity of the Earth and the existence of the antipodes, a situation which we will see arising also in other authors at around 1600 (Chapter 5). When Bacon published his very influential *Instauratio magna* (*The Great Instauration*), in the second part, *Novum Organum* 1.89, 1620, 107, he repeated the same mix-up, here for the first time publicly (Appendix, no. 5).[176]

The impact of new worlds on ancient texts

While traditional historiography considered the geographical discoveries c. 1500 as the major trigger of modernity, the last decades have seen a debate over the real impact of the New World on the Old. Historians such as John H. Elliott – although recognising the impact of the 16th-century European conquest of America, as we saw in p. 119 – indicated that it was "a blunted impact." It was considered that its effects were only harvested in the 17th–18th centuries, due to the resilience of the ancient and medieval authorities. The discussion was stirred in 1970 by Elliott's The *Old World and the New 1492-1650* (see also Elliott, 1976; Grafton, 1992[1995]).[177] Elliott softened his position later, "in common with many others, I overemphasized the degree of indifference shown by early modern Europeans to the new discoveries" (Elliott, 1995, 395),[178] and stablished a bridge to the Enlightenment:

> [...]. The authority of classical texts began to falter in the 16th century, given the growing understanding that in some things the Greeks and Romans were wrong, and not the least in their ignorance about the existence of America. [...] The moderns, at least in some respects, had proved superior to the ancients. Once the seed of doubt was sown, the road was open to the arrival of that idea of progress that was going to have such a profound influence on the Europe of the 18th and 19th centuries. [...].
>
> (Elliott, 2000, 42)[179]

According to the historian of Early Modernity Joan-Pau Rubiés, "there was a geography as well as a chronology of impact" (Rubiés, 2017, 64; 2006, 136). The voyages to America, but also to Africa and Asia, were a vital force "to understand how the Renaissance eventually led to the Enlightenment," even if their impact worked underground only to surface much later (Rubiés, 2006, 131, 132). Therefore, Rubiés affirms: "[a]t this point our two arguments tend to converge as Elliott emphasizes the importance of the 'work being undertaken behind the stockade' before 'any sustained break-out' could take place" (Rubiés, 2006, 139, note 17), in reference to Elliott (1970[1992], 29).

A space-time perspective fits the evidence of progressive diffusion of the awareness of the challenge of new discoveries first in Portugal, then in Spain, and in their Italian associates before these ideas spread further, even if some quickly became aware of them in certain places of northern Europe (such as Saint-Dié in the 1500s). In addition, to this *geographical* and *chronological* asymmetries of progressive impact across Europe, we could also suggest that there was a progression in the impact across disciplines, a *disciplinary asymmetry* in the impact of the discoveries; that is, there were topics much earlier affected by the new discoveries than others. Although he did not elaborate much on this, Elliott already noted that:

> The shock was perhaps greatest in the most specifically circumscribed area – Europe's inherited geographical knowledge and experience. Although the classical geographical tradition, as interpreted by fifteenth-century cosmographers, provided an inspiration and a springboard for overseas voyages of exploration, this tradition was itself the first and most decisive casualty of the voyages. […].
>
> (Elliott, 1976, 17)

Indeed, the Spanish Humanist Vives wrote, in a Dedicatory address to King D. João III of Portugal for his book *On Education*, that the "journeys of these [modern Portuguese] travellers […] give us a rude shock" (Vives, 1531, fol. Aij^{r}).[180] Even earlier, in 1523, Pomponazzi already felt that shock in his cosmographical teaching (see pp. 128, 159). The shock was not only felt in cosmographical topics (habitability of the Torrid Zone, existence of populated antipodes or evidence for the terraqueous model), but also in anthropology, in particular the early dispel of the old legendary accounts of monsters. Their existence had already been questioned by some late medieval missionaries who travelled to the east, Wilhelm of Ruybroeck, Giovanni de Marignolli, etc. (Relaño, 2002[2019], 38–39). Iberian explorers ended these legends (Relaño, 2002[2019], 39–40).

For our purposes, we are interested not in the period "after 1650" with the emergence of the "break-out" from the past, but with the previous work "behind the stockade" (Elliott, 1970[1992], 29), using Elliott's terminology; the period, as Grafton noted, when "between 1550 and 1650 Western thinkers ceased to believe that they could find all important truths in ancient books" (Grafton, 1992[1995], 1). However, as we have seen, in the field of cosmography (geography, the debate of the antipodes, the geological structure of the Earth, etc.) this chronology could be moved backwards half a century, to c. 1500, becoming actually one of the leading spearheaded fields of change and break-out. By the time of Acosta and Góis in the 1590s, this process was already consolidated, and ancient pagans and Christians were criticised alike regarding the discipline of cosmography.

Realisation of the importance of disciplinary asymmetry in the reception of the geographical discoveries explains why an author – such as Acosta – could attack Aristotelian cosmology, and at the same time, live happily in an overall Aristotelian framework. Acosta still recognised Aristotle's authority: "for I will not contradict *Ariſtotle*, but in that which is moſt certaine" (Acosta, *Historia* 2.12, 1590, 110).[181]

Moreover, he was happy to point to those topics on which Aristotle was right, as the sphericity of the Earth (Acosta, *Historia* 2.2, 1590, 16). Historian of science Reijer Hooykaas described this divide in Portugal several decades before Acosta:

> It was the Portuguese who had to cope as the first in Europe with these opposite influences. Their solution had two aspects: *to follow nature wherever it may lead, and, secondly, to respect tradition whenever this is possible* without violating the first directive.
>
> (Hooykaas, 1966[1983], 587)

As an example, we can mention Margalho, who said that just as Ptolemy corrected Marinus, the ability at sea of the Portuguese corrected Ptolemy (Margalho, 1520, fol. 7rb; Flórez, 1990, 358–359). Obviously, this was no obstacle for Margalho to agree with Ptolemy on other points, as on the terraqueous globe (see p. 110). Even earlier, Münzer and Columbus were able to recognise Aristotle's mistake in considering the Torrid Zone inhabitable and at the same time use him to justify the Atlantic crossing. The incorporation of new knowledge into the old framework also explains how natural philosophy could flourish in the Iberian empires in the 16th and 17th centuries, unaffected by the new sciences developed in other parts of Europe (see Gascoigne, 2013, and pp. 18–20).

The same situation also applied to Augustine: a significant blunder in geography does not mean that his overall authority was diminished. It just served as a warning that not absolutely everything had to be taken as the final truth. See, for example, Geraldini's comment in 1491 (p. 121), and similar ones that, as we have seen, also appeared in many books during the 16th century. Moreover, on this very topic, most authors fought to justify Augustine's error, while deeming it impossible to save Lactantius on cosmography, even though Lactantius himself, and Augustine, continued to enjoy a good reputation and authority in many other matters. In fact, in the case of Columbus, Augustine could be used both, to attack and to protect him (see p. 121). However, by the end of the 16th century, the impact of the discoveries forced some reformulations in theology (as we have seen) and was already reaching philosophy in a sceptical line advanced in 1523 by Pomponazzi (see p. 128). This was felt in particular among sceptical philosophers such as Francisco Sanches (1550–1623) who, based on the finding of new continents and the habitability of the equatorial and Polar Regions, asked:

> [...]. So how can you maintain that your propositions are eternally valid, incorruptible, infallible, and incapable of being otherwise – you miserable worm, who scarcely know, and are scarcely even capable of knowing, what you are and whence you come and whither you are going?
>
> (Sanches, *Quod Nihil Scitur*, 1581, 40)[182]

Other sceptical author is Michel de Montaigne (1533–1592), who in light of the recent overturn of Ptolemy's geography, wondered about the possibility of additional future challenges.[183] The rise of scepticism was not only triggered by the new discoveries refuting received wisdom, but also by the recovery of ancient sceptics's texts, showing the complex interplay between the old and the new, texts and experience, in the Renaissance. Of particular importance was the publication by Henricus Stephanus (Henri Estienne), in 1562, of the Latin translation of Sextus Empiricus's *Outlines of Pyrrhonism* (Floridi, 2010; see pp. 34–35 for ancient scepticism). The important point for us is to see how new geographical discoveries were soon incorporated in the argumentation of the 16th century, even for age-old discussions.

In connection with the debate on the degree of impact of the New World on the Old, there is a debate between those who see external or internal causes (experience from navigations or recovery of ancient texts) for the changes in European culture during the 16th and 17th centuries. However, Elliott has rightly affirmed that: "Rather than pointing Europe in totally new directions, the discovery of America emphasised and strengthened certain elements in European civilization at the expense of others. [...]" (Elliott, 1976, 17). More than providing radically new ideas, geographical experience allowed Europeans to decide between long conflicting bookish alternatives (the inhabitability of the tropics, the existence and inhabitability of the antipodes and, more importantly, the existence of actual inhabitants in these places). A good example is Acosta, who stated that the new geographical knowledge confirmed the views of those ancients in favour of the terraqueous model against other ancients, such Augustine (Acosta, *Historia* 1.6, 1590, 27). In fact, Elliott already pointed to that in 1970:

> Certain cosmographical ideas derived from classical antiquity were in fact vindicated by the discoveries. [...]. But other ideas-about uninhabitable regions or climatic zones-had to be abandoned or profoundly modified. [...]. The superiority of direct personal observation over traditional authority was proved time and time again in the new environment of America. And on each fresh occasion, another fragment was chipped away from the massive rock of authority.
>
> (Elliott, 1970[1992], 40)

> [...] the discovery of America was important, less because it gave birth to totally new ideas, than because it forced Europeans to come face to face with ideas and problems which were already to be found within their own cultural traditions.
>
> (Elliott, 1970[1992], 47)

Borrowing expressions from Grafton, the "contradictions between and tensions within the texts" and "their confrontation with external novelties" were what led to a revolution in European culture (Grafton, 1992[1995], 6).

The contradictions inside the western canon became more evident, in parallel with the geographical explorations, as more ancient books were recovered and as the printing press made them available for comparison (see pp. 105–106). As Blair and Grafton put it: "The New Science grew not only from the shock of the new worlds laid open by explorers and observers, but also from the shock of the old books recovered and explicated by scholars working patiently in their studies" (Blair and Grafton, 1992, 538). Grafton described the role of the external evidence of the discoveries as providing "a clinching piece of evidence to those who wished to argue for a new vision of history, for the superiority of modern to ancient culture," even though "the substance of that vision, ironically enough, often came from the very ancient writers whose supremacy it denied" (Grafton, 1992[1995], 157; on this paradox, see above in p. 136 a quotation from Elliott). However, as Grafton recognises, there was no other way to advance but on the shoulders of the ancients: "Ancient science, though evidently unstable, remained fruitful. It still offered the only moorings to which one could affix new facts and ideas" (Grafton, 1992[1995], 117). That is why many accommodated the new into the old.

The old geographies and maps were not thrown away; initially, they were 'supplemented' with new texts and labels, respectively (Hiatt, 2008, 184–223). In the last two decades of the 15th century, 'modern' maps (*Tabulae modernae*) were incorporated into the new editions of the *Geography* of Ptolemy (Akerman, 1995). Later, the solution was to separate the old historical maps from the new modern ones, leading to a new view that geography and cartography had a history and that the 'new' knowledge had to be distinguished from the 'old' outdated knowledge. This can be seen in the famous edition of the *Geography* in Strasbourg, 1513, which based the new maps on the "explorations of our age," (*ad saeculi nostri peragrationes*, title page, see Hiatt, 2008, 188). During the 16th century, the editions of Ptolemy's *Geography* drifted away from the classical model and eventually, the importance of progress in historical time was recognised, freeing the moderns from the yoke of the agreement with the past and giving rise to modern geographies. On the other hand, that also allowed recovering the old texts in their original purity as historical documents, purged from anachronistic later additions (Hiatt, 2008, 187–191).

That also explains why in the early 16th century, authors continued writing commentaries of ancient texts, even when they already had different views from them. On Vadianus's commentary on Pomponius Mela that we discussed in this chapter, Beer suggests that by that time (1518), ancient texts were still the "knowledge base in which to incorporate new insights" (Beer, 2013, 175) and that they functioned as "stepping stone[s] for adding and organizing knowledge" (Beer, 2013, 182). The situation changed by the end of the century.

It was in relation to geography and the voyages of discovery, that the topic of the antipodes was predominantly discussed, bringing with it the debate about the positions of Lactantius and Augustine. In this way, the discoveries impacted geography and almost immediately, theology (regarding such a delicate topic as the authority of the early Christians, including their highest authority, Augustine). The reason was that these two fields had already been linked for over a millennium. So even if in some topics the impact of the discoveries was delayed by decades or even centuries, on the topic here investigated, the impact was much quicker. As in geography, the theological implications of the antipodes were not brought about overnight by the discoveries; what the sailors brought with them were facts that finally allowed the resolution of old theological debates.

Some scholars had accepted the existence of the antipodes for centuries, but they were a minority; what the geographical discoveries in the late 15th and early 16th century did was to turn around the situation, sinking what for a millennium was the predominant view of one isolated Old World oikoumene surrounded by water.[184] It was only a matter of time until the new victors would look back in anger, confusing the nature of the real medieval debates and abusing the newly acquired availability of printed ancient Christian texts to create a new (distorted) interpretation of the past, as we will see in Chapter 5.

Conclusions

In previous chapters, we saw a rather uncritical acceptance of the ancients's geography (except on the topic of the antipodes), with only some criticisms of Lactantius in the 15th century; however, in this chapter, we have seen how things changed progressively as the revolutionary geographical discoveries of c. 1500 were absorbed in European culture. The moderns saw the ancients with increasing contempt. Ancient pagan geographers were criticised as well as ancient Christian authors, such as Lactantius, but in particular Augustine. This raises the critical question: who was challenged, i.e. who were those mistaken 'ancients'? We have observed that the strategy of the 15th century to justify Augustine's error in denying the antipodes became consolidated in the 16th century. However, as we approach the late 16th century, Augustine became part of a growing list of ancient Christian authors criticised for being mistaken in cosmographical matters (see pp. 130–133). Such lists resulted from increasing access to new ancient Christian texts, but also from a more careful study of already known authors. However, as seen in Table 4.1, there was not just one topic covered in these lists. The topic of the antipodes/antipodeans was prominent, but also the sphericity, generally referring not directly to the Earth, but to the heavens in general – although the sphericity of the heavens was connected to that of the Earth.

A key factor illustrated by the lists is the shift from criticising only Lactantius and Augustine to showing that there were *more* pagan and also early Christian authorities that were mistaken. Each individual list was, in fact, relatively modest, depending on the state of erudition of its respective author. However, the use in such lists of some generalising quantitative phrases such as "the other Christian writers" (*cæteri*, Senensis), "other people" (*alij*, Góis), and "many" ("plufieurs autres," La Popelinière, "molti," Mazzoni, "algunos y no pocos," Acosta[185]) could give the impression that these ideas were even more widespread than the authors explicitly mentioned in these lists, and that this was really a general trend among Christians overall. This impression was enhanced by the tendency not to mention those early Christians who held the opposite views – i.e. who were favourable to the sphericity and the antipodes – with the only exceptions being Acosta and Góis (for the sphericity of heavens) and Senensis, Mazzoni and Góis for the antipodes/antipodeans.

The lists are crucial evidence for the number of authors, but also for their importance and representativeness, including well-known saints and doctors of the

Church. For Mazzoni, the opposition to the antipodes was "the opinion of *some doctors* [of the Church], *based on the Holy Scripture*" (Mazzoni, 1587, 551),[186] and his whole argument was to provide a rationale to explain why Petrarch doubted the existence of the antipodes based on pagan and Christian authorities. According to Acosta, the deniers of the sphericity of the heavens were "yet many, (yea, of the *holy doctors* of *greateſt authoritie*)" (Acosta, *Historia* 1.1, 1590, 13).[187] And for Góis, "many *serious authors* have not agreed with the view that the heaven is round" (Góis, 1593, 202).[188]

Such generalisations could be seen as characteristic of Christian thinking *per se* and used in entirely different contexts, as in the early 17th-century authors discussed in pp. 134–135. An example is Rosweide's text, which was not only an early case of flat error, but also an example of how the flat error was already distorting the understanding of the past and influencing a misleading reading of ancient texts. In this way, Rosweide became an early victim of the flat error.

What nobody discussed was the chronological, geographical and intellectual location of these authors. That means that these lists forced artificial continuities of authorities belonging to separate chronological, geographical and theological milieus. Even where there was a general agreement – i.e. the denial of the antipodeans – this could be linked in some authors to a flat Earth cosmology, while others accepted a spherical Earth. In Góis and Wilkins, the problem would become more acute by mixing Late Antique authors with Abulensis, in the Renaissance.

Therefore, they did not discover that the denial of sphericity, except for Lactantius, was centred on Greek authors, mostly belonging to the Antiochian/Nestorian theological school; although Sixtus Senensis came close to realising this connection (see p. 130 and endnote 145). Not even the rediscovery of Cosmas Indicopleustes by Montfaucon (1706) triggered this connection until La Croze, 1724, 27–37. As this research is devoted to the origin of the flat error, which came into being in Western Europe, we have not devoted much attention to Eastern Christianity, where several true Christian flat-earthers flourished in Late Antiquity (although Cosmas's map is discussed in Chapter 2). What is crucial for us here is that, contrary to what was thought in the 19th and 20th centuries (see Chapter 1), these flat-earthers had no influence on the western medieval cosmology, as we saw in Chapter 2.

In any case, the authors surveyed in this chapter had the clear view that the debate on the sphericity of the heavens and the Earth was limited to the early church; for the denial of the antipodes, some of the late 16th-century authors pointed to Renaissance authors such as Petrarch (Mazzoni) or Abulensis (Góis). Therefore, in his lengthy introductory chapters, Acosta could insert the new discoveries in an historical sequence: the ancient debates on the sphericity of the Earth, the medieval/Early Renaissance discussions on the antipodeans, and the modern explorations finally reaching the antipodeans on the spherical surface of the globe. The seeds of the 17th-century *Querelle* between the ancients and the moderns, leading to the demise of the revered authority of the ancients and contributing to the idea of progress, were planted in the 16th century.[189]

The challenge to ancient authors was compounded by the first signs of misunderstanding about the past, as in Gómara's equivalence between denying the antipodes and denying the sphericity of the Earth. Other examples are Acosta's and Góis's discussions of Augustine as having seriously considered the possibility of a flat Earth.

Inevitably, as the criticism of Antiquity grew, an important issue was how to deal with ancient authorities – secular and Christian – increasingly under threat. All authors discussed in this chapter were Christians, conscious of the breakdown in the traditional medieval views on the antipodes, which questioned religious authorities. However, although these traditional authorities could be criticised regarding geographical matters, they were still esteemed as *theological* thinkers. The kind of comment that we saw in Geraldini (see p. 121) became a key to the reception of the early Christians.

The authority of Augustine, which – similar to the Bible – was supposed to extend to nearly every possible topic, now became challenged in relation to particular areas where he was shown clearly wrong. The solution found was carefully to split his authority into theological issues, where he maintained his unassailable position, and into other topics, where he was seen as reflecting the limited knowledge of his age (already in Vadianus). In this way, the general reputation of Augustine could be saved, while Lactantius was abandoned, as his error was deemed too great. Of course, the problematic question in dealing with this wedge was that of boundaries. As we have seen, Augustine could be used against and in favour of Columbus (we will see other ambivalent situations in Chapter 5). Furthermore, if Augustine was wrong in rejecting the antipodes, was he still right in defending monogenism? And, what about the movement of the Earth?[190] These became the next questions thrown at the 'Fathers' and – more dangerously – at the Bible itself in a post-Trent period, when a literalistic interpretation became fashionable among theologians.[191]

Geographical voyages in the 16th century provided some of the first examples of theology facing challenges from new discoveries. The adaptation to the new realities was facilitated by the diversity of views already present in the European baggage.[192] The finding of indigenous peoples in America, soon considered fully human antipodeans, led to several conclusions.[193]

First, rethinking the concept of monogenism led Christian authors to a dissemination model for the origins of humanity from a single place of creation to avoid accepting polygenism, which was increasingly fashionable. Various defenders of monogenism accommodated the indigenous peoples of American in one of the three branches of Noah's descend: Japheth, Cam and Sem (Gliozzi, 1977). This dissemination model had been rejected by Augustine, due to ancient geographical theories which negated the possibility of such human dispersion over the surface of the Earth across water barriers (see Chapter 2). However, the overcoming of these barriers in Early Modernity could lead to a new dissemination model, compatible with Augustine.

Biblical texts that considered evangelisation to be complete were reinterpreted (we saw this old debate in p. 45). These texts were now deemed hyperbolic or stemming from an original historical context with only limited geographical knowledge. A similar solution was found for some texts in the Bible that could hardly be reconciled with the new geographical knowledge: the view that Jerusalem, or at least the Holy Land, was the centre of the Earth was reinterpreted in spiritual, rather than geographical terms. To avoid challenging the divine authority of the Bible, the solution was found in Augustine's principle of accommodation, as we saw in Acosta in pp. 132–133. These solutions had already been proposed by those who, in the 15th century, explored the implications of real antipodeans *before* their discovery (e.g. Marzio, see pp. 75, 78).

There was a revival of the debate about the eternal destiny of the antipodeans. Authors, such as Baptistus of Mantua (book 3.11, 1501) accepted, like Marzio or Pulci (see Chapter 3), that those who lived according to natural law would find in God "justice and clemency" and thus had some hope for the afterlife (Marenbon, 2015, 286–287). However, this position was a minority one and difficult to accept for mainstream theologians at the time, such as John Major (book 4.3, 1509). Elliott pointed to an increasingly pessimistic view of the indigenous peoples of America over the 16th century, connected with the religious changes in Europe, emphasising the sinful human nature (Elliott, 1976, 15–16). The debate about the 'good savage' was launched.

Finally, there was a reignition of worldwide evangelisation -previously considered complete -that incentivised the Catholic Church to reach all the lands in the light of biblical texts such as Matthew 24:14. Only with the whole surface of the Earth explored could this prophecy be fulfilled. Of course, this programme of evangelisation needed to see all inhabitants of these distant lands as fully human and able to convert to Christianity, a position that was served better by Augustinian monogenism (which was the Catholic position, while Protestant views were more diverse). For John M. Headley, all that "placed a heavy burden on the missionary effort of the Catholic religious orders as they scrambled to effect this goal in what appeared to many as an eschatological twilight" (Headley, 1997a, 12). We already found this anticipated – and mixed with popular chivalric literature – in Pulci (see pp. 84–85). Columbus took these ideas very seriously – including this eschatological context – in the correspondence of his later years and his *Book of Prophecies*. This feeling was exacerbated after the Reformation in the 16th century, when many Catholics took territorial gains in America as *compensation* for losses in Central and northern Europe due to Protestantism, while a new crusade to reconquer Jerusalem was forgotten (Elliott, 1970[1992], 80; Segundo, 2012). This idea was also mixed with Catholic hopes and fears of *translatio ecclesiae* from the Old to the New World (Gil, 1989[2017], vol. 1, 284–295).

Without the 16th-century geographical revolution and the challenge of the traditional authorities that followed from the explorations of the late 15th century, the flat error would not have been able to appear and survive as a credible narrative of the past. It is my proposal that this was a necessary step for the flat error to appear.

Table 4.1 Table of ancient and medieval authors as referenced by early modern scholars (a colour version of this table is available at www.routledge.com/9781032893068)

Bold: Western Christian authors ***Bold Italics***: Eastern Christian authors *Italics*: Bible

The double vertical lines separate the authors involved in the Copernican debate (to the right)

		Senensis (1566)[a]	Mazzoni (1587)	Acosta (1588/90)	Góis (1593)	Foscarini (1615)[g]	Campanella (1622)[i]	Wilkins (1638)[k]
SPHERIC HEAVENS	**For**			*Ecclesiastes (and other Bible texts)* Aristotle Peripatetics Stoics ***Gregory of Nazianzus*** ***Basil*** **Jerome** **Ambrose**	*Ecclesiastes* Manilius ***Pseudo-Clement***[e] **Augustine?** **Jerome**			
	Against	**Lactantius** ***Chrysostom*** ***Procopius***		Epicurus **Lactantius** ***Chrysostom*** **Jerome?** **Agustine?** ***Theodoretus &*** ***Theophylactus*** ***Procopius***[d] "many … of the holy doctors"	**Lactantius** **Augustine?** ***Chrysostom*** ***Procopius*** Other people, referred to by ***St Damascene***	***Pseudo-Justin***[h] **Lactantius** ***Basil*** **Ambrose** ***Chrysostom*** ***Theodoretus &*** ***Theophylactus*** ***Acacius*** ***Procopius***	***Pseudo-Justin***[h] **Lactantius** ***Procopius*** ***Diodorus*** ***Eusebius of Emissa*** Others ***Chrysostom***	"…together with all those Fathers or other Authors who denied the Roundness of the Heavens" (1640)
SPHERIC EARTH	**For**							
	Against		**Lactantius**		**Lactantius**		[j]	
ANTIPODES / ANTIPODEANS	**For**	**Clements / Origen**	Pseudo-Aristotle[b] Cicero P. Mela Aelianus M. Capella **Virgilius**	Aristotle	Manilius **Pseudo-Bede**[f]			**Virgilius**

Against	**Lactantius** **Augustine** ***Procopius*** Other Christian writers	*Salomon (Bible)*[c] *Isaiah (Bible)*[c] Eratosthenes Polybius Strabo **Lactantius** ***Basil*** ***Caesarius of Nazianzus*** **Augustine** ***Chrysostom*** **Boniface** **Pope Zacharias** ***Elias of Crete*** ***John Tzetzes*** **Nicholas of Lyra** **Petrarch** "many of the ancient Fathers"	**Lactantius** **Augustine**	Lucretius **Lactantius** **Augustine** **Bede** **Abulensis**		Xenophanes **Lactantius** **Augustine** Others ***Procopius and his list of "all the Fathers"***	Herodotus Lucretius **Lactantius** **Augustine** ***Chrysostom (1640)***[l] ***Procopius*** **Bede** **Pope Zacharias** **Abulensis** "…with others" (only 1638)

a Augustine is quoted as criticising those who link their scientific views to their faith, putting the second at risk.
b The reference is to Pseudo-Aristotle's *De mundo*, of uncertain date, generally supposed to be Hellenistic.
c Despite listing these biblical authors as against the antipodes, according to the testimony of older theologians, Mazzoni defended the compatibility of the existence of the antipodes with the biblical texts (pp. 554–555).
d Reference taken from Sixtus Senensis.
e Pseudo-Clement's *Recognitiones* is a book dated around the 3rd–4th centuries.
f Pseudo-Bede's *De elementis philosophiae* has been attributed to William of Conches (c. 1100).
g This list is from the Letter from Foscarini to Bellarmine (dated Lent, 1615).
h Pseudo-Justin's *Questiones et responsiones ad orthodoxos* is a book dated in the late 5th century, attributed to the school of Antioch. First printed in the original Greek in Paris, 1551, and soon translated into French and Latin, becoming very popular. In his *Bibliotheca Sancta*, Senensis quickly recognised anachronisms in this text to be Justin's faithful 2nd-century work. The precise authorship is disputed (Toth, 2014; Voicu, 2015).
i Another list offered is that of those discussing the habitability under the equinox (against: Aquinas; favourable: Avicenna, Albert the Great). He also listed those who believe that the terrestrial paradise was located in the "other hemisphere of the earth": St Ephrem, Anastasius Sinaita, and Moses, Bishop of Syria (all eastern Christian authors).
j Campanella does not list flat-earthers; however, he lists authors who think that the Earth floats on water and believe that the sun circulates around the mountains in the north (a distinctive flat Earth belief): Xenophanes, Procopius, Eusebius "and others."
k I have not engaged in a complete study of this work and only checked the references in Russell, 1991 (see p. 13).
l Chrysostom does not appear in the 1st edition of 1638, but in the second of 1640.

Notes

1 For example, Crates view was considered antiquated already by Geminus in the 1st century BC and ignored by Ptolemy two centuries later. However, its appearance in Seneca's *Dream of Scipio* (1st century BC) implied that Macrobius explained it in a widely read medieval commentary. Similarly, the idea of an uninhabitable Equatorial Zone was rejected by Eratosthenes, Polybius, Posidonius, Geminus and Ptolemy, but transmitted to the Middle Ages (on these ancient views see Chapter 2).

2 A legend in Fra Mauro's map (1459) denied that the Indian Ocean was a closed sea (*stagnon*, Randles, 1986[2000], 3).

3 An example is the receding location of the Pillars of Hercules in Fra Mauro's map (1459); see Randles, 1992[2000], 5.

4 Examples of ancient authors's vindication of the Torrid Zone's habitability appear in the Humanists Élie Vinet and George Buchanan (Hooykaas, 1980, 236–239).

5 Strabo accepted the reports of ancient circumnavigations of Africa that Ptolemy dismissed (see Chapter 2, endnote 22). An example of the use of Strabo is Vinet (see endnote above). Columbus frequently referred to Strabo in the margins of his copy of the *Historia rerum ubique gestarum* and his son and biographer, Ferdinand Columbus (1571), indicated that Strabo was an essential authority for the two most important lines of evidence in favour of his father's project of crossing to America: first, a large eastward extension of India leaving water covering only a small part of the surface of the Earth; and second, that the Atlantic was of not a great extension, and could even contain other landmasses, with a clear encouragement to find India sailing from Western Europe (Anastos, 1952[1979]).

6 As with other marginal notes in Columbus's books, there are doubts about whether the note in Ailly's *Imago Mundi* describing Dias's arrival to Lisbon was written by Columbus himself or his brother Bartholomew (Las Casas, *Historia de las Indias* 1.28; on that, Wey Gómez, 2008, 134, 154–156; however, Gil, 1989[2017], vol. 1, 144, does not give credit to Las Casas).

7 Geraldini, *Itinerarium ad regiones sub æqvinoctiali* 14, 1631, 204–205. See p. 121.

8 Randles (1994[2000], 47–8; 1990b[2000], 18). This is based on the favourable attitude of the Cardinal Pedro González de Mendoza as described by Oviedo, *La historia general delas Indias* 2.4, 1535, fol. v[v]. It was towards Cardinal Mendoza to whom the words of Geraldini were directed. In any case, the final success of Columbus was the result of the lobbying of different friends and supporters.

9 Also in marginal notes to Ailly's chapter 8 (for the full-text of Ailly and Columbus's notes, see Ailly, 1930). The terraqueous model is defended with similar arguments in a letter from Columbus to the Catholic Monarchs on his third voyage to America in 1498, where he first referred to Ptolemy, and several pages later mentioned a mix of pagan, Muslim and Christian authorities: Pliny, Peter Comestor, Lyra, Aristotle, Averroes, Ailly, Seneca and even the pseudo-epigraphical 4 Esdras (Spanish-English edition in Columbus, 1847[1870], 134, 144–146; this passage is commented in Wey Gómez, 2008, 147–154). In the same letter, Columbus speculated that the earth was not wholly spherical but pear-shaped, with the protrusion being the Paradise which he located in South America (Columbus, 1847[1870], 134–147, 150; on that identification, see p. 118).

10 "salvo tierra, e agua, abrasada en redondez" (first published in Bernaldez, 1856, vol. 1, 269; my translation). In the following lines, Bernaldez indicated that Columbus believed it was possible to move around the earth and water as described by Mandeville. References to Mandeville reappear in chapters 123, 126 and 127 (Bernaldez, 1856, 305–306, 313 and 317, respectively, and other references in chapters unrelated to Columbus).

11 Basurto (1494, fol. avii[r]); Latin-Spanish edition in Flórez et al. (1989, 262–263). I thank Prof. Miguel Ángel Granada (University of Barcelona, Spain) for his help in understanding this complex text.

12 Fernández-Armesto (1991, 127–128). See also Columbus's letters quoted in pp. 117–118.

13 "[C]he diconvi quello che dice li philosophi, che la terra ha forma spherica. Et che cierto lo admirante ha havuto grande animo et ingiegno havere discoperto l'altro mondo opposito al nostro, con tante fatiche et sudori, et visto la mutatione che fa la tramontana per esser ito di là dalla linea del Equinoctiale, che mai lo harei creduto che ci potese habitare gli homeni, che stimavo fussi tuta aqua et non terra" (my translation using a Spanish version – which I compared with the original Italian – in Vannini, 1974, 39).

14 His publications have been very controversial because although based on authentic letters by Vespucci, they also contain added imaginary material and are considered forgeries, but in any case, they were considered authentic at the time of publication by most readers. Nowadays, some authors refer to the author as Pseudo-Vespucci. The *Mundus novus* is considered a modified version of the real letters from Vespucci to Lorenzo de Pierfrancesco de Medici in 1501–1502, in particular the so called Bartolozzi letter, 1502 (Randles, 1994[2000], 51, and Hiatt, 2008, 191–192). It was first printed in Florence in Latin translation (Vespucci, 1502–1503) and then went through no less than 60 editions in six languages by 1529 (Randles, 1992[2000], 22). *The four voyages of Amerigo Vespucci,* although presented as a letter to Soderini, dated in 1504, it is considered a mix of authentic letters and additions extolling Vespucci. In general, the first voyage, and often also the fourth, are not considered authentic. The text was first printed in Florence in Italian (Vespucci, c. 1505) and in Latin at Saint-Dié (Ringmann and Waldseemüller, 1507). See Randles, 1994[2000], and Hiatt, 2008, 191–192.

15 Pacheco, *Esmeraldo de Situ Orbis* 1.2, 1892, 6–7; Castro, *Tratado da sphaera* 1, 1968a, vol. 1, 50–51. The date c. 1535 for the *Tratado da sphaera* has been disputed; also dated into the 1540s and even into the late 16th century. João de Castro has been considered the author of this text, although it has also been attributed to other authors (such as André de Avelar). Another book attributed to him is *Da Geografia por modo de Diálogo* (Castro, 1940 and 1968a1). For details, Vargas, 2013 (particularly, pp. 85–86) and Santos, 2016 (see p. 471).

16 According to Randles (1994[2000], 11, 56–57), the first to use the Cratesian term *perioeci* was Volterra (1506) from recently rediscovered Cleomedes (1497).

17 Critics were numerous enough to gain the disdain of the Aristotelian Gaetan de Tiene (published in 1476, eleven years after his death); see Randles (1994[2000], 65).

18 For Mayor, see Randles (1993b[2000], 22), Grant (1984, 16); for Celaya, see Navarro (1994, 361); and for Lax, see Randles (1993b[2000], 21).

19 The same idea is repeated and linked to Ptolemy in fol. 7va (see Randles, 1993b [2000], 21).

20 A hugely popular miscellaneous book, with around a hundred editions in one century in many European languages, including English (London 1571, 1576 and 1651).

21 "o mar oceano nom he outra cousa senom huma muito grande halaguoa metida dentro na concauidade da terra e ha mesma terra e ho mar ambos juntamente fazem huma Redondeza" (Published in Pacheco, 1892, 7; translated in Pacheco, 1937, 13).

22 English translation in Erasmus (1977, 381). Erasmus also criticised the eccentric model in a letter to Theobaldus Fetichius in 1533, published this year in the first Greek edition of the *Geography* of Ptolemy (Ptolemy, 1533, fol. 3^{r}); quoted by Vogel (1995, 457; 2006a, 469).

23 "el agua 7 la tierra es todo juntamẽ vn cuerpo" (translated by Randles, 1994[2000], 66). Enciso continued with what can be read as criticisms of the Ptolemaic and eccentric models: "the earth does not surround the water nor does the water surround the earth" ("la tierra no cerca al agua ni el agua a la tierra"). A little later in the same page there is a sentence not translated by Randles: "the earth together with the water is the centre" ("la tierra juntamẽte con el agua es el cẽtro"). This is repeated with slightly different wording twice more in the same page. The partial English translation of 1578 did not include the cosmographical section where this was mentioned.

24 *Terremarisque mensores quod ex aqua et terra constat pro una habent sphera* (my translation).

25 "whiche yerth and the water to gyder with all / So Ioynyd make a rounde fygure ſperycall" (Rastell, c. 1520, fol. A.vi^{v}).

26 *Qvom terrae et aque ſuperficies ſit vna/ et ſphaerica*… (Apian, *Cosmographicus Liber* 1.4, 1524, fol. 10).

27 *mihi tamẽ admodũ ſit veriſimile eam orbis partem non integi atqʒ abduci pelago* (translated by Hiatt, 2008, 216). A copy at the National Library of France (Res.G.2926) has in the last page the date 1524, but this is considered a mistake and it is dated after the last geographical discovery mentioned (1526), see King (2019a, 34, note 6). While King dates it to 1526–1529, Hiatt does it to 1527–1530.

28 *Quum igitur perſpicuum ſit tum eruditorum virorum monimentis, tum fideli nauigantium teſtimonio* (translated by Randles, 1994[2000], 68).

29 "a agoa e a ter*r*a juntamente fazem hu*m*a soo sphera" and "ambos juntos fazem hu*m*a soo sphera" (my translation). Also "ambos iuntos fação hu*m*a soo esphera […] ambos fazere*m* hu*m*a soo figura spherical" (Castro, *Da Geographia por modo de dialogo,* c. 1545; quoted from Castro, *Da Geographia por modo de dialogo*, 1968a1, vol. 1, 112).

30 "Estes dous elementos agoa e terra fazem ambos juntos huma sphera" (my translation; see discussion in Hooykaas, 1980, 308, 320). The same idea appears in an undated printed booklet of Nunes (considered from 1537 to 1542), *Astronomici introductorii de spaera epitome* 1, section *Quod terra cum aqua sit Sphærica* (from its first modern republication, Guimarães, 1916, 272–273).

31 "El agua y tierra eſtan juntos, y el agua poꝛ vnas partes y la tierra poꝛ otras: ambos juntamente debaxo de vna ſuꝑficie hazẽ vn cuerpo redõdo" (my translation; see also Faleiro, 1535, fol. bvii^{r}–bviii^{v}).

32 "Baſta che à queſto tutti ſi accordano, che la Terra inſieme con l'Aqua fecino vn corpo Sferico, & rotondo" (my translation).

33 The book came with additions by Vadianus defending the terraqueous model.

34 "Como la tr̃ra y el agua ambos elemẽtos haʒẽ vn cuerpo redõdo. Notar ſea q̃ eſtos dos elemẽtos Tierra y Agua ãbos jũtamẽte haʒen vn cuerpo redõdo" (my translation). He already mentioned that in his 1543 unpublished *Coloquio de cosmographia* 37 ("el agua y la tierra ſon vn cuerpo Redondo" fol. 13^{r}) and 38 (fol. 13$^{r–v}$). There, question 38 is directed specifically against the eccentric model (see also question 40, fol. 14$^{r–v}$).

35 "The Torrid Zone is not uninhabitable, because today the Portuguese navigate through it; indeed it is very populated. Below the equator there is the fort of Mine belonging to the Most Serene King of Portugal and which we have seen" (Grant, 1974, 635; note to chapter 6 of Ailly's *Imago Mundi*, at the margin of a portion borrowed from Oresme's *Traitié de la sphère* on the uninhabitability of the Torrid Zone; Latin-French edition in Ailly, 1930, vol. 1, 196–197).

36 Geraldini, *Itinerarium* 14, 1631, 204. See p. 121.

37 See discussion in pp. 135–140. Grafton has rightly been criticised for overlooking this evidence (Almeida, 1995, 359, note 5; 2012, 390–392; see also here Chapter 1, endnote 16).

38 The dates refer to first editions; many had several editions and translations. In some cases, these works were unpublished until later (in these cases, I give the date of composition). See bibliography for details. The list is reduced to the books I have consulted.

39 Published in 1841.

40 Published in 1847.

41 Published in 1509.

42 Published in Kunstmann (1854).

43 *Decades* was the title of a book that Anghiera wrote over the years and published in stages. The original texts were letters sent by Anghiera to Cardinal Ascanius Sforza

from 1493 onwards, which were published as the first decade in an abbreviated pirated edition (Venice, 1504) and later included in Fracanzio da Montalboddo's *Paesi novamenti retrovati* (Vicenza, 1507). An apparently unofficial version was later printed (Seville, 1511). The first three decades were finally published in an official way (Alcalá de Henares, 1516), followed by the fourth (Basel, 1521), and the whole eight decades (Alcalá de Henares, 1530b).

44 Published in Marcondes de Souza (1958).

45 Published in Pacheco (1892).

46 Published in Pires (1944).

47 Published in Erasmus (1533).

48 Published in Nardi (1965).

49 The relevant portion was published in Verrazzano (1841) and again in Verrazzano (1970).

50 Published in Pérez de Oliva (1985).

51 Published in Las Casas (1875–1876).

52 Published in Italian in Columbus (1571).

53 Castro (1940; 1968a).

54 Castro, *Roteiro do Mar Roxo*; extracts related to the topics discussed here were published in English in Purchas (1625, vol. 2); the complete original text was published in Castro (1833).

55 Castro (1940; 1968a1).

56 Published in Rheticus (1876).

57 Published in Gorlaeus and Rheticus (1651) and Rheticus (1984).

58 Although this book was composed in the late 1530s.

59 "rudos marinheiros," "longa experiencia," "Iulgão por falſos, ou mal entendidos" (my translation).

60 "Eſto que he dicho no ſe puede apꝛender en Salamanca / ni en Boloña / ni en Paris" (translated by Elliott, 1970[1992], 40).

61 See Galateus, p. 109. Hooykaas notes this situation: "to the Renaissance scholar *all* ancient authors seemed venerable, [...]. Consequently, by several artifices [...] it was possible to use the authority of the Ancients for widely divergent and even opposite purposes" (Hooykaas, 1979, 419). An example was to remember Posidonius's defence of the habitability of the tropics, while most ancient authors rejected such an idea (Hooykaas, 1979, 20; 1980, 238–239; On Posidonius, see Chapter 2, p. 39 and endnote 21).

62 "que por tempo de mil & quinhentos annos ou mais ſoube de todo eſquecida & morta" (Pacheco, 1892, 1).

63 *Laetatus est igitur Infans, quia homines ibi habitari expertus est contra cosmographos, qui sub torrida zona negant habitationem hominum* (Münzer [1493b], first published by Kunstmann [1854, 350]; the translation is mine from the Portuguese translation of Vasconcelos [1932, 146], and the French of Randles [1980, 36]).

64 Anghiera, in his letter nº 165, already reported that Columbus believed that the Hispaniola Island was Ophir (To Cardinal Bernardino, Tortosa, 9 August 1495, in Anghiera, *Opus Epistolarum* 8, 1530a, fol. xl^{r}; translated by Thacher, 1903, vol. 1, 75–76). Columbus already made this identification in a letter to the Catholic Monarchs in 1498, describing his third voyage to America (Spanish-English edition in Columbus, 1847[1870], 112–113) and in another in 1503 on his fourth voyage (Columbus, 1847[1870], 203–204). It reappeared in a letter to the Pope in February, 1502 (translated by Fernández-Armesto, 2010, 210). He also dedicated an important space of his unpublished *Book of Prophecies* (c. 1501–1506) to collect biblical texts related to Tarshish and Ophir, and commentaries upon those texts (Columbus, *Book of Prophecies*, fol. 77^{v}–83^{r}, 1997, 320–347). Acosta rejected these ideas in *Historia* 1.13–15). Acosta's criticisms were preceded by Santaella, 1503, fol. 2^{v}. On those topographical identifications, see Gil (1989[2017], vol. 1, 49–57, 269–284, 295–301).

65 Several authors rejected that, explaining that there was nothing specific regarding America, apart from the speculations about the antipodes (Acosta, *Historia* 1.11, 1590, 43–48) or Plato's Atlantis (Acosta, *Historia*, 1.12, 22, 1590, 48–49, 75–78).

66 Acosta attacked Aristotle for considering the Torrid Zone as uninhabitable, but then mentioned how other ancients knew better (Plato, Pliny, and moreover Ptolemy and Avicenna, Acosta, *Historia* 2.9, 1590, 103). Anghiera noticed that the ancients considered uninhabitable the Torrid Zone "a fewe excepted" (Dedicatory letter to Charles V, 1516, fol. a.i^{v}; translated by Eden, 1555, fol. d.iv^{v}). Eden pointed to Albert the Great as a correct precedent defending the habitability of the Torrid Zone, but he recognised that it was a minority position (Eden, 1553, fol. A.iii^{v}).

67 "e nas outras todas sem nehuma pejo somos vencidos" (Castro, 1833, ii).

68 Randles (1992[2000], 7; 1961[2000]; 1993a[2000], 86–88). Centuries earlier Dio Cassius, in his *Historia Romana* 62.4.2 (c. 220s), referred to Britons as dwelling "on a different earth and under a different sky" (Cassius, 1925, vol. 8, 89).

69 "Nam ha duuida que as nauegações deſte reyno de cem ãnos a eſta parte: ſam as mayoꝛes: mais marauilhoſas: de mais altas 7 mais diſcretas conjeyturas: que as de nenhũa outra gente do mundo" (My translation).

70 "diguo q̃ ſe sabe mais ẽ hũ dia aguora pellos Purtugueſes do q̃ ſe ſabia ẽ 100. ãnos pollos Romanos" (translated in Orta, 1913, 127–8).

71 "la maior cosa despues de la criacion del mundo, sacando la encarnacion, y muerte, del que lo crio, es el descubrimiento de Indias. Y assi las llaman mundo nuevo" (translated by Elliott, 1970[1992], 10).

72 *nouas ꝓuintias noua regna nouas inſulas 7 quaſi nouos 7 incognitos oꝛbes* (translated by Rogers in Lucena, 1958, 45).

73 *Etenim ꝗd tu aliud obſecro Rex nobis, ꝗ̃ terras alias, mare aliud, alios mundos, aliaq́; poſtremo ſidera, nõ magis ĩueniſti, ꝗ̃ ab æternis tenebris, & à ueteri penè dixerĩ chao rurſus ĩ hanc publicã lucem protuliſti?* (translated by Randles, 1992[2000], 26; see also Hooykaas, 1979, 9–10).

74 Rosen, "Nicholas Copernicus. A Biography," in Copernicus and Rheticus, 1971, 372. In the 17th century, the vocabulary of the 'new worlds' was applied to telescopic discoveries by Wilkins (Wilkins, 1638 and 1640) and to microscopic ones by Huygens (cited in Pagden, 1986, 3).

75 *alium orbem omnibus auctoribus incognitum* (my translation).

76 On the terminology used after the discovery: Randles (1993a[2000], 84–88; 1961[2000], 2–4).

77 Translated by Eden (1555, fol. 48^{v}–49^{r}). Also, "almoſt weſt *Antipodes*, fownde of late by the Spanyardes" (Anghiera, *Decades* 2.1, 1516, fol. d.ii^{v}; translated by Eden, 1555, fol. 51^{r}). Italics in the English original in both cases.

78 In the original, the meaning is that these new lands were unknown to the inhabitants of the Old World; see the Italian-English edition in Harrisse (1888, 9–11).

79 Téllez (1508a, fol. 2^{r}; 1508b, fol. 2^{v}–3^{r}; 1516, fol. a.iiij^{r}); on this text, Miralles (2016, 158–162).

80 Sebastiano del Piombo (Sebastiano Luciani), "Portrait of a Man, Said to be Christopher Columbus," Metropolitan Museum of Art, New York (accession number: MMA 00.18.2; available at: https://www.metmuseum.org/art/collection/search/437645).

81 On the comparison of the Portuguese and Roman Empires by Portuguese authors of the early 16th century, see Hooykaas (1979, 53–63). For Spain, see Maravall (1966[1986], 438–446).

82 "A mi pareſce que con mas raȝon conuenia a eſte tiempo o ſiglo llamarſe la era doꝛada, q̃ la que los antiguos llamarõ de Saturno" (my translation).

83 "q̃ ni Griegos, ni Romanos, ni los Cartagineſes, ni otra nacion alguna tanta parte del mundo poſſeyeron ni poſſeen" (my translation).

84 "Si los ãtiguos alcãçarã lo q̃ los modernos alcançã, no eſtuuieran poꝛ đſcubꝛir las ĩdias" (my translation).

85 João Teixeira's discourse before King D. João II (1489, published in Texeira, 1562), compared him with the Queen of Saba (Randles, 1961[2000], 11).

86 "luz devinal" (Zurara, 1841, 10).

87 See, for example, references in the prologue to the Catholic Monarchs to Columbus's *Diario* (log-book) of his first voyage in 1492–1493 (preserved by Bartolomé de Las Casas in a manuscript published in Fernández de Navarrete 1825, vol. 1, 2; English translation by Thacher, 1903, vol. 1, 514); also Columbus's first letter announcing the arrival to America, addressed to Luis de Santangel on 15 February 1493, printed in April in Barcelona (Spanish-English edition in Columbus, 1847[1870], 16–17); Columbus's letter to the Catholic Monarchs describing his first voyage to America on 4 March, 1493 (English translation in Zamora, 1993, 190, 194–195); another letter to the Catholic Monarchs of 1498 (Columbus, 1847[1870], 108–109); a letter to the nurse of Prince Juan of Castile in 1500 (Columbus, 1847[1870], 153), a letter to the Pope in February, 1502 (translated by Fernández-Armesto, 2010, 211); a letter to the Catholic Monarchs of 7 July, 1503 (Columbus, 1847[1870], 191, 192, 208), etc. His missionary concern appears in the mysterious signature that he started to use in early 1493 (during his first voyage), including a series of capital letters, to which from 1502 he added "XPO FERENS" or Christ-bearer (Varela, 2010, 260–261). The first map depicting the new American lands by Juan de la Cosa (1500), part of the crew of Columbus during the first voyage, contains in place of Central America an image of Saint Christopher bearing Christ on his shoulders across the waters (Museo Naval of Madrid, N. 257), see at http://bibliotecavirtualdefensa.es/BVMDefensa/i18n/consulta/registro.cmd?id=16822. This idea of Columbus bringing Christ to America was mentioned in a sermon in Rome in 1497 (see quotation in p. 117).

88 For more details, see Delaney (2006; 2011), León Azcárate (2006; 2007), Gil (1989[2017], vol. 1, 49–57, 236, 231–260), see in particular note 138 in pp. 63–64 for examples of these ideas in different peoples in the years immediately preceding the arrival of Columbus in America, as some were part of the propaganda of the Spanish monarchy.

89 For example, in his letter to the nurse of Prince Juan of Castile (1500) where he affirmed: "God made me the messenger of the new heaven and the new earth, of which He spoke in the Apocalypse by St. John, after having spoken of it by the mouth of Isaiah; and He showed me the spot where to find it" (Columbus, 1847[1870], 153, the references are Isaiah 65:17ff and Revelation 21:1ff). See also p. 117.

90 On Columbus as guided by God, see the letter to the Catholic Monarchs dated in 1501 and inserted in *Book of Prophecies* (fol. 4[r]) and the letter to the Bank of Saint George in Genoa of 2 April 1502 (Spanish-English edition in Harrisse, 1888, 5–6). Some of his contemporaries endorsed this view of himself, as in the quotations referred to in pp. 118, 128 and endnotes 87, 89, 90 and 137. Las Casas considered Columbus chosen by God in his *Historia de las Indias* (1527–c. 1560, unpublished until 1875), an idea endorsed by Pérez de Oliva in his *Historia de la invención de las Yndias* (written in 1528, which the author gave to Columbus's son Ferdinand and was then lost until recovered and published in Pérez de Oliva, 1965). On the preaching of the Gospel in new lands, see the same letter in the *Book of Prophecies*, fol. 6[r]. Of course, this was an implicit reference to the text of Matthew 24:14, explicitly mentioned in fol. 20[r] (also indicating references to the parallels in the Synoptics) with a discussion about its fulfilment in the past (fall of Jerusalem in 70 AD) or at the Parousia (Columbus mentioned the commentaries by Lyra and Abulensis in fol. 21[r]), topics which we already saw discussed by Augustine in p. 45. Columbus also mentioned there Psalm 18/19:4/4-5 (fol. 14[v], fol. 21[v]). For the *Book of Prophecies*, see Columbus (1997).

91 I have picked up this term from Delaney (2006, 280), and created the equivalent temporal term 'chronoeschatological.'

92 See Columbus's letter to the Catholic Monarchs, dated in 1501 and inserted in *Book of Prophecies*, fol. 5[r]; see also fol. 25[r-v].

93 Joachinism was a radical religious movement inspired by the eschatological ideas of the Italian theologian Joachim of Fiore (c. 1135–1202).
94 On that, see also Vogel (2006b, 836–839), and the quotation from 30 years later of Elliott in p. 135. Also Maravall (1966[1986], 435, 442, 446, 452, 453) for examples in 16th-century Spain of the feeling of superiority derived from the geographical expansion.
95 *parua quędam ĩſula ẽ uti libro de ſomnio Scipioniſ auctor ẽ Cicero* (my translation).
96 As Nothaft (2017) remarked, nothing in Lilio shows him defending Lactantius's flat Earth views, as has been suggested by Randles (1980, 31), Lecoq (1996, 32) and Allegro (2017).
97 In *Natural History* 2.65.162 Pliny mentioned how the measurements of the height of mountains by Dicaearchus led to the conclusion that they were negligible protrusions in comparison with the size of the Earth (Keyser, 2001[2017], 361–365, who dates these measurements c. 305 BC). See also in Marzio, Chapter 3, p. 77 and endnote 44.
98 Specialists are unsure if these notes were written before his first oceanic crossing of 1492 or in later times, seeking to make sense of his discoveries and to put forward the kind of argumentation that we find in his later writings (Wey Gómez, 2008, 143–144).
99 On this letter, see endnote 9.
100 See marginal note to chapter 8 of Ailly's *Imago Mundi* (Latin-French edition in Ailly, 1930, vol. 1, 208, 213; on Ailly's discussion of this topic, see Chapter 3, endnote 91). The same idea – only one-seventh of the Earth covered with water from 4 Esdras – appears in the 1498 letter to the Catholic Monarchs about his third voyage (Spanish-English bilingual edition in Columbus, 1847[1870], 146) and in another letter to them describing his fourth voyage, 7 July 1503 (Columbus, 1847[1870], 183). While in the letters he mentioned 4 Esdras correctly as defending that six parts of the Earth were uncovered, in the note to Ailly, he went further and rendered them as inhabited (*Sex partes sunt habitate*...; quoted in Ailly, 1930, vol. 1, 208).
101 Although Columbus referred to Ambrose's *Hexaemeron* in his 1498 letter to the Catholic Monarchs, mentioned in endnote above, I have not found a fitting reference in Ambrose. However, Ambrose was very fond of 4 Esdras and recommended it to Horontianus (*Epistle* 34.1-2), placed it over Plato (*De bono mortis* 10.45, 11.51) and mentioned it in several of his books.
102 Geraldini mentions Diego of Mendoza, but it should rather be his powerful brother, Pedro Gonzalez de Mendoza (1428–1495), called the "Great Cardinal."
103 *cui cum referrem Nicolaum à Lyra virum ſacrę Theologiae exponendae egregium fuiſſe, & Aurelium Auguſtinum doctrina, & ſanctitate magnum, tamen Coſmographia caruiſſe* (translated in Wey Gómez, 2008, 113, slightly modified).
104 Geraldini wrote between 1520 and 1522, but his book was only published in Rome in 1631. This has led to some suspicions regarding its authenticity. However, that seems unfounded, as several 16th-century manuscripts witness its transmission before it was printed (Oliva, 1993 and González Vázquez in Geraldini, 2009, 79–93). He proudly recalled his friendship with Columbus and how the navigator acknowledged it by giving the name of Geraldini's mother to a Caribbean island: Gratiosa. Geraldini affirmed that in this way her name would remain immortal in the antipodes (*Nomen ad Antipodes retinet tamen illa perenne,*... *Itinerarium* 12, 1631, 192).
105 On the combination of contradictory cosmographies in the book, see Iommi (2017).
106 However, Reisch affirmed that southern antipodes had not been found to the northern lands.
107 *Hos tamen diuus Auguſtinus negat libro. xvi. de ciuitate dei c.ix quia fortaſſe nõdum tanta longitudo haditabilis* [sic, *habitabilis*] *in onĩm noticiam venerat.* (I thank Harry Prof. Harry Hine (University of St Andrews, UK) for this translation).
108 Born in St Gallen (Switzerland), Vadianus studied at the University of Vienna, becoming chair of Poetry (1512), dean (1516) and doctor of medicine (1517). In 1518 Vadianus returned home, married in 1519, became a member of the city council in 1520,

joined the Reformation in 1522 and led the conversion of St Gallen to Protestantism after elected mayor in 1526.

109 I thank Dr Christa Gray (University of Reading, UK) for this translation. In addition to the texts of the Appendix, no. 17, other shorter translations of this correspondence, also by Dr Christa Gray, appear in the following endnotes.

110 A brief criticism appears already in Vadianus (1515, fol. B.ii[r]): "In the meantime, to prevent you from turning against Augustine, who denies [the existence of the Antipodes], or grant belief to the dreamer Lactantius" (*Interĩ aũt, ne ue ab Auguſtino abhorreas negante, uel Lactantio fidem prebeas ſomnianti*).

111 *terram rotundã eſſe negare noluit.*

112 *multa pro Auguſtino, & cõtra Lactantiũ locuti.* See endnote 115.

113 On this commentary, see Beer (2013, 170–177).

114 On this debate, see Vogel (1992), who summarised it in English in Vogel (2006a).

115 Vadianus not only had a copy of Marzio's *De incognitis vulgo* (*DIV*), which he tried to get published (see Chapter 3, endnote 60), but also mentioned Marzio and his *DIV*, chapter 14, in his 1518 commented edition of Mela's *Cosmographia* (fol. 5[r]). This mention appeared in the context of a reference of Vadianus to his own 1515 letter to Agricola (which did not mention Marzio, and was annexed to the 1518 edition of Mela). There were other references to Marzio in Vadianus's commentary (fol. 6[r] and 9[r]). In the reedition of the commentary in 1522, a reference to Marzio and *DIV* was inserted by Vadianus in the very text of his 1515 letter to Agricola, as an example of an author who had voiced objections against Lactantius on the antipodes (Mela, 1522, fol. Gg3[r]).

116 From 1529 this text appeared with the revision of Gemma Frisius, who in 1533 made an enlarged version that was reprinted many times and translated into several languages.

117 However, it was not included in the simplified version, Apian, *Cosmographiae introductio* 19, 1529, no pagination.

118 This addition appears under the name of Gemma Frisius, but as explained by Ortroy (1920, 207), this text is not found in any of Frisius's works.

119 This is part of an addition in the book taken from Gemma Frisius, *De Principiis* 1.7, 1530, fol. B.1[v]–B.2[r].

120 *ipſi huius partis notitia caruere* (my translation).

121 Pacheco, *Esmeraldo*... 1.21, 1892, 35–36; translated in Pacheco (1937, 58–59).

122 Nunes (1537d2, fol. B.i[r–v]).

123 *cum enim falsum ... consequuntur incurrere* (see this text in Lactantius's quotation in Appendix, no. 12, which has the wording as *consecuntur*).

124 Gómara mentions Marzio in support of Scandinavian inhabitants (Gómara, *Historia* 3, 1552, fol. iv[r]–v[r]). A discussion of northern regions also appears in Mejía (see above).

125 "y los que niegan los antipodes" (for the context, see endnote 128; my translation here and in all fragments from Gómara).

126 Mayaud (2005, 84–85), affirms that this reference in Clement was unknown in his enormous collection of 16th and 17th texts debating the relation of Copernicanism and the Bible. However, it is mentioned in well-known authors of this time (but not part of the astronomical debate) such as Gómara (*Historia* 4, 1552, fol. v[v]), Richard Eden (1555, fol. 321[r]) and Sixtus Senensis (1566, 92). On this text of Clement, see p. 33.

127 "Latanҫio e Iſidro no tuuieron cauſa para negarlos. San Aguſtin tuuo las que dixe."

128 "Mas Anaximandro, y Anaximenes, y Lactancio, y los que niegan los antipodes, afirman ſer llano eſte cuerpo redondo, que haʒen agua y tierra." Incidentally, we can see there Gómara advancing the terraqueous view.

129 "Dexando aparte autoꝛes gentiles, digo que tambien ay chꝛiſtianos q̃ niegan auer antipodas. Los que tenian a la tierra poꝛ llana los negaron. Y Latanҫio Firmiano los contradiʒe ..."

130 Eden added a marginal note "he meaneth S. Auguſtine and Lactantius" to a generic criticism by Anghiera to "men of ſingular witte and great lernyng, which haue denyed that there is *Antipodes,*" italics in the original). His essay "That the Spaniards have ſailed to the Antipodes…," in his translation of the *Decades*, criticised Clement for accepting the uncrossability of the ocean, without pointing that, at the same time, Clement seemed to accept antipodes beyond the ocean, as noted by Gómara (Eden, 1555, fol. 321^{r}).

131 However, according to Grant (1984, 22, 27–32), there was a sophisticated scholastic resistance in the late 16th century, in particular from the conimbricenses (this was an influential group of Jesuits from Coimbra, Portugal; see, for example, Manuel de Góis in p. 133). Wootton (2015, 129–135), discusses the survival of other old cosmographical ideas in the late 16th century (such as in Antonio Berga or Jean Bodin).

132 "Sainct Auguſtin notamment, Lactance & pluſieurs autres" (my translation).

133 "Platon meſmes a confeſſé les Antipodes: mais ils ne nous en ont laiſſé les demonſtrations. Qui fut occaſion à ſainct Auguſtin de croire bien la rondeur de la terre, mais de nier qu'il y euſt des Antipodes ſous nous […]. Mais L'actance a bien plus hardiment nyé la terre ronde, l'affermant plate afin qu'il peuſt mieux probablement confuter l'opinion des Gẽntils, […]" (my translation).

134 "Enquoy nous devons faire noſtre profit, remarquãt en ces bons docteurs la fragilité de la nature humaine. Car pouſſez d'vn ardant deſir d'aneantir la doctrine payenne pour pluſtoſt avancer la notre: ils ſe ſont ſi aheurtez à ſoudain cõdanner & contredire les opinions des Gentils, qu'ils n'ont à l'avãture trop bien regardé comme ils aſſeuroient les leurs. Et nous en voyõs auiourdhuy qui n'ayans employé un bon an aux eſtudes, condennent neantmoins comme fauces & impiez les opinions qu'ils ne ſçauroient bien entendre" (my translation).

135 Original Latin texts, from preserved students's notes, are published in Nardi, 1965, 43: *Ad hoc respondeo dicendo quod hoc solvant fratres, quoniam ego credo ecclesiae sanctae* and *quare nescio quid dicendum est, nisi forte quod Christus fecit se crucifigere etiam in alio polo!* (I thank Prof. Harry Hine (University of St Andrews, UK) for this translation).

136 The Latin original and an English translation are available in Thacher (1903, vol. 1, 206–212).

137 "And indeed inasmuch as Columbus frequently declared himself to have been selected by God that through him this prophecy might be fulfilled" (Thacher, 1903, 209). For more on that, see endnotes 87, 89, 90.

138 "que ate os Sanctos duuidaram" (my translation).

139 "Finalmente tutti coloro, tenendoſi al prouerbio Castigliano, il quale in ciò, che non par ragioneuole, ſuol dire, Dubita Santo Agostino, per cioche detto Santo al IX Capitolo del XXI libro de Ciuitate Dei reproba, & ha per impoſſibile che ci ſiano Antipodi, & ſi poſſa paſſar dall'vno Emiſperio all'altro" (my translation, see Appendix, no. 6).

140 "Otros alegaban a San Agustín, el cual, como tocamos arriba, negaba que hubiese antípodas, que son los que decimos que andan contrarios de nuestros pies, y así traían por refrán, 'duda San Augustín'" (my translation).

141 "ſiempre dudoſo" (translated in Acosta, 1604, 3).

142 *Et qui Antipodes abnuunt, in qua hœreſi fuit alio quin, cœleſti ingenij acumine Diuus Auguſtinus, at experientia, & oculorum ſenſus manifeſto contrarium demonſtrat.* I thank Prof. Harry Hine (University of St Andrews, UK) for the following translation: "and those who deny the antipodes; among others Saint Augustine, with his divine sharpness of intellect, belonged to that heresy, but experience and eyesight clearly demonstrate the opposite."

143 See here below and Table 4.1 (on this text by Procopius see Petit, 1991, vol. 1, xxxiii ff and Ceulemans, 2017). As with Cosmas Indicopleustes (under the name of *Christianus historiographus*, see Herren, "Introduction" in Aethicus, 2011, l–li; Bischoff and Lapidge, 1994, 25; Lapidge, 2005, 32), traces of Procopius seem to have been

known in the west before, in the context of the school of Archbishop Theodore and Abbot Hadrian in Canterbury c. 700 (Bischoff and Lapidge, 1994, 227–229 and Lapidge, 2005, 32, note 7; but more doubtful Ceulemans, 2017, 208–210).

144 It had been translated into Latin by Mutianus at the request of Cassiodorus in the 6th century. A contemporary independent translation by the African Bishop Facundus was less well-known.

145 *Hanc ſententiam iiſdem penè uerbis uidetur Procopius accepiſſe ex 14. & 27. homilia Chryſoſtomi in epiſtolam ad Hebræos* (I thank Prof. Harry Hine (University of St Andrews, UK) for this translation).

146 We will not discuss here if the inclusion/exclusion criteria as flat-earthers or deniers of the antipodes/antipodeans were then, or are today, justified for each ancient Christian author, as this is another topic outside our scope. An example of a recent study of the history of how early and medieval Christians interpreted the biblical text in relation to ancient cosmology is Randles (1999), and a recent collection of texts of those authors on Genesis 1-11 is Louth (2001).

147 *mercatores noſtri ... Quibus conſideratis prudentisſimi patres Antipodas negarunt* (I thank Prof. Harry Hine (University of St Andrews, UK) for this translation).

148 Mazzoni displays here a list of ancient authors negating the accessibility of the Southern Hemisphere "as Macrobius showed, and before him Eratosthenes, Polybius, Strabo and Tullio" to which he surprisingly adds "and after all Galeotto da Narni" (Mazzoni, *Della difesa della Comedia di Dante* 3.33, 1587, 555; for the translations of Mazzoni's Italian text, here and below, I am grateful for the help of Teresa Canale (Scuola Materna Italiana Madrid, Spain) and Juan Sánchez (Facultad SEUT, Madrid, Spain). As Miggiano has pointed out, that "suggests that he quoted by ear, without having direct knowledge of the original *De incognitis vulgo*" (Miggiano, 2004, 191, note 3; my translation).

149 Books 1 and 2 were written in America (1582) and published in Latin to introduce the treatise on missiology *De Procuranda Indorum salute* (Acosta, 1588). Back to Spain, he translated both to Spanish, and composed five more books in Spanish on American natural philosophy. This complete book was published in 1590 (from which I quote) and soon translated into many languages becoming very influential (Pino, 1978).

150 "Y aun parece tirar algo a eſto lo que dize ſant Hieronymo" (translated in Acosta, 1604, 3).

151 We will see below an early reference to Augustine as a flat-earther in a letter from 1589 by Brahe (Brahe, 1596, 156–157; see pp. 169–170). For the recent debate, see p. 40.

152 "vaſe con el vulgo", "la fuerça de la razon" (translated in Acosta, 1604, 22, 23).

153 "muy otra fue" (my translation; Acosta, 1604, 25 translates that as "was other", missing the intensity of the original); "tomado de las entrañas dela ſagrada Theologia" (translated in Acosta, 1604, 25).

154 "q̃ aunque tan gran Philoſofo, ſe engañò en eſta parte" (translated in Acosta, 1604, 28) The ambiguous judgement of Augustine is applied to Aristotle.

155 Brackets in the English texts are my corrections in the light of the 1590 Spanish original.

156 "Que Plinio y los mas de los antiguos ſintieron lo miſmo que Ariſtoteles" (translated in Acosta (1604, 33). Acosta discussed that again in a chapter entitled "For what reaſon the Ancient held, that the burning Zone was inhabitable [*sic*, it should be uninhabitable]" ("Que les mouio a los antiguos, a tener por coſa ſin duda, que la Torrida era inhabitable", Acosta, *Historia* 2.2, 1590, 86; translated in Acosta, 1604, 83).

157 In the 17th century, the debate on the antipodes and polygenism was recast into the pre-adamites debate, connected with the origins of the indigenous peoples of America (Livingstone, 1992).

158 Already Juan López de Velasco, in a manuscript of 1574, suggested that humans could have arrived to America due a present or past connection with Asia in the north (*Descripción universal de las Indias*, first published in López de Velasco, 1894, 3–4).

159 He also wondered about endemic animals of America (Acosta, *Historia* 4.36, 1590).
160 "Mas habla la eſcritura conforme a nueſtro modo de imaginar y hablar" (translated in Acosta, 1604, 11).
161 "Y no ſolamente las letras profanas, mas aun las ſagradas tambien hablan eneſſa ſorma, accõmodandoſe a nueſtro lenguaje" (translated in Acosta, 1604, 27).
162 See above pp. 131–132. Góis points to one of the references mentioned by Acosta, *De genesi ad litteram* 2.9.
163 *id noſtri ſeculi nauigationibus* and *conſtituentibus Geographis vltimos terræ fines Herculeas columnas* (I thank Prof. Harry Hine (University of St Andrews, UK) for this translation).
164 Abulensis's commentary on Genesis 1:24–31 (written in 1436), published in Fernández de Madrigal, *In librum Genesis comentarii* 1, 1596, fol. 6[v], left column, I-K.
165 See pp. 128–130 and also references to these Bible texts above in Chapters 2 and 3.
166 *Verum quia res eſt prorſus theologica, paucis reſpondemus ſiue aliquis Apoſtolorũ Æquinoctialem circulũ pertranſierit, ſiue non: tamen in omnem terram id eſt, in præcipuas mundi oras, quæ tunc notæ erant, id eſt in Aſiæ, Africæ, & Europæ partes ipſis adhuc Apoſtolis viuentibus Euangelium perueniſſe, quod nunc indies in alias atque alias nationes particulatim diſſeminatur* (I thank Prof. Harry Hine (University of St Andrews, UK) for this translation).
167 Krüger dated it 1627, as well as Vassiliev (1893, XXXV), who used Migne's edition (1849, vol. 73, cols. 415–428), which was a reproduction of the 1628 edition.
168 *vbi cælum terræ ſe cõiungit* (online English translation by Benedict Baker available at: http://www.vitae-patrum.org.uk/page33.html, last updated 12 July 2004).
169 *Fuit veterum multorum ſententia, planam eſſe terram, nec Antipodes [n]ullos eſſe; atque ita cælum alibi terræ iungeretur. Quin multi paradiſum certâ terræ regione ſtatuerunt, sublimiore, ubi cælum terræ iungitur* (I thank Prof. Karla Pollmann (University of Tübingen, Germany) for this translation).
170 The objective is similarly worded: *vbi cælum terræ ſe iungit.*
171 Although it could have been written in the 11th century, date of the oldest manuscripts. See Vassiliev (1893, XXXV–XXXVIII), Kazhdan (1991, vol. 2, 1270–1271), Acerbi (2009, 119).
172 The problematic word would be "many" (*multorum*, see endnote 169; for a discussion of the usage of this and similar terms in the context of the flat error, see pp. 140–141). This is a subjective term that can give the impression that flat-earthers were dominant at that time, and in this case, it would become a misleading quantification of the real impact of the flat-earthers in early Christianity. Nicolaidis indicates that in the Eastern Orthodoxy, during the Middle Ages, the Alexandrian spherical Earth was embraced by scholars, the royal court and the higher clergy, while the Antiochian flat Earth "became the popular cosmology [...], propagated both orally and in such written works as the 'Lives of the Saints'" (Nicolaidis, 2011, 24, 39). The text here discussed is indeed one of those "Lives of the Saints."
173 An old tradition exists of placing the earthly Paradise at the top of a mountain (Caudano, 2016, 187). The idea of reaching the edge of the skydome touching the flat Earth is a recent fancy.
174 Many medieval maps indeed located Paradise at the eastern edge of the oikoumene, to intensify its remoteness following Genesis 2:8, but that did not mean that they imagined Paradise compressed in an angle formed by the sky coming down to a flat Earth. As we saw in Chapter 2, these maps were compatible with a spherical Earth, and the mountain of Paradise was imagined emerging from the spherical Earth (as later in Dante in the early 14th century, see p. 81, and Columbus in 1498, see endnote 9). The two ideas mixed by Rosweide did appear in Eastern Christianity, as in Cosmas, in the 6th century, who imagined Paradise as a rectangular land against the vertical eastern edge of the sky descending on the Earth (Kominko, 2013, 52–53, 58–65). However,

there is not a hint at these cosmological ideas in the *Vita Sancti Macarii*. I thank Rev. Dr Robert D. Keay (Belfast School of Theology, UK) for a helpful discussion about the theological reading of this text.

175 Translated in Farrington (1964, 77–78).

176 Translated in Bacon (1858, vol. 4, 87–88).

177 On the history of the discussion regarding the intellectual impact of the discoveries over the centuries, see Pagden (1986). The term "blunted impact" was in the title of Elliott (1976); and already mentioned in Elliott (1970[1992], 104).

178 In an interesting criticism to Elliott, the historian of science Pamela H. Smith mentions that "In 1530, corn was already growing in Avila, Spain. [...]. [That] indicates that exchange and assimilation of information, materials, and practices was already much livelier at an early stage than the texts studied by Elliott lead us to believe" (Smith, 2009, 375, note 89).

179 "[...]. La autoridad de los textos clásicos comenzó a tambalearse en el siglo XVI, ante la creciente comprensión de que en algunas cosas los griegos y los romanos estaban equivocados, y no por último en su ignorancia sobre la existencia de América. [...] los modernos, al menos en algunos aspectos, habían demostrado ser superiores a los antiguos. Una vez sembrada la semilla de la duda, el camino estaba abierto a la llegada de aquella idea de progreso que iba a tener tan profunda influencia en la Europa de los siglos XVIII y XIX. [...]" (my translation).

180 ... *ſiue prolatione rerum inauditarum, quæ tam vehementer facie, conſuetudine, more a noſtris abhorrerent*. The above translation from Vives, 1913, 3, is rather simplified. Vives is referring to the shocking (abhorrent) impression that the new cultures found in the travels produced to Europeans.

181 "porque no me determine a contradezir a Ariſtoteles, ſi no es coſa muy cierta" (translated in Acosta, 1604, 110). On the modification and survival of authorities such as Aristotle and Ptolemy, see Pagden (1993, 91–92).

182 Translation from Sanches (1988, 222); on Sanches, see Almeida (2001).

183 Montaigne, *Essais* 2.12 [An Apology of Raymond Sebond], 1580, vol. 2, 348–349; a reference to the antipodes was inserted in the edition of 1588, 241v; with a further addition in 1595, 376; he also had a parallel reflection concerning the Copernican challenge to geocentrism, Montaigne (1580, 344–345).

184 As we have seen before, the contradictions among ancient authors themselves on the antipodes debate were noticed by Petrarch (p. 82) and Agricola (p. 122), who agonised over such 'internal' contradictions. However, Vadianus could reply to Agricola referring to recent geographical discoveries to settle the issue. See also endnote 66.

185 The English translation of 1604 rendered these Spanish words as "many." A more literal translation would be "some and not few" (see endnote 187).

186 "che fù opinione d'alcuni dottori, fondata nella scrittura Sacra" (my italics in the translation).

187 "algunos y no pocos, ni de los de menos autoridad entre los ſagrados Doctores" (translated in Acosta, 1604, 1–2), which could be more literally translated as "some and not few, *nor those of lower authority* among *holy doctors*" (my translation and my italics).

188 *Mvlti tamen graues Auctores ſententiæ de cœli rotunditatem* [sic, *rotunditate*] *non acquieuerunt* (I thank Prof. Harry Hine (University of St Andrews, UK) for this translation; my italics).

189 On this periodisation scheme, see pp. 180–183. A large study on the topic of ancients and moderns was undertaken by historian José Antonio Maravall (1966[1986]). A specific monograph on the idea of progress was published by sociologist Robert Nisbet (1980[1994]), although missing the component of the geographical revolution in the 15th–16th centuries, and – typically – with this topic only mentioned in relation to the 17th-century England (p. 145).

190 Pomponazzi posed similar questions to the Aristotelian tradition (see p. 159).

191 See pp. 95–96.
192 The complexity and contradictions in the European cultural canon made it more adaptable, a fact that accounted for its resilience up to the Enlightenment and even after, as indicated by Grafton (1992[1995], 6, 10, 48, 148, 152, 157, 237, 242) (see also pp. 135–140). As seen in Chapter 3, geographical and theological speculations about the antipodeans existed even before they were found.
193 Augustine's problem could be avoided by rejecting the human nature – and Adamic descent – of indigenous peoples of America. However, their human character was already affirmed by Columbus. The vigorous defence of the humanness of the indigenous peoples, mostly by Spanish clerics, was endorsed by Pope Paul III in the bull *Sublimis Deus* (1537). I have not seen the human nature of the indigenous peoples of America disputed so as to justify that they were not human antipodeans.

5 The early Copernicans and the flat-earthers (1541–1616)

Copernicus, Lactantius and the rise of the concept of a science and Christianity conflict

On 18 March 1523, after explaining the geographical ideas of Aristotle and Averroes (see p. 128), philosopher Pietro Pomponazzi asked his students: "But what should we today think?" (Pagden, 1993, 90; translated from Nardi, 1965, 42, who does not provide an actual Latin quotation). He then referred to a letter from Antonio Pigafetta, one of the survivors of the first circumnavigation of the Earth (1519–1522) and questioned the authority of the philosophers with these words: "If we are ignorant about things on Earth which we can see, how will we know about the heavens?" (Pagden, 1993, 91).[1] At that time he could not know that Nicolaus Copernicus (1473–1543) was working on a new astronomy which would indeed challenge the ancient 'knowledge' of the heavens. Copernicus not only defied Aristotle but also the more technically challenging astronomy of Ptolemy,[2] as Owen Gingerich noted: "If Ptolemy as geographer-cosmographer could be challenged, what about Ptolemy as astronomer-cosmologer?" (Gingerich, 2001, 29).

Copernicus against the scholastic eccentric model

Although Copernicus's *De revolutionibus* (Nuremberg, 1543; Basel, 1566; Amsterdam, 1617) was primarily an astronomic work with the Sun at the centre of the new proposal. And at the heart of such new ideas was the new view of the Earth as a terraqueous globe, ironically, inspired by Ptolemy's geography. Of course, the terraqueous globe was a handy topic for weakening the authority of scholastic geocentrist philosophers. However, it was also necessary for heliocentrism, because only a single, compact, terraqueous sphere could be envisioned as realistically rotating at high speed around its axis, something unthinkable with a multi-spherical body with several centres.[3] In fact, a single-sphere planet Earth could more convincingly be defended as naturally having a circular motion (*De revolutionibus* 1.5). It could also move with its nearby air to meet the objection that the Earth's movement would fling objects on its surface into space (*De revolutionibus*, 1.8). The connection of the geographical advances with Copernicus has been already discussed by several authors,[4] and Vogel considers the terraqueous sphere as "crucial, even indispensable, for Copernicus's cosmological innovation" (Vogel, 2006a, 480). This

DOI: 10.4324/9781003545682-6

was something Copernicus may already have had in mind when he first drafted his heliocentric ideas in the *Commentariolus* (c. 1510, before 1514).[5]

Therefore, Copernicus inherited the entire terraqueous tradition of debates we have seen in Chapter 4 and which developed in parallel with his own life (he studied at Italian universities during the critical period of 1496–1503, with its vital geographical explorations).[6] It could then be expected to see Copernicus rejecting the eccentric model and referring to the recent navigations, and indeed he did. He started *De revolutionibus* book 1 with chapter 1, "The universe is spherical," followed by chapter 2, "The earth too is spherical" (Copernicus, 1543, fol. 1ʳ; 1978, 8). This was in the same order as in Ptolemy's *Almagest* and later in Sacrobosco's *De sphera*, from the heavens to the Earth, which Lactantius criticised.[7] It should be noted that, in passing and in a very quiet way, Copernicus dismissed the theologically supported defence of the idea that the waters were higher than the earth – which was part of the eccentric model – in the very last sentence of chapter 2. In chapter 3: "How earth forms a single sphere with water" (Copernicus, 1543, fol. 1ᵛ;[8] 1978, 9), Copernicus defended a single globe using a wording already common in treatises of navigation and cosmography during the first half of the 16th century, as we saw in our previous chapter, although it had precedents in ancient literature (see pp. 31–32). He also criticised there the Peripatetics's 10:1 water to earth relation, rejecting the eccentric model of our planet and the differentiation between the centre of gravity and of magnitude, mentioning the limited oikoumene of Ptolemy and its expansion by "the moderns," specifically the explorations of "Spain and Portugal." Indeed he referred to "the existence of antipodes or antichthones," considering that America (in fact, South America, betraying a dependence of Ringmann and Waldseemüller, 1507, as discussed by Rosen, 1943) was "diametrically opposite the Ganges district of India." "From all these facts," Copernicus defended the terraqueous model and attacked all theories of a non-spherical shape of the Earth, starting with the flat Earth of Empedocles and Anaximenes (Copernicus, 1543, fol. 1ᵛ–2ʳ; 1978, 9–10).[9] All those points were already well known by 1543. In this initial chapter, Copernicus could have criticised Lactantius and Augustine, as we have also seen in many authors; however, as we will discuss below, he left that outside his book and only decided to mention Lactantius in the last moment in the dedicatory to the Pope.

Copernicus's disciple, the Protestant Wittenberg mathematician Georg Joachim von Lauchen (Rheticus, 1514–1574), endorsed the connection of the new Copernican astronomy with the criticism of the scholastic eccentric Earth and the denunciation of the limitation of the Ptolemaic oikoumene (while accepting a terraqueous model also of Ptolemaic origin). These new ideas benefited from the recent navigations, as discussed by historians Burmeister (1969) and Hooykaas (in Rheticus, 1984, 123–132). Both Copernicus and Rheticus mentioned geography, as they had a specific interest and engagement in geographical activities, particularly cartography. In the dedicatory letter of his *Chorographia* (1541b, published in 1876) to Duke Albrecht of Prussia, Rheticus – at that time helping Copernicus to finish *De revolutionibus* – affirmed that "as to geography Ptolemy is of little use" and that

Iberian sailors "have brought us another and as it were a new world" in the place expected to be covered by the ocean (in Rheticus, 1984, 125). Similar words were found in Rheticus's *Treatise on the Holy Scripture and the Motion of the Earth* (1541c, published in 1651)[10] in the context of criticisms of the eccentric model (Rheticus, *Treatise* 38–46, 1984, 86–92), for example: "Ptolemy believed that there were waters where the Portuguese have discovered in their maritime voyages that very fertile territories exist" (Rheticus, *Treatise* 38, 1984, 86). Interestingly, Rheticus "was personally acquainted with Vadianus" (Hooykaas in Rheticus, 1984, 130, note 48), who had defended the terraqueous globe in print since the 1510s.

Copernicus against the flat-earther Lactantius

We have seen in Chapter 4 that by the mid-16th century it was routine to mention Lactantius in works related to geographical topics (as the shape of the Earth was a central concern for geography). However, there was no obvious reason to mention him in an astronomical work devoted to the *movement* of the Earth, a topic Lactantius had not discussed.

A connection that might trigger a reference to him could be the first book of *De revolutionibus*, in particular the first three chapters, which were devoted to the shape of the universe and the Earth, and to geographical issues. However, neither Lactantius nor other Christian authors were mentioned there, as could be expected, and it was in his letter of dedication to Pope Paul III (1468–1549) that Copernicus attacked Lactantius as a religious flat-earther.[11]

Surprisingly, Lactantius was mentioned by Copernicus in relation to a theological – not a scientific – concern and for a specific strategic purpose. A delicate piece of humanistic literature, the preface-dedication to the Pope was crafted to present Copernicus's revolutionary proposal in the most acceptable way, including the framing of it as a recovery of ancient 'Pythagorean' wisdom. However, near the end of it, there was a surprising twist. Copernicus, who had expressed himself in moderate terms – as he also did in the book itself – suddenly rose to the attack.

Interestingly, this seems a later move. In *De revolutionibus* 1.3 Copernicus only listed ancient Greek Pre-Socratics as flat-earthers, without mentioning Christian flat-earthers,[12] while in the later preface-dedication to the Pope, he only mentioned as a flat-earther one Christian, Lactantius. The original preface, preserved in the autograph of Copernicus, but finally replaced by the preface-dedication to the Pope, did not mention the flat Earth or Lactantius; the same can be said of his early unpublished essay *Commentariolus* (c. 1510, before 1514). Why this radical difference? Why introducing the Christian element in the geographical/astronomical discussion? In addition, Copernicus, "this usually mild man," –in the words of Hooykaas – employed at this point unusually aggressive language (Hooykaas in Rheticus, 1984, 30). The text of the book was written many years before the preface-dedication,[13] which was the last part to be written, shortly after the beginning of printing in 1542. What happened then between the composition of book 1 and the preface-dedication?

The theological threat and Osiander's correspondence with Rheticus and Copernicus

In June 1539, a month after Rheticus arrived at Frauenburg, Martin Luther publicly criticised the idea of the movement of the Earth.[14] Furthermore, after Rheticus first disseminated Copernicus's ideas in his *Narratio prima*, finished in September 1539 and published in March 1540 in Danzig, Philip Melanchthon named and attacked Copernicus in a letter to Mithobius (Burkhard Mithoff) in October 1541, a month after Rheticus left Copernicus to print *De revolutionibus* and returned to Wittenberg. This negative attitude of the leading Reformers was probably well known in Wittenberg, a place where it seems that Rheticus returned briefly to lecture in the winter of 1540, returning afterwards to Copernicus (Kraai, 1998; 2000).

The theological threat was also perceived by Rheticus's friend Achilles Pirmin Gasser. In March 1540, Rheticus sent him a copy of *Narratio Prima*. In his reply, printed as the foreword of the second edition of *Narratio Prima* (1541a), Gasser affirmed that it "could be considered heretical (as the monks would say)" (1541a, 4).[15] The historian of Copernicanism Michel-Pierre Lerner pointed that this was "the first printed mention of the term 'heretical' associated with the new astronomy," which was reprinted in 1566 and 1596 (Lerner, 2005, 12). This ironic remark is not surprising if we keep in mind that Gasser was a Protestant.

In parallel, shortly after arriving at Frauenburg in 1539, Rheticus began corresponding with the Protestant Reformer Andreas Osiander (1498–1552). Copernicus also took part in this correspondence.[16] They discussed how to deal with the foreseeable criticisms from theologians. Could the exchange with Osiander have reinforced Copernicus's suspicions about the theological danger of his views and the need to pre-empt the attacks?

If we look at the chronology of this exchange, we see that Rheticus sent a complimentary copy of his *Narratio Prima* to Osiander in 1540. Upon receiving it, Osiander asked Rheticus to put him in contact with Copernicus. Copernicus's letter from 1 July 1540, which for unknown reasons only reached Osiander in March 1541, is lost. What is preserved, thanks to a manuscript of Kepler (composed in 1600–1601, published in 1858)[17] is a fragment of Osiander's reply on 20 April 1541, to Copernicus, proposing an hypothetical presentation to "pacify the Peripatetics and the theologians whom you fear to be about to raise objections" (Kepler, 1858, vol. 1, 246).[18] The same day Osiander also wrote to Rheticus, of which a portion was preserved by Kepler, affirming:

> The peripatetics and theologians will be easily placated if they hear that there can be diverse hypotheses about the same apparent motion and that they are not advanced as being certainly so, but rather as governing the calculation of apparent and composite motion as expediently as possible; [...].
>
> (Kepler, 1858, vol. 1, 246)[19]

The coupling of "theologians and Peripatetics" in both letters from Osiander has been interpreted as a reminiscence of the terminology that Copernicus (and

Rheticus?) could have employed in sharing his fears with Osiander (Lerner, 2006).[20] In 1615, Cardinal Bellarmine repeated Osiander's proposal of solution to also calm "philosophers and scholastic theologians" (Blackwell, 1991, 265; on this letter, from Bellarmine to Foscarini, see p. 174).

Copernicus's final strategy

It was after Rheticus started with the printing in Nuremberg (May 1542), that Copernicus prepared the preface-dedication to the Pope (June 1542),[21] one year after the proposal of Osiander, which Copernicus rejected:

> Although we do not have Copernicus's [reply] letter, it is evident that he did not share his correspondent's opinions. As Kepler put it, "Strengthened by a stoical firmness of mind, Copernicus believed that he should publish his convictions openly, even though the science should be damaged."[Note: Kepler, 1858, vol. 1, 246].
>
> (Rosen, in Copernicus and Rheticus, 1939[1971], 23)[22]

This rejection is not only inferred from the correspondence with Osiander, it also appears in the works by Rheticus, while staying with Copernicus, such as the *Narratio Prima* (1540) and the *Treatise on Holy Scripture and the motion of the Earth* (composed in 1541). Copernicus's response to Osiander appears in *De revolutionibus*. By 1542, the astronomer had a clear view of what to expect from theologians, which would explain his final strategy.

As a first strategic step, he inserted at the front of the book a letter from Cardinal Nicolaus Schönberg, dated in 1536, which showed vivid interest in the work of the astronomer and requested a copy of it at his own expense (Copernicus, 1543, fol. ij^{r}; 1978, xvii). Then, in the preface-dedication to the Pope (Copernicus, 1543, fols ij^{v}–iiij^{v}; 1978, 3–6) is where Copernicus response was fully displayed; in particular in its later part (Copernicus, 1543, fol. iiij^{r-v}; 1978, 5–6; see Appendix, no. 7, for a different translation of the whole relevant portion, available at www.routledge.com/9781032893068).

On the one hand, the defensive step. Instead of placating the theologians and philosophers as Osiander had suggested, Copernicus appealed to the highest authority, the Pope, for protection. Copernicus sequenced his strategy in an interesting way. First of all, before addressing the Pope directly, Copernicus asserted that his was a technical science book by saying upfront that "I have no doubt that acute and learned astronomers will agree with me" if they "thoroughly" consider the argumentation – the statement in the next paragraph, that "Astronomy is written for astronomers" would reinforce that view. Then Copernicus appeals to the Pope as one of those learned people, because he is known for his "love for all literature and astronomy too." Finally, he requested the Pope's protection from a diffuse enemy: "you can easily supress calumnious attacks although, as the proverb has it, there is no remedy for a backbite" (Copernicus, 1543, fols iiij^{r-v}; 1978, 5).[23]

He could have stopped there, but Copernicus developed his second line of response to Osiander's proposal, the offensive step, as if filled with fury. It was in the following sentence, opening the last paragraph, that Copernicus became more specific, pointing to "babblers,"[24] who would pretend to judge astronomy although being ignorant (as distinct from the Pope, whose astronomical knowledge had been praised a couple of sentences before). However, ignorance of astronomy was not the only problem, as Copernicus accused his enemies on theology grounds of "badly distorting some passage of Scripture to their purpose," that is, to "censor" Copernicus. He then repeated his message again in reverse order, first attacking babblers who *distorted* the Bible in the person of Lactantius (avoiding the terms theologians' or 'theology', and calling him "illustrious writer"). Afterwards, Copernicus reminded them of their ignorance of astronomy ("Astronomy is written for astronomers"), before once more addressing the Pope and turning to the topic of the calendar reform (Copernicus, 1543, fols iiij^{v}; 1978, 5).

It was in this second charge against babblers, that he deployed a pre-emptive attack on those abusing the Bible to reject heliocentrism by drawing a parallel between the flat-earther Lactantius and the geocentric rejection of heliocentrism. This parallelism was a clever propagandistic move, as nobody denied the sphericity of the Earth at that time. For this parallel he chose the most vulnerable target, Lactantius, of whom he only highlighted his worst blunder, the denial of the spherical Earth. Copernicus simplified matters and ignored the antipodes debate. In any case, both debates were already closed in favour of the new terraqueous globe by the 1540s, with little opposition (Hooykaas, 1984, 127–128). Lactantius was an ideal, already weakened "easy target" and "scapegoat" for literalism, according to Hooykaas (Hooykaas, 1984, 30, although he did not reject the Earth's sphericity with the Bible). Randles also noted that Lactantius was considered since the 16th century as a "perfect example of an obtuse and reactionary mind" (Randles, 1980, 14; translated by Russell, 1991, 91, note 92). Still in Copernicus we find Lactantius treated formally with respect, using the usual formula that he was an "illustrious writer"; note that he is not considered a theologian, something we will discuss in p. 166.

Copernicus must have thought that the best defence of his revolutionary ideas from theological criticisms was to attack. Therefore, he charged against Lactantius as a general warning to theologians to keep their hands off astronomy and to avoid making in astronomy the same mistake that Lactantius made in geography, i.e. denying a scientific idea without proper acquaintance with its technical content. Of the two threats discussed with Osiander, that of the philosophers was already addressed in *De revolutionibus* itself, but the theological question was conspicuously absent, so the preface-dedication to the Pope was the last chance to mention it without modifying the text of the book itself.[25]

The reference to Lactantius is interesting in itself for several details. First, it is couched in an already familiar rhetoric of praise before attack. Second, not mentioning that Lactantius also denied the antipodes, not only simplified the message, it also meant that Copernicus could avoid the need to mention Augustine, a Doctor of the Church on whose theology of accommodation his disciple Rheticus had just built in 1541 a theological defence of heliocentrism (see pp. 168–169). Indeed, the language used, in particular, the reference to those ignorant of astronomy and "badly *distorting* some passage of Scripture to their purpose…" (my italics), that is, to criticise Copernicus, could remind the reader of the language of Augustine referring to those who *abused* the authority of the Bible for scientific issues in *De genesi ad litteram*, which was explicitly quoted by Rheticus and later by the Italian Catholic physicist and astronomer Galileo Galilei (1564–1642) for his own defence in his *Letter to the Duchess of Lorraine* (1615).[26] A third interesting detail is that Lactantius is mentioned without quoting and without a specific reference, apparently giving the impression of being already well-known.[27] In fact, the source of Copernicus's knowledge about Lactantius is still unknown. For Rosen, Copernicus's information about the discovery of America (in *De revolutionibus* 1.3) derived from Ringmann and Waldseemüller's *Cosmographiae introductio* (1507).[28] However, Lactantius is not mentioned there, but as is shown in Chapters 3 and 4, this story was so well-known, in particular for those into geography and recent navigations, that Copernicus could have found it in many places.

Copernicus's role in the origin of the flat error

Copernicus's plans failed. On the one hand, although he rejected Osiander's "placating" strategy, it was that framing of Copernicus which Osiander finally imposed on *De revolutionibus* with an anonymous note to the reader, taking advantage of being entrusted with supervising the final stages of its printing in the absence of Rheticus, who left to take up a post at the University of Leipzig in October 1542 (Rosen, 1984, 127). Osiander introduced his "Address to the Reader" at some point after October 1542, probably close to the finishing of the printing before 21 March 1543, when a copy was sent to Emperor Charles V (Rosen, 1984, 204–205).[29]

On the other hand, Copernicus also failed in his own defensive strategy, as it did not deter the attacks. Incidentally, Pope Paul III established the Roman Inquisition on 21 July 1542, which would condemn Copernicus's teaching in the next century. As historians of science Hooykaas (in Rheticus, 1984, 20–27) and Granada and Tessicini (2005), have shown, the Erasmian spirit of Copernicus and his circle was out of step by this time with the rising counter-reformation in Rome. In fact, an early censure of Copernicus by the Dominican theologian Giovanni Maria Tolosani was only stopped by its author's death in 1549 (Tolosani, 1973; see Lerner, 2002; 2005, and here endnote 31).

However, Copernicus opened a new avenue in the reception of Lactantius's *Divine Institutes* 3.24, which his followers took up. No longer part of a cosmographical debate on the antipodes or the shape of the Earth, it was now the text employed in the context of a science and Christianity debate, as an example of all that could go wrong when someone with religious convictions, but without scientific knowledge, pontificated against a scientific development. Thus, this text became an iconic story to explain the origin of science and Christianity conflicts, connecting an old discussion on the shape of the Earth (and of the antipodes) and a contemporary debate on the movement of the Earth. In this way, a temporal line of conflicts was drawn, looming large after the Galileo trial a century later. This chain of conflicts became the backbone of the growing ideology of a perennial science and Christianity conflict, which will search for new examples to flesh out this timeline, as we will see with the rediscovery of Virgilius of Salzburg (see pp. 170, 172).

Copernicus also contributed to the flat error with a specific detail overlooked by Jeffrey Burton Russell. It is certain that the preface-dedication to the Pope cannot be seen as an example of flat error, as Russell recognised in 2001/2002, after a more critical view in 1991 that we discussed in Chapter 1, because Copernicus correctly attacked a single person, Lactantius, who indeed was a flat-earther. However, by the juxtaposition of sentences, a dangerous inference was facilitated. Copernicus first mentioned those who distorted Scripture. Two sentences later, he mentioned Lactantius. Copernicus indeed described him as an "illustrious writer," not identifying him as a theologian, not even as a Christian! However, many readers would infer that Lactantius was an example of distorted use of Scripture to support the flat Earth. This is factually wrong, as Lactantius never mentioned the Bible in discussing the shape or habitability of the Earth, and Copernicus never said so explicitly. However, it was very easy to make that erroneous connection for anyone not directly acquainted with Lactantius's text, as can be seen even today. A similar discussion arrangement can be seen in Rothmann's manuscript *Observationum stellarum fixarum liber primus* and Kepler's *Epitome* vol. 1 (see pp. 167–168). This is, in my view, Copernicus's indirect, specific contribution to the flat error.

Finally, regarding Copernicus's reference to Lactantius, it is important to mention that this passage, quoted by Galileo in his *Letter to the Duchess of Lorraine* (1615), was one of those portions ordered to be censored from *De revolutionibus* in 1616.[30]

The flat Earth and the antipodeans in early Copernican literature

If Copernicus needed to defend himself, his disciples needed no less reinforcement. And Lactantius was now available to be deployed to shut the mouths of those willing to attack from a *theological* angle. Soon other ancient and medieval authors enlarged the list of examples of improper theological intervention in scientific affairs, as we will see in the next pages.

However, although I found interesting examples of this usage of older geographical polemics in many Copernican works up to the first two decades of the 17th century, there are further Copernican works where I found no reference to the topics of the sphericity and the antipodes.[31]

Lactantius among the early Copernicans

The first Copernican to mention Lactantius was Rheticus, even before the publication of *De revolutionibus*, in his *Treatise on the Holy Scripture and the Motion of the Earth* (1541c, published in 1651). The attack had the usual rhetorical 'praise' and, as Copernicus, focussed only on the denial of the sphericity: "Lactantius, otherwise a man of great learning and eloquence, ridicules those who claim that the earth is round" (Gorlaeus and Rheticus, 1651, 35).[32]

Christoph Rothmann (1560/70–c. 1600), a German Protestant astronomer and one of the few 16th century Copernicans, in his manuscript text *Observationum* 23, fol. 70ᵛ–71ʳ (c. 1589),[33] had a discussion of the relation between the Bible and astronomy in connection with the fluidity of the heavens (published in Granada, 1996, 827 and Rothmann, 2003, 203; see Appendix, no. 16 from where I quote below).[34] He criticised those who "misuse the authority of Sacred Scripture," not "learned Theologians," but "babblers" (note the same contrast in Copernicus, see p. 164). He exemplified this with Lactantius, borrowing several words from Copernicus and even including – in one of the draft versions – Copernicus's rhetorical 'praise' of the ancient Christian author. In fact, Rothmann – not without professional pride – warned in a letter to Tycho Brahe on 19 September 1588 regarding the debate over the fluidity of the celestial matter:

> Unless this question is decided by us [astronomers/mathematicians], it will not be decided by anyone, whether theologian or physicist [philosopher]. [...]. For we know and understand about the heights and the matters discussed by us only as much as we discover mathematical demonstrations through trigonometry.
>
> (Quoted in Howell, 2002, 93–94)[35]

The English Protestant physician and pioneer of magnetism, William Gilbert (1544–1601), was well disposed toward Copernicus; however, he only accepted the rotation of the Earth in *De magnete* (1600). While explaining that the *movement* of the Earth is not felt because everything on it moves at the same time, he ridiculed Lactantius and his double denial of the antipodes and the Earth's sphericity (in contrast to the traditional rhetorical praise, he labelled Lactantius a "weak mind," *puſillanimi*):

> Wherefore empty and ſuperditious is the feat that ſome weak minds have of a ſhock of bodies (like Lucius Lactantius, who, in the faſhion of the unlettered rabble and of the moſt unreaſonable men ſcoffs at an Antipodes and at the ſphærick ordering of the Earth all round).
>
> (Gilbert, *De magnete* 6.3, 1600, 219)[36]

Some pages later, he compared the "fooliſh folk" *(ſtupidi)* fearing that everything "muſt neceſſarily be ſhaken and overthrown by the Earth's motion" with it fearing the antipodeans to "fall into the oppoſite region of the ſky" (Gilbert, *De magnete* 6.5, 1600, 227).[37]

The German Protestant astronomer David Origanus (or Tost, 1558–1628) developed a geo-heliocentric model with a rotating Earth, and in the Letter-Preface to the Elector of Brandenburg, in his *Novae motuum coelestium ephemerides*

Brandenburgicae (1609), he listed several ancient flat-earther philosophers, followed by Christian flat-earthers, where Lactantius was not forgotten (Origanus, 1609, fol. a3^{r-v}; Appendix, no. 14).[38]

Johannes Kepler (1571–1630), the outspoken German Protestant Copernican astronomer, became fond of mentioning Lactantius's denial of the sphericity of the Earth in several of his works where he rejected the (ab)use of Scripture against heliocentrism, as in the introduction to *Astronomia nova*, 1609, fol. 6^{r} (Appendix, no. 9)[39] and the *Epitome astronomiae Copernicanae* 1.5, 1618, 138 (Appendix, no. 10).[40] As other authors, Kepler also treated Lactantius with apparent respect, calling him "Sacred" (*Sanctus*) and "revered Church Father" (*sanctos Ecclesiae Patres*), respectively, before attacking him as a flat-earther. These epithets are oddly applied to Lactantius, and they would instead fit Augustine. In any case, after dealing with several biblical texts, in *Astronomia nova* Kepler discusses "the opinions of the Saints," starting by affirming the traditional disciplinary boundary between theology and philosophy. While the former is based on "Authority," the latter depends on "Reaſon." On that basis, Kepler disputes venerable Christian authorities on cosmographical matters, opposing the "Truth" that he can "demoſtrate from Philoſophy." Clearly, for Kepler cosmographical matters were no longer the competence of the theologian, but rather belonged to the natural philosopher who should use reason, in line with the previous quotation from Rothmann's letter to Brahe of 1588.

Kepler differentiated between Lactantius and Augustine, and used this to construct an historical progression of theological errors regarding cosmographical topics, such as the sphericity of the Earth (Lactantius), the antipodes (Augustine), and the motion of the Earth (Kepler's theologian contemporaries).[41] From the 4th to the 17th centuries, he developed a three-stage model, which expanded the previous historical two-stage theory of Copernicus (sphericity and motion of the Earth). Kepler's new model played into the hands of later authors (even today) who wished to portray science and Christianity as being locked in a perennial conflict (Felipe and Jeeves, 2017).

Augustine among the early Copernicans

In contrast to the geographical/cosmographical literature about the antipodes, Augustine was hardly discussed by some early Copernicans/geokineticists who concentrated their fire on Lactantius, rather than Augustine, and mainly on the topic of the sphericity of the Earth.[42] Lactantius provided a clear and *easy* target to attack (in comparison with Augustine), someone whom nobody by that time wanted to defend after decades of scorn.

However, there was more in the Copernican agenda. Very early Copernicans identified Augustine's theology (in particular his 'principle of accommodation')[43] as an ally, rather than as an enemy, in their effort to provide a science and Christianity model of relations which could avoid the clash over Copernicanism. That was something that both Catholics and Protestants could do, as both revered Augustine[44] (e.g. Rheticus, Rothmann, Galileo, etc.), despite his specific mistake regarding the antipodeans.[45] Augustine became a safe alternative to Lactantius, and therefore Copernicans had every reason to concentrate their fire on Lactantius while restraining their criticisms of Augustine.

A notable exception was the geo-heliocentrist Danish Protestant astronomer Tycho Brahe (1546–1601), who obviously was in less need of Augustine's accommodation, as in his system the Earth remains in the centre and the planets move around a Sun that revolves around the Earth. In the late 1580s, there was an exchange of letters between him and Rothmann about the fluidity of the heavens (where both agreed) with references to the Copernican system (where both strongly disagreed), which were published by Brahe in 1596. In a letter of 19 September 1588, Rothmann defended that "God speaks accommodating Himself to the capacity of the Hebrews" (Granada, 2008, 571), therefore, "the prophets did not know about this topic more than the uneducated people and even Lactantius" as that was not part of what the Holy Spirit revealed to them (my translation from the Italian in Granada, 1996, 810).

Brahe reacted in his letter of 21 February 1589, criticising Rothmann as one who:

> [...] detracts too much from the prophets by saying that they did not understand more about the nature of things than other vulgar people. Although they were not physicists by profession, indeed this was not the nature of their gift, nonetheless they mixed many physical propositions in with their prophecies, which no one, however deeply imbued in natural philosophy, could deny.
> (Blair, 1990, 363)

On 22 August 1589, Rothmann replied that he did not detract from the authority of the Scriptures, appealing to Augustine: "If you read Augustine, you will find that he speaks much more freely than I do about the Scriptures" (my translation from the Italian in Granada, 1996, 811).

Finally, Brahe wrote back on 24 November 1589, and seemed not to understand why Rothmann had mentioned Augustine, as this Church Father did not discuss the movement of the Earth. Brahe challenged Rothmann to produce any quotation from the Bible, Augustine or other early Christian authorities supporting Copernicus. It is clear that Rothmann did not intend to enlist Augustine in support of Copernicus, but to use his principle of accommodation to counter the Bible-based rejection of Copernicanism. In the following sentence, Brahe, apparently not satisfied with having rejected the usage of Augustine in the discussion about Copernicus, produced a surprising and unnecessary attack on the Bishop of Hippo using the commonplace of the ignorance of mathematics (but without the praise, we saw in others):

> But I know perfectly well that Augustine, the only person you name, nowhere allowed any movement of the Earth, either annual or diurnal, once he – not being much of a mathematician – undermined its roundness by denying the antipodes. He can be forgiven for that. So since he did not realise

> that it [sc. the Earth] was round, far less could he allow a motion that went along with roundness.
>
> (Brahe, 1596, 156–157)[46]

In the first sentence, Brahe just seemed to state his belief that by denying the antipodes, Augustine undermined the sphericity of the Earth (not realising that someone can deny the antipodes accepting a spherical Earth). However, in the third sentence, he seems to imply more clearly that Augustine was a flat-earther. Rothmann never replied, leaving at his death many manuscripts unpublished, including the *Observationum*. As far as I can see, in these few sentences of Brahe, there is the first case of accusing Augustine of being a flat-earther, even though it was worded indirectly. It is also an early example of conflating the sphericity of the Earth and the antipodes. These are ideas that became critical elements of the flat error (Table 5.1).

Eventually, some geokineticists and Copernicans – in particular Protestants – also ended up accusing Augustine of being a flat-earther (together with "Lactantius and others," as part of "many of our ecclesiastical writers") as in David Origanus (1558–1629) treatise of 1609 (Appendix, no. 14). More precise, but no less harsh, were the words of Kepler, the first Copernican I know to criticise Augustine, in this case correctly, for his denial of the antipodes. One example is his 1609 *Astronomia nova*, in a text where he also attacked Lactantius (see p. 168, Appendix, no. 9), and another one in the preface to book 4 in the second volume of the *Epitome astronomiae Copernicanae*, 1620, 428–429 (Appendix, no. 11).[47] The reference in this second text started with a startling affirmation: "Antiquity teaches us by examples how vainly people set up boundary posts where God has not set them up: How severely were *all* the astronomers blamed by the first Christians" (my italics). However, the examples following this absolute generalisation are an anecdotal reference from Eusebius to an unnamed astronomer, the mention of Augustine and Virgilius of Salzburg. From there, he jumped to his present time, with a series of assertions that reminds us of the text of *Astronomia nova*: "Yet today we…" As discussed in p. 168, such a depiction of the situation aligned with his portrayal of continuous science and Christianity conflicts.

Virgilius of Salzburg among the early Copernicans

In Chapter 2, we explained how the old forgotten 8th-century debate, between Virgilius and Boniface, was recovered by the Protestant and strong critic of the papacy Johannes Turmair (Aventinus, 1477–1534). It appears in a book composed in 1517–1521, and published as *Annalivm Boiorvm libri septem*, 1554, 297, taking the information directly from the Pope's letter (Appendix, no. 2).[48] In putting the conflict into context, Aventinus described Virgilius as an erudite man who rejected Augustine's denial of the antipodes. However, Augustine was not mentioned in the actual letter by Pope Zacharias I to Boniface, which is the only available record of the debate and that Aventinus paraphrased with his comments. Furthermore, Aventinus depicted that age in wholly negative terms: "Virgilius was also educated, more highly than Christian custom of the time allowed, in the disciplines

Table 5.1 Table illustrating how the ancient and medieval ages became devaluated among early Copernicans[a]

	Lactantius's denial of Earth's sphericity based on the Bible	Augustine as flat-earther	Boniface and Pope Zacharias I against Virgilius	Medieval Catholic clergy as flat-earthers	Middle ages as barbaric	Indiscriminate generalisations to disregard the past	Lists of mistaken authors[b]	Sequential linear progression of conflicts
Copernicus (1543)	X							
Ramus (1567 & 1569)			X					
Brahe (1589; published 1596)		X						
Rothmann (1589)[c]	X		X					
Origanus (1609)		X	X	**(implied)**		X	X	
Kepler's *Astronomia nova* (1609)								X
Foscarini's *Lettera* (1615)						X		
Foscarini's *Leter to Bellarmine* (1615)[c]							X	
Galileo (1615)	X[d]							
Kepler's *Epitome* I (1618)	X							
Kepler's *Epitome* II (1620)			X			X		
Campanella (1616/1622)		X				X	X	X
Maestlin (1621)[c]		X	X	X	X	X	X	X
Campanella (1636–1637)[e]			X				X	X

a I understand Copernicans loosely to include geokineticists like Origanus or geocentrists friendly towards Copernicans such as Campanella
b I consider it a list when three or more authors are referred to (see Table 4.1).
c Unpublished in the 17th century.
d This is just a quotation from Copernicus.
e Based on the partial study of this book as described in pp. 176–177.

we call mathematics and in secular philosophy." Without trying to understand the medieval debates on the antipodes, he presented Virgilius as a misunderstood precursor ahead of his time: "he had taught ideas that in our age do not have to be investigated by argument but have been discovered by experiment." Aventinus then mentioned how the Pope condemned Virgilius's ideas and ordered him to be expelled from the church. In this way, a brief unclear notice in one medieval letter was utilised by Aventinus to foster the idea of a science and Christianity conflict as far back as the 8th century.[49]

Apparently this story entered the scientific discussion by the French Protestant philosopher Peter Ramus (1515–1572), a critic of Aristotle interested in Copernican astronomy.[50] Ramus referred to Virgilius, mentioning Aventinus in his *Prooemium mathematicum*, 1567, 187–188 (Appendix, no. 15)[51] which became later books 1–3 of *Scholarum mathematicarum* (1569). He attacked Boniface, and his modern counterparts, for slandering mathematics and astronomy "as not just useless but impious." Although this was not in the context of a Copernican polemic, here we have the first example I know of this story used in defence of modern science, similarly to how Copernicus used Lactantius. Copernicans happily incorporated it into their historical repertoire.

So far, the first Copernican whom I was able to find using Virgilius's antipodes story was Rothmann, in his unpublished *Observationum* (c. 1589, published in 2003). A first brief mention is found in chapter 1:

> And one would hope not to see theologians showing contempt for this art [Astronomy], since it brings into clearer view the greatest and best God's immeasurable wisdom, power, and ineffable majesty, and his providence in the administration of the universe. For it is a long time since the demolition of the obtuse and quite asinine ignorance of Boniface, when he condemned Virgilius bishop of Salzburg for saying before an assembly that the antipodes exist.
>
> (Rothmann, *Observationum* 1, c. 1589, fol. 5^r, from Rothmann, 2003, 66)[52]

A more detailed discussion appears in *Observationum* 23, fols 70^v–71^r, where as examples of the "babblers" of Copernicus, he mentions Lactantius and Virgilius, depicted as condemned for impiety (Appendix, no. 16).[53] Nevertheless, this second reference was used in debating the fluidity of the heavens, not Copernicanism. Rothmann's reference to this story seems borrowed from Ramus, according to Granada *et al.* (Rothmann, 2003, 66).

The story of Virgilius was reported after a list of ancient pagan and Christian flat-earthers in Origanus's *Novae motuum coelestium ephemerides Brandenburgicae* (1609), who took it – again – from Aventinus, assuming that Virgilius was condemned and expelled from the Church (Appendix, no. 14). He also made a strong parallel between these old conflicts and the debate over the movement of the Earth. In the preface to book 4 of his *Epitome* (1620), Kepler referred briefly to Virgilius (Appendix, no. 11). Finally, in a manuscript by Maestlin (c. 1621), Virgilius was paralleled with Copernicus (see pp. 178–179).[54]

Copernican defences and the path to the flat error

Foscarini's and Campanella's lists of early Christian authorities

Unsurprisingly all the authors mentioned after Copernicus in previous sections were Protestants. The freedom of Copernicus, and the old Erasmian-oriented Catholic circle around him, was not possible after Trent. In particular, it is not surprising that Catholics refrained from criticising Augustine or the troubles of Virgilius. Instead, they would try to defend their church against these accusations, as can already be seen in the mathematician and Minim priest Marin Mersenne (1623), who discussed Origanus.

In the critical year from early 1615 to early 1616, leading to the (in)famous condemnation of Copernicanism by the Inquisition, three Catholic authors did enter this debate supporting Copernicanism. They fell out with their own church, while being embraced by Protestants. Galileo briefly mentioned Lactantius in his *Letter to the Grand Duchess Christina* (summer of 1615),[55] which is a detailed theological defence of Copernicus. In quoting the preface-dedication to the Pope from Copernicus, Galileo reproduced the reference to Lactantius but without glossing it. After circulating in manuscript form, it was published in Strasbourg in the original Italian and translated into Latin (Galilei, 1636).

A few months before Galileo's letter, the Italian Carmelite theologian Paolo Antonio Foscarini (c. 1565–1616) published in Naples a theological defence of Copernicanism, dated 6 January 1615 (*Lettera sopra l'opinione de' Pittagorici, e del Copernico*). Foscarini made more ambitious arguments than Copernicus, and in a similar historical spirit to Kepler. He compared the "many" thinkers through the ages who denied the antipodes and "the unanimous consent of all" who denied the habitability of the Torrid Zone with those who in his time considered "our opinion of the Earth's motion" to be absurd.[56] Not one but two errors of the past were used to criticise the "Venerable Antiquity" in contrast with the discoveries ("Experiments") of the moderns. The problem was that by not giving names and using the terms "many" and "all," the net result was the oversimplification and generalisation of the far more nuanced early Christian and medieval debate.

When he learnt of the negative evaluation of the censor of his printed *Lettera*, based on his disagreement with "the common explication of the Holy Fathers,"[57] Foscarini quickly sent a copy of the Italian *Lettera* with a Latin letter directly to Cardinal Robert Bellarmine, the most influential theologian of the time in Rome and member of the Inquisition since 1599, requesting his opinion. While the *Lettera* was packed with biblical texts, the new personal letter was filled with impressive lists of references to theologians and early Christian authorities.[58] It was composed with the help of patristic texts and anthologies from the library of the Carmelite convent of Traspontina at Rome where Foscarini had been invited to preach for the Lent of 1615, the time when the letter is dated. The restrained tone of the printed *Lettera* was in contrast to his letter to Bellarmine. In the latter Foscarini deployed all the documentation he had researched painstakingly in the library of the convent. He listed three different errors of the early Christian authors on natural topics:

(1) that the Earth was founded above the waters (Justin, in fact, Pseudo-Justin), (2) that the Moon was the second largest heavenly body after the Sun (Basil and Augustine), and (3) that the sky was a dome instead of a sphere (Pseudo-Justin, Basil, Ambrose, John Chrysostom, Theophylactus, Theodoretus, Lactantius and Procopius). We can see how, correctly, Foscarini did not include Augustine in this last group. This latest list of references from Foscarini's letter is longer than any other outlined in Table 4.1.

Bellarmine's reply to Foscarini on 12 April 1615 was negative, appealing to the consensus of the "Holy Fathers,"[59] and foreshadowing the condemnation of Foscarini's printed *Lettera* in 1616. In the meantime, Foscarini, far from being intimidated, worked on his defence of the *Lettera*, based on his Latin letter to Bellarmine. Neither this new text, entitled *Defensio Epistolae Pauli Antonii Foscareni ... super mobilitate terrae*, nor the original letter to Bellarmine were printed, as Foscarini died on 10 June 1616, shortly after the condemnation of his printed *Lettera* on 5 March 1616.[60] Although Foscarini's letter was read only by Bellarmine and a few other people, and not published until recently, it illustrates the possibility to gather such information at the time, which we can also observe in the long published lists of Acosta and Campanella (Table 4.1).

In early 1616 the Italian Dominican theologian and philosopher Thomas Campanella (1568–1639) wrote, in a Neapolitan dungeon, his *Apologia pro Galilaeo* (1622, published by his Lutheran friend Tobias Adami).[61] There two of the key traditions here researched merged: on the one hand that of the theologians investigating the history of the interpretation of the Bible (particularly Genesis) on cosmographical matters, and on the other hand that of the Copernicans using the ancient/medieval flat Earth and antipodes debates as precedents to deflate theological criticism of their heliocentric proposal. Both trains of thought were strongly connected with the navigational discoveries, as seen in Campanella's references to "navigators," "Columbus" and "Magellan" (see below).

Although a geocentrist himself,[62] Campanella strongly defended Galileo's academic freedom, while tirelessly attacking Aristotle. His short book was mostly devoted to the history of theology, particularly the cosmographical speculations of the early Christian theologians and the scholastics, of which Campanella had an encyclopaedic knowledge. Although he mentioned Senensis,[63] his display of authorities was even more significant, as shown in Table 4.1. His knowledge of available sources allowed him to point to many errors and contradictions among theologians, in addition to those errors about the shape and habitability of the Earth, such as the composition of the heavenly matter or their ideas about the Torrid Zone.[64] He also addressed Bellarmine twice directly in his text, pointing out that, if the Catholic Church were to condemn Copernicanism, this would allow the Protestants to criticise the Catholic Church of hostility against the sciences (Lerner, 2005, 23–24).

Campanella referred to ancient Christian authors who denied the sphericity of the heavens: Lactantius, Procopius, Diodore of Tarsus, Eusebius of Emissa, Pseudo-Justin (*Quaestiones ad orthodoxos*), John Chrysostom *& alij* ("and others," Campanella, *Apologia* 4, 1622, 35; 1994, 88). Although this is not strictly a list of flat-earthers, Campanella mentioned several times those who thought that the Earth was floating on water: Procopius, Eusebius "and others" (Campanella, *Apologia* 4, 1622, 35; 1994, 88). Those ancient authors were not really supporting something like the late medieval scholastic eccentric spherical model of the Earth. What Campanella rather had in mind was the floating flat Earth, similar to the pre-Socratic model of Xenophanes with whom he associated them (Campanella, *Apologia* 3, 1622, 28–29; 1994, 77).[65]

The anti-antipodeans also had a prominent place in Campanella's book: besides the pagan Xenophanes, the Christians Lactantius, Augustine, and "others." As usual, Campanella introduced the observation that Lactantius and Augustine "were holy and learned men," before accusing them of denying the antipodes (Campanella, *Apologia* 3, 1622, 12; 1994, 52–53). The linkage of Lactantius and Augustine culminates in the implication that because they both rejected the antipodes, they both believed that the Sun rotated around a northern mountain and not below the Earth, which Campanella linked with the beliefs of Xenophanes.[66] In this way, he practically accuses Augustine of believing the Earth to be flat (Campanella, *Apologia* 4, 1622, 40; 1994, 95–96). In addition, Campanella mentioned that Procopius compiled rejections of the antipodes, "drawing a list of comments on the Sacred Scripture taken from the writings of *all* the Fathers" (Campanella, *Apologia* 3, 1622, 28; 1994, 77, my italics).[67]

As Campanella attacked these erroneous early Christian cosmographical views to criticise the use of theology against Copernicans, no care was taken in providing lists of early Christian authorities with alternative views, which inevitably gave the impression that most if not all shared these errors. Exceptions made things even worse, as Campanella affirmed without providing any data, contrary to his usual custom of erudite referencing: "Indeed some of the Fathers who have held that the antipodes do exist have even been taken to be heretics" (Campanella, *Apologia* 3, 1622, 28; 1994, 77). None of those 'heretics' were mentioned, and the only reference given is: "Procopius says it is heresy to deny that the earth floats on water" (Campanella, *Apologia* 4, 1622, 41; 1994, 96).

Sadly, what Campanella was not able to mention was the vigorous defence of the sphericity of the Earth by the 6th-century Alexandrian Christian philosopher John Philoponus, an author he mentioned with respect for his early rejection of the Aristotelian fifth element (Campanella, *Apologia* 4, 1622, 34–35; 1994, 86–87). Philoponus's fundamental anti-flat Earth treatise, *De opificio mundi*, would only be published in print for the first time by Corderius in Vienna (Philoponus, 1630).[68]

Campanella's assessment was more balanced on the habitability of the Torrid Zone, noting that Aquinas followed Aristotle "rather than the arguments of Albert the Great and Avicenna" (Campanella, *Apologia* 3, 1622, 12–13; 1994, 52–53). Those authors were closer to him chronologically, and readily available. When touching on those who believe "that the earth floats on water," he pointed out that "others who held the contrary were not condemned" in order to argue in Galileo's favour that the movement of the Earth should not be condemned (Campanella, *Apologia* 4, 1622, 35; 1994, 88). This is complementary to what he affirmed in the above text that "some" Fathers who accepted the antipodes were "taken to be heretics." Although it seems he knew they existed, Campanella did not list those who accepted the Earth's sphericity and the antipodes.

Similar cosmological discussions are mentioned in a very different book by Campanella, *De gentilismo non retinendo*, composed earlier, in 1609–1610, but published in 1636, and republished with additions in 1637. This was a general introduction to his large *Philosophia realis*, of which the fourth part was already published in 1623, including the famed *Civitas solis* (*City of the Sun*). In *De gentilismo non retinendo* Campanella attacked the pagan philosophy, proposing the creation of a new Christian one, for which he was not ashamed of having Lactantius on his side. He considered the famed text of Daniel 12:4: "But you, Daniel, roll up and seal the words of the scroll until the time of the end. Many will go here and there to increase knowledge" (NIV), in the same way as in Bacon's *Instauration Magna* (1620), as fulfilled in his own time by the advances in the knowledge of nature. Campanella listed the finding of a new world by Columbus against theologians and philosophers, an inhabitable Equator against Aristotle, Vergil and many others, as well as the discovery of new stars by Galileo, also noting the new astronomical advances by Copernicus and Peuerbach, the Portuguese navigation, etc. (Campanella, 1636, 5–6; 1637, fol. bij^{r}).[69]

Some of these stories became almost proverbial tales in Campanella. When mentioning examples of known mistakes in judgment, Campanella referenced the denial of the antipodes by Augustine and others, and when rhetorically referring to a negation of truth, he compared that with an hypothetical condemnation of Columbus and other discoverers of truth (Campanella, 1636, 10; 1637, fol. biij^{r}). Similarly, Virgilius's condemnation for affirming the antipodes (Campanella, 1637, fol. d^{r}, d^{v})[70] or Augustine's denial of the antipodes are used as an example of ignorance (Campanella, 1636, 52, 55; 1637, fols eij^{r}, eij^{v}). Another example is Thomas [Aquinas] denial of the habitability of the Torrid Zone (Campanella, 1636, 62; 1637, fol. eiiij^{v}). In fact, Campanella denounced the alliances of theology and sciences, noting that this could in time lead to exposing faith to ridicule (Campanella, 1636, 13; 1637, biiij^{r}),[71] later pointing to the example of the denial of the antipodes by Augustine and Lactantius (Campanella, 1636, 30; 1637, ciiij^{v}). Closer to the end Campanella indicated that it was better no to list more errors, as it was shameful for the [Church] Fathers: "I do not want to expose other obvious errors: for the shame of the Fathers must be covered over" (Campanella, 1636, 55; 1637, eij^{v}).[72]

In fact in *De gentilismo non retinendo* Campanella collected several embarrassing examples of religious opposition to important historical figures, such as Jerome and Cardinal Baronius, ending with the opposition to Columbus for proposing to discover the antipodes, which Campanella mentioned were denied by Augustine, Justin (in fact, Pseudo-Justin) and Lactantius, and accepted by Virgilius. He finished retelling the misfortune of Virgilius with an explicit reference to Aventinus in a marginal note (Campanella, 1636, 13; 1637, biiij^{r}). Later Campanella retold the same story of the discovery of a new world by Columbus, pointing to the denial of the antipodes by philosophers and theologians based on Augustine, Lactantius, Chrysostom and others *(& aliis)*, with the favourable opinion of Virgilius, "deposed" as "heretic" (Campanella, 1636, 48; 1637, e^{r-v}). In the next sentence he also linked this story to that of Galileo's discoveries of new constellations with the telescope and recent inventions (Campanella, 1636, 48; 1637, e^{v}).

That leads us to realise that, as in the case of Kepler (see p. 168), Campanella also began to detect a chronological line connecting science and Christianity conflicts. Already in his 1616 *Apologia pro Galilaeo* Columbus is first contrasted with Christian authors (Campanella, *Apologia* 3, 1622, 28; 1994, 77). Later, Columbus and Magellan, two closely related navigators – both chronologically and in their activity, i.e. geographical exploration – appear connected and opposed to ecclesiastical authors (Campanella, *Apologia* 4, 34; 1994, 86). However, a linear chronological connection is not as straightforward as in Kepler; they appear instead linked as a pair on a single topic, in contrast with ecclesiastical authors.

A more suggestive parallelism, between Columbus and Galileo, appeared in *De gentilismo non retinendo* (1636) and in his *Quaestiones ... de politicis*. Those *Quaestiones* were initially also composed in 1609 (as *De gentilismo non retinendo*), completed in 1613 and published with the second editions of *De gentilismo non retinendo* and *Civitas solis* in his large *Philosophia realis* (1637).[73] As a defensive colophon to *Civitas solis*, in the *Quaestio IV*, Columbus and Galileo were cited as examples of discoverers rejected by monkish bookish learning:

> Point 8. It is natural for people to investigate the works of God, to travel about the world, to acquire everywhere the sciences and to experience all things. Living however in such a republic they are as monks who study only their own books, and when they hear what is not written in them they are scandalized and disturbed as now scarcely do they believe the observations of Galileo, just as formerly they did not believe Columbus, the discoverer of a new hemisphere, because St Augustine had denied it.
>
> (Campanella, *Quaestio IV de Optima Republica* 1 in Campanella, 1637, 101)[74]

This was a more ambitious historical alignment, as in Kepler, connecting the 15th and the 17th centuries and two very different disciplines: geography and astronomy. However, the context in which this linkage appeared is a list of objections against Campanella's *Civitas solis*, formulated and answered by Campanella himself. It is therefore difficult to know to what extent Campanella was really endorsing this connection. In his reply to this particular objection, he defended the erudition of cultivated monks devoted to science (Campanella, *Quaestio IV de Optima Republica* 1 in Campanella, 1637, 103–104), in line with Copernicus, who also praised science-friendly clerics, in particular the Pope to whom he dedicated *De revolutionibus*. Campanella did not discuss objections against his work involving Columbus or Galileo.

In any case, this text shows that the parallelism between Columbus and Galileo, as opposed to religious bookish learning, was already in the air by 1609–1610, the date of the original composition of those two texts by Campanella.[75] This was the time of the first telescopic discoveries of Galileo, which became known for challenging Aristotelianism and supporting Copernicanism. It was very tempting to compare the challenging nature of the new worlds on Earth of Columbus with the new worlds in the sky of Galileo, fulfilling the prescient words of Pomponazzi.

Maestlin's fury in the wake of the Vatican condemnation of Copernicanism and Descartes's fear after the condemnation of Galileo

None of the above stopped the 1616 condemnation of Copernicanism by the Roman Inquisition.[76] Furious, Michael Maestlin (1550–1631), the German Protestant astronomer and enthusiastic Copernican teacher of Kepler, reacted producing a harsh text (c. 1621) which was meant to accompany the second edition of the *Mysterium Cosmographicum* by Kepler (together with the *Narratio prima* by Rheticus). However, the volume was published in 1621 without Maestlin's text, which was found among his papers and published in the 19th century by Christian Frisch, published in Kepler (1858, vol. 1, 56–58; Appendix, no. 13).[77]

Maestlin began recalling the recent decree of the Cardinals condemning Copernicanism. He attacked their ignorance of astronomical matters, retraced the history of the two editions of *De revolutionibus*, and the support from Catholic authorities that led to its publication. After a reference to the precedents among the Pythagoreans and Aristarchus's heliocentrism, he drew parallels between Copernicus and the medieval science and Christianity cosmographical conflicts, starting in this way: "So the same misfortune is afflicting Copernicus before those cardinals as also faced Virgilius, Bishop of Salzburg, in about AD 743."

Maestlin devoted the rest of his text to glossing in detail the conflict between Virgilius and Boniface. Maestlin deprecated the state of erudition at the time of Virgilius, affirming that he "was more learned in [...] mathematics and secular philosophy than [...] the barbarism of that age allowed." Here Maestlin was strongly influenced by Aventinus, borrowing from him most of the words that described the conflict (in Appendix, no. 13, Maestlin's words from Aventinus are underlined). However, Maestlin embellished Aventinus's narrative, connecting Virgilius's erudition with the rejection of "Augustine, Lactantius and others" and his teaching "that the earth is spherical, and that human beings are distributed all over it. From this it follows that on the earth there are antipodes." Here we see Maestlin linking Lactantius and Augustine in the negation of both, the sphericity of the Earth and the antipodeans. The mention of Lactantius and of the Earth's sphericity were not part of Aventinus's text. While Aventinus had added the comparison with Augustine to use such an obscure and hardly known letter in the context of the antipodeans debate,[78] Maestlin adds a second reference to Lactantius. As a result, he thus shifts the focus to the flat Earth.

After he recounted the whole episode, Maestlin – no longer quoting Aventinus – addressed his readers in order to reflect on "these holders of the apostolic see and of the archbishoprics in that age – plus the cardinals too, and whoever they employed as advisers in their councils when it came to decisions on difficult cases." He continued attacking the ignorance that prevented them from realising "that the earth's surface is not flat, and what necessarily follows from that fact." This accusation of the entire medieval Catholic hierarchy of being flat-earthers is what lies at the heart of the flat error, and here is, as far as I know, the first vigorous example of it. Maestlin also scorned the "wise ignorance" of "those wise priests," which made them reject "things that long ago were ably proved by philosophers, and taught in ordinary schools," and that later were known to everyone after the sailings to America.[79] Maestlin finished with the unfounded statement that "Virgilius was condemned for heresy" (see p. 180), and a reinforcement of the parallel with Copernicus similar to the beginning of his discussion: "The same is now happening to Copernicus and the hypotheses of his astronomy."

After the Inquisition condemned Galileo in 1633, the philosopher, mathematician and scientist René Descartes (1596–1650) wrote from Amsterdam to the mathematician and Minim priest Marin Mersenne in Paris in April 1634. A Copernican himself and a Catholic, Descartes was very concerned about Galileo's condemnation and established this parallelism with the topic of the antipodes:

> However, seeing that this censure has not yet been authorized by the pope or by a Council, but only by a particular Congregation of Cardinal Inquisitors, I am not entirely giving up hope that it will not happen as it did with the antipodes, which were once condemned almost in the same way; thus, my *World* might see the light of day in due course, [...].
>
> (Quoted in Finocchiaro, 2005, 46)

Descartes's fear was sadly confirmed and his *De mundo* (*World*) was only published posthumously (1664). From his wording it is not very difficult to discover that he was referring – without naming him – to Virgilius's condemnation. Such parallelism, as well as the other one with Columbus, will become recurrent during the 17th and 18th centuries.

Was the belief in the antipodeans heretical in the Middle Ages?

While the sphericity of the Earth was the core of medieval natural philosophy in essential medieval treatises such as those by Bede or Sacrobosco, the antipodes and the antipodeans were topics of intense debates. A.D. White, a major proponent of the flat error in the late 19th century, was convinced that to believe in the antipodes (and antipodeans) was heresy in the Middle Ages (White, 1896, vol. 1, 106–107). He even identified some scholars apparently accused of adhering to this heresy and persecuted for it: Pietro d'Abano (c. 1257–c. 1315) and Cecco d'Ascoli (1257–1327). However, the antipodes were not the reason why they were persecuted by the Inquisition (see Thorndike, 1923, vol. 2, 874–947 [Abano] and 948–968 [Ascoli]). White's error-laden stories about these two scholars are just part of the modern flat error.

As seen with other components of the flat error, the idea that belief in the antipodes or the antipodeans was considered a heresy in the medieval period was also born in early modern times. We find this view in Montaigne (1588),[80] Bacon (1607, published in 1653),[81] Maestlin (c. 1621)[82] and Campanella (1622 and 1636–1637).[83] In some cases, even the condemnation of specific individuals was affirmed, particularly Virgilius (8th century),[84] who, mentioned or not, seems to be the real inspiration for this alleged medieval heresy after his rediscovery by Aventinus in 1554. As discussed in Chapter 2, it is not easy to assess what really happened. The extant letter from the Pope indicates that he condemned the acceptance of the antipodes, but it is not known if Virgilius was actually called to trial, and there is no record that he was formally condemned as heretic (see Mayaud, 2005, 94–99). In any case, he was later promoted to Bishop and canonised. The only other testimony of this alleged condemnation of the belief in the existence of the antipodes as 'heresy' could be the recollections by Geraldini of the *Junta* in Santa Fe (1491), where he affirmed that "many Spanish bishops plainly alleged that he [Columbus] was manifestly guilty of heresy" (Geraldini, *Itinerarium* 14, 1631, 204).[85] Campanella, without mentioning Geraldini, also noted that Columbus was attacked as an heretic because he proposed to reach the antipodes, *Alij contra Columbum tanquam hæreticum inuehebantur, quod propoſuiſſet inuenire Antipodas* (Campanella, 1636, 13; 1637, fol. biiij^{r}).

Against that, we could mention the different authors who discussed the topic of the antipodes and antipodeans freely for centuries, in some cases accepting their existence. In addition, we can also mention the medieval preservation and copying of ancient texts which defended the existence of antipodes and antipodeans, such as those by Macrobius and Capella, even if some strongly criticised these authors, e.g. Manegold's rejection of Macrobius. Even more, the *Travels* of Mandeville (c. 1360), with its acceptance of the antipodes, became a bestseller in the Late Middle Ages without known opposition (see p. 58). While some people thought of this belief as heretical, no real persecution and real victims ever existed. Even the letter of Pope Zacharias I condemning the antipodes was never turned into an official condemnation. As Cardinal Fillastre (1414–1418) carefully affirmed, belief in the antipodeans "*seems* to lead to heresy" (see p. 70 and Appendix, no. 8, my italics). The argumentation by Augustine was quite reasonable and persuasive in itself, and while no hard data was available there was no strong motivation for anyone to defend as real what was only speculation. In addition, as we have seen, there were ways to discuss the existence of antipodeans without the thorny problem of their genealogical origin, mainly by positing them on extreme edges of the known oikoumene (i.e. Spain and India), as Fillastre himself did. Ironically, once the antipodeans were found outside the ancient oikoumene, it was their denial which could be regarded as 'heresy', as in the case of Monachus (c. 1526–1529), which was discussed in p. 129.

The periodisation of history and the invention of the Dark Ages

Light and darkness have long been used as metaphors, particularly in a religious context, as seen in the initial verses of the Gospel of John. Ancient and medieval Christians saw the pagan past as a time of darkness until the birth of Christ. From this dichotomy arose the idea of a two-stage 'progress' in the recent history of salvation. Lactantius divided history prior to his own time in three periods, adding an initial idealised stage with biblical and pagan roots: a Golden Age of justice and worshipping of God, a period of injustice away from God, and the third age after Christ's incarnation when true knowledge of God has been restored (Kendeffy, 2015, 66–67, note 53).[86]

In the 15th century, Renaissance authors secularised this metaphor and used it to contrast the previous centuries of intellectual decline with their own age of the rebirth of arts and letters, creating a new tripartite scheme already present in the mid-14th century with Petrarch: luminous antiquity, medieval dark ages, and their own modern renaissance as rebirth and restoration of light (Mommsen, 1942).[87] In itself, this scheme suggested not so much a progress story but rather a nearly circular return to the greatness of the past. This scorn of the recent medieval past went hand in hand with the exaltation of the accomplishments of the present: salvaging classical culture, restoring the learning of Latin and even Greek, emulating classical architecture and art, and for the particular topic that concerns us here, even surpassing the medievals and the ancients in geographical exploration and knowledge – as well as imperial conquests. Indeed, it was the 16th century and its novelties, among them the geographical discoveries discussed in Chapter 4, which moved modern history from the idea of rebirth towards an ideological overarching narrative of progress from the Renaissance onwards – while keeping the medieval past in the dark closet of history.[88]

We can see how the history of discoveries became inserted into this periodisation narrative that denigrated the Middle Ages. In the first volume of *Navigazioni e viaggi* (1550), Ramusio wrote a *Discorso sulle spezie* praising the ancient Roman commerce with the East Indies which was lost due to the barbarians, bringing on 400 years of darkness which was only recently overcome by the Portuguese (Ramusio, 1550, vol. 1, fol. 398^r, also fol. 400^r).[89] In 1553, Ramusio composed a *Discourse* (published in the third volume, 1556), with an exaltation of Columbus who had brought half the world out of darkness into the light. Over the second half of the 16th century, the mythification of Columbus steadily continued to grow.

This tripartite history became theologised again when Protestantism linked the Renaissance of culture in the 15th century with their Reformation of the Church in the early 16th century, eclipsing almost a millennium between Augustine and Luther.[90] In 1518, Luther used against the scholastic Johann Eck in Leipzig not only Scripture, but also history, to show the contradictions of Catholic hierarchy and tradition.[91]

In his *On the Babylonian Captivity of the Church* (1520), Luther would parallel the present situation of Christianity with Israel's Babylonian exile (Ungureanu, 2019c, 108). It is difficult to escape from the impression of a tripartite history consisting of a glorious past, followed by a gloomy present captivity, and a hopeful future liberation. Such periodisation had strong links with Augustine's view in *City of God* of the restless struggle between the two kingdoms, the City of God and the Earthly City of its enemies. Protestant church history also linked the periodisation of history with eschatology, as the third period of the Reformation was seen as the "latter days" before the final apocalyptic days predicted in Revelation (Williamson, 2008, 40–49). Linking church history and eschatology was less original than it might seem, as it can also be found in patristic and medieval apocalypticism, such as Joachinism (see Chapter 4, endnote 93).

Later, in an enthusiastic preface to Robert Barnes's *Vitae romanorum pontificum* (1536), Luther stressed the importance of history to complement Scriptural attack on the Pope.[92] Luther recognised in the preface that in his initial works he had not much expertise in history. In the 1510s there were "limited resources available to support a comprehensive revision of the prevailing historical narrative" which was pro-papal, that changed by the late 1520s (Headley, 1963, 51; Lotito, 2019, 142–143).

With the division of the European Church in the second half of the 16th century, history – and this periodisation – became a battlefield for Catholics and Protestants "in the service of confessional propaganda and polemic," according to Renaissance historian Wallace Klippert Ferguson.[93] For Protestant historiography, the dark ages were not only propaganda; they were also part of their doctrine to justify the need to reform the church and recover the original purity of the Gospel, spurring historians to search for fresh data with an anti-Catholic agenda.[94] In contrast to Catholic historical continuity, Protestants enshrined the need for reformation, restoration, purification and the return *ad fontes*, leading inevitably to a periodisation of church history (Ungureanu, 2019c, 107–111). That the objective was this return to the original purity can also be connected with "the principle of the sufficiency of the Scriptures," leading to "assert that the Early Church was to serve as paradigm" (Norelli, 1997, vol. 2, 745). Historian of science and religion, James C. Ungureanu notes that in the 17th and 18th centuries, this strategy "had unintended consequences. A new generation of Protestants used similar strategies against other Protestants, opposing and accusing one another of irrationality and obstructing religious progress" (Ungureanu, 2019c, 257).

An example of this periodisation can be seen in the division of church history, up to the 16th century, in three periods of 500 years found in Simon Goulart's rewritten *Catalogus testium veritatis qui ante nostram aetatem Pontifici Romano atque papismi erroribus reclamarunt* (1597);[95] division also found in Carion's *Chronicle*, created in the circle of Melanchthon (Peucer, 1562).[96] Goulart explains that during the first 500 years, the Church was faithful to true doctrine, in the next 500 years that truth was gradually replaced by falsehood, and in the last 500 years, a reign of error trumped the true Christian doctrine under the tyranny of the Popes, with only a small minority of critical voices (Goulart, 1597, vol. 1, 629–630; see partial translation and discussion in Backus, 2011). As the historian of the Reformation Irena Backus notes, "this method put doctrine at the centre of ecclesiastical history" (Backus, 2011).

More famous was the Protestant periodisation of Flaccius Illyricus in his influential ecclesiastical history *The Magdeburg Centuries* (1559–1574), divided arbitrarily by centuries in plans developed by him in 1556 (Lyon, 2003, 260). For some time Flaccius considered a division into five ages: "The end-points of these five ages – the Nicene Council, Charlemagne, Henry IV, Wycliff and Huss, and the present – represent major turning points in the history of the church" (Lyon, 2003, 260).

Aventinus's anti-Catholic *Annalivm Boiorvm libri septem* (1517–1521, published in 1554) became influential as a source of information. A key example was the story of Virgilius, borrowed from Aventinus by Flaccius in the *Octava Centuria* (1564).[97] Virgilius and the antipodes also appeared in Goulart's *Catalogus* (1597, vol. 2, 71), although this story was not mentioned in the first two editions of the *Catalogus* by Flaccius (1556 and 1562).[98] From Aventinus, some Protestant scientists borrowed the story of Virgilius, and they may have also acquired Aventinus's darkened depiction of the Middle Ages, as we have seen by comparing his text with Maestlin's. Not only did they adopt a timeline of conflicts from Protestant historiography, but these very conflicts became evidence for the existence of a chain of witnesses of "truth," such as Virgilius.

Against Catholic accusations that the Reformers were innovators without a tradition connecting them to the early Church, the Protestant historiography created the idea of a faithful remnant of true believers throughout the medieval darkness. This appeared in Flaccius connected with 1 Kings 19:18, mentioned in Romans 11:4–6 (Lyon, 2003, 264). Luther's use of the idea of the Babylonian captivity (above) already brings to mind such a concept of a remnant.

When we arrive at Kepler, Bacon or Maestlin in the early 17th century, the periodisation of both the secular and church histories was well established. What we see then is how this view was also read back into the history of science,[99] which in turn reinforced the idea of the Dark Ages, which has become entrenched in Western culture (regardless of nationality or religious confession) until the present – despite the revolt of medievalists since the early 20th century.[100] The soil was ready to nurture the seed of the flat error.

The crisis of Augustine's authority

In Chapters 3 and 4, we discussed the reluctance of Renaissance scholars to attack Augustine over the topic of the antipodeans. Reasons for this could include Augustine's overall authority, which surpassed that of the less well-known Lactantius, and that Augustine's theology of accommodation was key for incorporating new discoveries.

During the 15th and 16th centuries, the scholars' reaction to the discovery of the antipodes and antipodeans was to defend Augustine by reducing his authority to theology. By the early 17th century, Protestants (and as seen in pp. 173–178, some Catholics engaged in the Copernican debates) would have fewer reservations in criticising Augustine's authority *even* in theological matters. Although a more detailed analysis would be necessary to study how each of the authors here discussed saw Augustine and his authority in general terms, Augustine's authority was increasingly discussed during the 16th and 17th centuries in light of the religious debates of the time.

While Erasmus was not very fond of Augustine,[101] according to Backus both Catholics and Protestants "wanted to claim him as their doctor" (Backus, 2013, 74–75). This led to ambiguity regarding the authority of Augustine, who, according to Backus, "had been viewed during the Middle Ages as the universal doctor of one universal Church" (Backus, 2013, 74). Such clashes gave way to criticising Augustine's ideas as confessional divisions increased over time. Backus gives an interesting example from Illyricus's *Centuries*, "which draw the reader's attention to Augustine's 'pure' doctrine and his 'uncommon gifts' (col. 1117). However, the same article in the *Centuries* also mentions the 'blemishes' (*naevos*) which Augustine inherited 'in considerable number' from his predecessors" (Backus, 2013, 78). The offending topics related to justification and good works, as well as Augustine's views on free will. There were internal disputes in each confession, with Augustine being appropriated by Lutherans, Calvinists and Zwinglians among Protestants, and by Jesuits and Dominicans among Catholics (Backus, 2013, 80–81; 2003, 337). Thus, unsurprisingly we can find in Kepler or Campanella criticisms of Augustine regarding topics of secondary importance, such as the antipodes. We could then agree with the historian of early modern literature, Arnoud S. Q. Visser, that "in many ways confessionalism actually promoted new scholarship, sharpened critical awareness, and refined philological and historical method." (Visser, 2011, 137).

Investigating William Laud (1573–1645), Visser noted that historicising Augustine "as an author who lived in a remote historical world" (Visser, 2011, 111) was a strategy "to limit Augustine's theological authority" (Visser, 2011, 112) during the theological debates of the early 17th century on grace and predestination. I would like to argue that Augustine being historicised in the context of the antipodes's debate long preceded the 17th-century theological disputes. In fact, we already saw such an historicising approach in Chapters 3 and 4, starting in the 15th century, when Augustine and other ancient and medieval deniers of the antipodes were criticised for a geographical ignorance associated with the old ages in which they lived.

Conclusions

We arrive here at the end of this investigation of the reception of two key texts by Lactantius and Augustine in an effort to detect the origin of the flat error. In Chapter 4 we saw the rise of an essential change of attitude in the first half of the 16th century, as the 'moderns' saw themselves in possession of a knowledge and perspective that allowed them first to equal and then to surpass the ancients. The new knowledge afforded them the opportunity to see the past from an historical perspective and appreciate better its shortcomings.

If there was a discontinuity in geography because of modern navigations, the proposal of Copernican astronomy was yet more challenging, even if it was couched as ancient Pythagorean wisdom. Difficulties from theology and philosophy were discussed even before Copernicus published his landmark book in 1543. Therefore, the new heliocentric astronomy needed a protection strategy. While Osiander attempted with his note to the reader to conceal the real intention of Copernicus, negating his claims to apprehend nature's reality, Copernicus took a risky step forward by addressing the Pope in the preface-dedication of *De revolutionibus*. As the book itself was directed at the experts in secular knowledge (astronomers and philosophers), the theological issues were not covered. Instead, Copernicus's disciple Rheticus elaborated a detailed theological defence inspired by Augustine's principle of accommodation in a separate specific treatise. However, Copernicus did not ignore theology and deployed in his preface-dedication to the Pope a simpler pre-emptive attack on theological objections by putting forward an example of the worst blunder in the relationship between Christianity and 'science' (in this case, geography/cosmography), i.e. Lactantius's attack on the sphericity of the Earth.

Copernicus's reference to Lactantius was hardly a novelty. As shown in Chapter 4 this topic was well-known, particularly in geographical texts likely familiar to Copernicus (as he embraced the modern concept of a terraqueous globe). In *De revolutionibus* Copernicus reminded his readers of something that was already commonplace, not even needing a reference. In this way, the mention of the mistaken views of the ancients regarding geography was re-deployed in a new context, not to correct an old geographical error, but to warn contemporaries to avoid theological interference in science.[102] The novelty of Copernicus's reference to Lactantius was the parallel established between the rejections of the sphericity and the movement of the Earth, geography and astronomy, and their theological implications: a reminder of the power of Pomponazzi's suspicion, quoted at the beginning of this Chapter. Sailors had already opened the door to criticism in the 15th century, and now disturbing novelties would creep in.

Over the following decades, many Copernicans followed his example. They cited other well-known blunders associated with early Christians, in particular Augustine's denial of the antipodes. If, however, astronomers had difficulties finding more examples of ancient theologians to criticise for their past mistakes in science, theologians sympathetic with Copernicanism came to the rescue. In particular, two catholic authors may serve here as examples, viz., Foscarini (1615) and Campanella (1616/1622), see Table 4.1.

Lists of early Christian authorities, or later theologians, appeared in other contexts such as theology or geography texts (see pp. 130–133). Among Catholics, after the Council of Trent there was a pressing need to interpret the Bible in ways not contrary to the "unanimous agreement of the Fathers," as indicated in the *Decree on the edition and on the interpretation of the Sacred Scriptures* (*Decrees of the Council of Trent*, session IV, 8 April 1546, translated by Blackwell, 1991, 183). However, in line with the Copernican strategy, Foscarini and Campanella did not rest with simply providing lists of authorities to buttress their own views. They also twisted the purpose of compiling lists of 'orthodoxy' and provided lists of cosmographical errors by early Christian authorities: e.g. denying the antipodes, denying the sphericity of the heavens, flat-earthers, etc. Regarding this rhetorical strategy, philosopher and historian Richard J. Blackwell explained that its purpose is:

> [...] to determine the accepted meaning and the orthodoxy of various points under dispute. Also at times the purpose is the show that the authorities do not agree on a given point, and thus the matter is still an open question. In either case, the larger the number of authorities quoted, the stronger is the argument using this method. Such argumentation from authority had had a very long history before Campanella used it. Further, this method was reemphasized by

> the decree of the Fourth Session of the Council of Trent when it insisted that the acceptable interpretation of Scripture was the one given by the Church and found in the unanimous agreement of the Fathers.
>
> (Blackwell, in Campanella, 1994, 25–26)

This was common in other discussions in the Renaissance. For Lerner, showing the contradictions among Christian authorities "denounced the idea, held in his [Campanella's] time by the great majority of Catholic theologians, that there existed a unity of doctrine among the Church fathers and the scholastics concerning the question of the structure of the universe" (Lerner, 2005, 23). It is important to note that the use of authorities was already outdated in geography by the beginning of the 17th century (see pp. 135–140), but was still used in natural philosophy up to the end of the 17th century (and even later in theology). We already saw Rothmann and Kepler rejecting authorities in scientific topics.

As discussed in pp. 140–143, lists like these, without context or alternative views from other early Christian authorities, would ironically be later used in denigrating the past and supporting the flat error. There were indeed other authorities who believed in the existence of the antipodes and the sphericity of the Earth without being considered heretics. Campanella mentioned these authors as a general group, but without providing names. This attitude reinforced the view that all, or most, past Christian authorities were either flat-earthers or condemned as heretics. As the objective of Campanella and the Copernicans was to weaken the authority of the early Christian authors, no attempt was made to compile lists of those with 'correct' scientific views, heretics or not, or at least to contextualise the 'errors' of the wrong views (in fact, there is a contrast between the 17th-century Copernican lists in Table 4.1 and those of previous authors, with the Copernicans only giving one-side examples). This was the consequence of the habit of that age of either relying on or demolishing traditional authorities.

The Tridentine and Bellarmine's overconfidence in the superiority of the early Christian authorities on any topic led to their demise. As their real views were recovered and compared with the latest progress of cosmography, their shortcomings became apparent. Foscarini tried to warn the Catholic Church about that focussing on two topics: the population of the antipodes and of the Torrid Zone, which had been simultaneously demonstrated by modern navigators. He was not heard and his book was condemned.

In parallel to the reflection on recent events, other authors developed a complex view of the history of science and Christianity, focussing on instances of conflict to highlight the problem of theological interference in science. In 1609 an astronomer imbued with a strong sense of the historical development of science, such as Kepler, went beyond a two-stage history of conflicts (past/present, as Copernicus did), to describe a three-stage historical conflict in cosmography: sphericity, antipodes, and heliocentrism (referencing Lactantius, Augustine, and anti-Copernicans). In 1618, he pointed to an even earlier example of conflict in ancient Greece: the religious attack against the heliocentrist Aristarchus of Samos by Cleanthes, which Kepler compared with the contemporary geocentrist attacks on Copernicus.

The hostility of Pope Zacharias I towards Virgilius, the cleric who allegedly defended the antipodes in the mid-8th century, was unearthed by mid-16th century Protestant historians. This conflict became a tremendous tool in the hands of Protestants to criticise the hated medieval Catholic hierarchy, particularly the papacy. Scientists took this narrative from historians and theologians, and in 1620 Kepler referred to Virgilius as having been attacked for his belief in the antipodes. This affirmed the idea of a dark medieval period for science, which had been employed for the periodisation of history in the previous three centuries. This added an intermedium point of conflict – between the early Christian authorities and the contemporary Copernican debate – in the three-stage historical periodisation of Kepler in 1609.[103]

I found the earliest example of an aggressive usage of this episode for the flat error in Maestlin (c. 1621). He is well-known for his anti-Catholic bent, but his grossly distorted historical discussion of the Virgilius's episode -accusing the high clergy of that time of flat-earthism, which is one of the clearest early cases of flat error- can only be understood in the light of the dramatic condemnation of Copernicanism in 1616 to which it is compared by Maestlin himself. This was the culmination of half a century of Protestant usage of the story of Virgilius to reinforce the examples of Lactantius and Augustine in their denunciation of the errors of the past. Indeed, the words of Maestlin and other Copernicans -mainly Protestants- need to be viewed in the context of the increasing periodisation of history, initiated by humanists and continued by Protestants in their campaign to denigrate medieval Catholicism. Although Maestlin's text was left unpublished at the time, and appeared too late to influence the early stages of the flat error, it is an interesting example of the state of thought felt among Protestant (and probably also some Catholic) astronomers, and intellectuals in general, after 1616, as foreshadowed by Campanella. Sadly, these scholars' anger did not help mitigate the 1616 condemnation, and was reinforced by the condemnation of Galileo Galilei in 1633, which ended up becoming the paramount example of religious intolerance against science (Finocchiaro, 2005).

An interesting example of the same state of thought is Francis Bacon, who we discussed in Chapter 4.[104] A geocentrist, but on the side of the moderns for his defence of experimental science,[105] in 1606–1607 he drafted two similar texts, and published a similar third one in 1620 (*Novum Organum*). In these he not only conflated the denial of the sphericity of the Earth with disbelief in the antipodes, but he also made a deeper historical connection in the field of science and religion. Bacon drew a parallel between problems ancient philosophers had with religious authorities in the pre-Christian Greek era and the attacks of early Christian authors on cosmographers, mentioning their trial for impiety. In reality, only the 8th-century Virgilius fits into that pattern. Finally, in a third stage, he drew a second parallel with how "things now are" to attack the resistance of the scholastics to modern ideas (see Appendix, no. 5).

A more specific attack came from Campanella, who in a book written in 1616 praised Columbus and Magellan as more trustworthy first-hand witnesses than late ancient and medieval Christian authors. In 1609–1610, Campanella had compared the religious opposition to Columbus (inspired by Augustine) to attacks against Galileo, while Descartes was comparing in 1634 the condemnation of Virgilius with that of Galileo. In this way, Columbus became included in the historical list of conflicts between science and Christianity: a key piece in the development of the flat error.

The connection between Columbus and Galileo discovering new worlds had been already pointed to Galileo himself on the occasion of his new telescopic discoveries by Lorenzo Pignoria in 1611 (Pagden, 1986, 3; 1993, 97) and Giovanni Battista Manso in 1610 (Pagden, 1986, 4; 1993, 98). Columbus was transformed into a romantic hero in the second half of the 16th century.[106] Columbus's statue was the only one named among those of the "principal inventors" in the *Salomon's House* described in Bacon's *New Atlantis* (Bacon, 1626[1627], 45). By the 18th century, it was a "commonplace [the] connection between Columbus, Galileo and Copernicus" (Pagden, 1993, 196, note 39). Therefore, "Columbus had been successfully integrated into a genealogy of modern science" (Pagden, 1993, 102). And we can add that he was also integrated into a genealogy of modern science and Christianity conflicts.

If Galileo was seen by Manso in 1610 in the light of Columbus as "almost a new Columbus," Columbus was reinterpreted as a predecessor of Galileo, wrongly seen as intentionally *wishing to discover a New World*, and as *wishing to demonstrate that the Earth was round* (in cases of fully-fledged flat error).[107] This historical 'transfer' and re-contextualisation is quite common. For example, we mentioned Maestlin's interpretation of Virgilius's condemnation in light of the recent condemnation of Copernicanism.

The research on which this Chapter has drawn confirms that it is impossible to point to a particular author as the single originator or 'culprit' of the flat error. What we can see is a slow change of focus regarding the past during the 16th and early 17th centuries, which led to an increased perception of the late ancient and medieval periods as dark and anti-scientific (Table 5.1). We will discuss this in more detail in the following General conclusions.

Notes

1 *Si nescimus de istis rebus quae sunt in terra et possunt videri a nobis, quomodo sciemus de celo?* (Nardi, 1965, 42).

2 The same Ptolemy whose *Geography* the navigators demolished was the author of the *Almagest*, the most elaborate text of geocentric astronomy that Copernicus needed to undermine/replace (Hooykaas in Rheticus, 1984, 126–128).

3 The eccentric model placed the huge sphere of water not symmetrically around the Earth's axis, but on one side (the Western Hemisphere), which complicated the viability of its rotation.

4 Thorndike (1941, vol. 5, 427–428), Goldstein (1972), Randles (1993b[2000], 23; 1994[2000], 69–70). Reviewed with other earlier authors by Vogel, 1995, 18–30.

5 As can be seen in his identification of "the center of the earth" and the centre "of gravity and of the lunar sphere" (Copernicus and Rheticus, 1939[1971], 58). Later on, in the fifth assumption in the *Commentariolus*, he was even more explicit: "The earth together with its circumjacent elements performs a complete rotation on its fixed poles in a daily motion" (Copernicus and Rheticus, 1939[1971], 58) and "Thus does the earth

rotate together with its circumjacent waters and encircling atmosphere" (Copernicus and Rheticus, 1939[1971], 63). The word used is "circumjacent" (*circumfluis*) which "is equally applicable to both *aqua* and *aere*" (Rosen in Copernicus, 1985, 96, note 67, who translated the later text as "Thus does the earth rotate together with its circumjacent waters and nearby air." Copernicus, 1985, 83, a point that will also reappear in *De revolutionibus* 1.8). The connection of geography and topics related to the Earth, atmosphere and astronomy was part of the long tradition of 'cosmography' (see p. 7).

6 Copernicus owned a copy of Ptolemy's *Geography* (Lerner et al. in Copernic, 2015, vol. 1, 811–814), which he quoted in *De revolutionibus* 1.3, 1543, fol. 2ʳ. This chapter of *De revolutionibus* shows how Copernicus was well aware of the contemporary cosmographical debates.

7 See pp. 38–39 for Ptolemy and Lactantius and p. 57 for Sacrobosco.

8 *Quomodo terra cũ aqua unum globũ perficiat.*

9 See Rosen (1943), Hooykaas (1980, 315–317) .

10 On dating it to 1541, see Hooykaas in Rheticus (1984, 144–145). Rheticus's text was the second treatise of a single volume where only the author of the first treatise was identified: David van Goorle (Gorlaeus and Rheticus, 1651). Rheticus's authorship in 1541 of the second text, the anonymous *Treatise on the Holy Scripture and the Motion of the Earth*, was defended by Hooykaas (Rheticus, 1984). This has been recently questioned, proposing an unidentified 17th-century authorship closer to its publication date (Higgins, 2013). However, the presence of the topics discussed in this paragraph is more fitting for a mid-16th century context. See, for example, the similarity of quotations from *Chorographia* with that from the *Treatise* in this paragraph.

11 This letter was studied in detail by Granada and Tessicini (2005; with references); however, they did not discuss this portion.

12 Besides Aristotle's discussions on the shape of the Earth, Copernicus mentioned the books by Cicero (*On Academic Scepticism* 2.123, in *De revolutionibus,* dedication and 1.5) and Pseudo-Plutarch (*The Opinion of the Philosophers* 3.13, in *De revolutionibus,* dedication and 1.3, at that time falsely attributed to Plutarch, see Rosen in Copernicus, 1978, 342, 348) where he found ancient references to the motion of the Earth. Copernicus also owned Pliny's *Natural History* (Lerner et al. in Copernic, 2015, vol. 1, 811–814). In these books he could find references to the ancient debates on the sphericity that we discussed in pp. 30–33.

13 Rosen (1977), dates the beginning of the *De revolutionibus* in 1515, with most work done in the 1520s, and finished by 9 June 1541, shortly before Rheticus returned to Germany in September 1541 with a copy for printing. The first chapters seem to be completed by 1524 (see endnote 28, dating *De revolutionibus* 1.3 before 1522), with Rheticus helping Copernicus mainly with the two final books of *De revolutionibus* (Segonds and Luna, 2013, show the usage in *De revolutionibus's* final chapter of the Greek edition of the *Almagest*, published in Basel in 1538 and brought by Rheticus to Copernicus in 1539).

14 Although it is commonly assumed that this was an attack against Copernicus, it has been suggested that it was against Celio Calcagnini, a humanist scholar who had proposed the rotation of the Earth 20 years before, although published in 1544 (Barker, 2005; Lerner, 2006; 2009; on Calcagnini, Omodeo, 2014, 209–213). In any case, it was an attack on geokinetism, also relevant for Copernicus. Lerner (2009), mentions previous research (Elert, 1962, 418 [original text from 1931]; Kleinert, 2003) pointing to the identification of Copernicus as the target of Luther as a late 19th-century propagandistic invention in the context of Catholic/Protestant polemics, with some earlier references in the first half of the 19th century. This process runs parallel with the popularisation of the flat error.

15 *...Scholarũ theoricis contrarius, et (ut Monachi dicerent) hęreticus exiſtimari...* (translated in Rheticus, 1939[1971], 394; Rosen, 1984, 184).

16 Osiander was a key figure in Nuremberg, where Rheticus had been on his scientific tour (1538–1539) before visiting Copernicus. In fact, it is believed that he was part

of the circle of people Rheticus met in Nuremberg (also including Petreius, later the printer of *De revolutionibus*) and who encouraged him to engage in a new trip to visit Copernicus. The leading role of Osiander in Nuremberg and his connections with Duke Albrecht of Prussia explain the interest of Rheticus and Copernicus in knowing Osiander's reactions to the new astronomy and his advice on how to gain the favour of intellectuals there. Rheticus deployed a successful campaign to attract the important patronage of Duke Albrecht (see in pp. 160–161 a reference to the *Chorographia*, which was part of his plan). For a detailed discussion of the patronage context surrounding the publication of *De revolutionibus*, see Westman, 1990; Barker and Goldstein, 2003; Granada and Tessicini, 2005; Barker, 2005.

17 *Apologia Tychonis contra Ursum* (*A Defense of Tycho against Ursus*, English translation in Kepler, 1984). For the dating, Kepler (1984, 9), and Rosen (1946). First published by Frisch in Kepler (1858, vol. 1, 215–287).

18 *Sic enim placidiores redderes peripatheticos et theologos, quos contradicturos metuis.* (Translated in Kepler, 1984, 152; also Rosen, 1984, 193).

19 *Peripathetici et theologi facile placabuntur, si audierint, ejusdem apparentis motus varias esse posse hypotheses, nec eas afferri quod certo ita sint, sed quod calculum apparentis et compositi motus quam commodissime gubernet, et fieri posse*... (Translated in Kepler, 1984, 153; also Rosen, 1984, 194).

20 I believe there was a genuine concern in Copernicus and Rheticus about a negative theological response to heliocentrism. Of course the theologians were not alone, but were here helped by the Peripatetic (Aristotelian) philosophers. Still on 26 March 1598, Kepler wrote to the Babarian Chancellor, Hans Georg Hewart von Hohenburg: "there is not an astronomer who puts these new hypotheses [of Copernicus] one bit behind those of antiquity; the struggle against Copernicus is waged exclusively and entirely by natural philosophers, metaphysicians, and theologians" (quoted in Rosen, 1975, 330). The two sides appear again: astronomers (surely not all of them were Copernicans, Kepler was again optimistic in his *Epitome* of 1618, Appendix, no. 10) and an 'alliance' of natural philosophers, philosophers and theologians. The same sides clashed around Galileo, and surely the astronomers also had theologians on their side (as Foscarini and Campanella). There is not a clean disciplinary divide in the debate.

21 Rosen, in Copernicus and Rheticus (1939[1971], 400–402), Rosen (1984, 184), Barker and Goldstein (2003).

22 Fuller text in Kepler (1984, 152). On Copernicus's rejection of Osiander's 'solution,' see Lerner (2006, 431–432).

23 Lerner (2005, 28, note 14) explains that this proverb was taken from Aristophanes (*Ploutos*, line 885) by Erasmus, who translated it into Latin (Erasmus, 1508). I found the word translated by "backbite," *sycophanta*, attached to Lactantius a century before (1444), in Antonio da Rho's *Dialogues* against Lactantius (Rutherford and Schulten, 2023, 348–349, 648–649 and 660–661), and in Niccolò Arcimboldi's annotations to the same work (Rutherford and Schulten, 2023, 784–785, 882–883).

24 See also in Rothmann, pp. 167, 172. This is a rare word Copernicus used in Greek *ματαιόλογοι*. Lerner (2005, 28, note 15) suggests that it was taken from the New Testament (Titus 1:10, where Paul denounced false religious teachers; this word was translated in the Vulgate as *vaniloqui*).

25 Gasser's reference to the heretical implications of Copernicanism in the presentation of *Narratio prima* in 1541a, could be seen as addressing this point by dismissing it instead of hiding it. The same approach was used by Copernicus in his preface-dedication to the Pope in 1542.

26 I have not had the chance to study this in detail. A careful intertextuality study should be carried out to confirm this speculation.

27 Rheticus's contemporaneous reference in *Treatise on the Holy Scripture and the motion of the Earth* was equally tangential (see pp. 166–167).

28 Discussed in Rosen (1943). See also Goldstein (1972, 38–42), who argues that Copernicus's geographical description in *De revolutionibus* 1.3 predates the first circumnavigation of the Earth in 1522, probably shortly after his first conception of heliocentrism in *Commentariolus* (before 1514), which he did not update afterwards. Other authors date the beginning of the writing of book 1 later in the 1520s, although without ruling out completely Rosen's view of an earlier start before 1515 (Goddu, 2006, 49, note 29).

29 Several people knew he was the author of this text, but it was only made public by Kepler in *Astronomia Nova* (1609, on the back of the cover page), as his earlier detailed account in *Apologia Tychonis contra Ursum* remained unpublished (see endnote 17; on the history of this text, see Lerner and Segonds, 2008).

30 *Monitum Sacrae Congregationis ad Nicholai Copernici lectorem* (1620): "In the preface, toward the end, delete everything from the beginning of the last paragraph up to the words 'this work of mine,' and substitute: 'For the rest, this work of mine ...'" (Finochiaro, 1989, 200–201). The Latin text after the censoring was: *Cœterum hi nostri labores...* (see Appendix, no. 7).

31 Rheticus (1540), Digges (1576), Bruno (1584a, 1584b, 1584c), Zúñiga (1584), Galileo Galilei (1610, and his correspondence with Castelli [1613] and Dini [1615], as well as his unpublished notes on Copernicus [1615]; I checked the correspondence and notes from the Spanish translation in Galileo (1987), Kepler (1596 and 1610, although in this last book there is a passing reference to Columbus and the antipodes irrelevant for of our work here; see Pagden, 1986, 4) and Lansbergen (1619). These references are absent in the anti-Copernican text of Tolosani, *De coelo supremo immobili et terra infima stabili, ceterisque coelis et elementis intermediis mobilibus* (c. 1547–1548, published in 1973; see p. 165).

32 *Lactantius, vir alias eruditiſſimus & eloquentiſſimus irridet eos qui profitentur, terram eſſe rotundam* (reproduced in Rheticus, 1984, 54 [Latin], 84 [English translation]).

33 This manuscript was written between the letters to Brahe in September 1588 and August 1589 (Granada, 1996, 814).

34 I thank Prof. Harry Hine (University of St Andrews, UK) for this translation.

35 For a longer quotation and discussion, see Felipe (2018, 53–54), and the 2019 erratum.

36 *Quare vanè, ſuperſtitioſè, concuſſionem corporum timent puſillanimi quidam (Lucij Lactantij more, qui Antipodes, & globoſum vndiq; terræ ornatum, indoctiſſimi vulgi, & importuniſſimorum hominum more deridet:)* (translation in Gilbert, 1900, 219).

37 *ex telluris motu concuti, & dirui neceſſariò exiſtimant* and *in contrariam cœli partem ruant* (translation in Gilbert, 1900, 227).

38 I thank Prof. Harry Hine (University of St Andrews, UK) for this translation.

39 Translated in Salusbury (1661, vol. 1, 467).

40 Translated in Rosen (1975, 334).

41 The third stage is not attached to individuals (as the first two linked to Lactantius and Augustine), but to the *Officium* of the moderns. As discussed in the Appendix, no. 9, endnote 4, this word was translated as "Liturgy" already in 1661, which in fact is *Officium Divinum*, or more commonly translated as a reference to the Inquisition. Although *Sanctum Officium* is used to name the Inquisition, it is doubtful that Kepler refers here to this Catholic institution exclusively in 1609, i.e. before the condemnation of Copernicanism by the Inquisition in 1616. In addition, I cannot think of any part of the Liturgy discussing these topics. In a personal communication, Prof. Miguel Ángel Granada (University of Barcelona, Spain) pointed out that the true grammatical antecedent of *qui* (masculine plural) is not *Officium* (neutral singular), but *hodiernorum* (the denials of the Earth's movement). Taking this into account, and the fact that the context is ironic and the word *Sanctus* is used before to refer to Lactantius (who was not a Saint) and to Augustine (who was indeed a Saint of the Catholic Church), I believe that the whole string of connected references is ironic, not referring to formal

institutional nomenclature. In addition, Kepler knew by 1609 that the problem of Copernicanism was not only the Catholic Church, as this introductory text to the *Astronomia nova* (1609) was initially the chapter 5 of the *Mysterium* (1596), but was censored at that time by Lutheran authorities (on Kepler's problems with Lutheran censorship in the 1590s, see Rosen, 1975). My own hunch is that this is not specifically an anti-Catholic text (as we see in Maestlin's post-1616 text in pp. 178–179), and that *Officium* here denotes the religious authorities, both Catholic and Lutheran, which resisted Copernicanism and the "philosophical" (scientific) inquiry into the truth. The final critical reference to the *Doctorum Ecclesiæ* also seems to point in this direction.

42 However, there were exceptions, such as the geokineticist Gilbert, who mentioned Lactantius and both topics (see p. 167 and endnote 36). Later, in the early 17th century, a committed Copernican, such as Kepler, attacked both authors and topics (see Appendix, no. 9).

43 See Chapter 3, endnote 55. For its impact on Copernicanism, see Howell (2002), and Hooykaas (in Rheticus, 1984). On the principle of accommodation, in connection with some ancient and Renaissance Christian scholars (including Copernicans), see Felipe (2018).

44 On that, see Felipe (2012). Hooykaas points to this explanation in his discussion of Rheticus's criticism of Lactantius, which ignores the antipodes (Rheticus, 1984, 122).

45 Campanella noted how Augustine violated his own principles by denying the antipodeans: "But it is surprising that both Augustine himself and the Fathers made a similar mistake, namely, an error in the particular premise, but not in the universal premise which they were teaching, in the following syllogism" (*Sed quid mirum, quando & ipſe Auguſtinus & alij patres ſic errauerunt, non in vniuerſali quam docuerunt, ſed in particulari huius ſyllogiſmi propoſitione deficientes.* Campanella, *Apologia* 3, 1622, 28; translated in Campanella, 1994, 76).

46 *Id ſatis ſcio, Auguſtinum, quem ſolum nominas, motum Terræ, neq́; annuum neq́; diurnum vſquam conceſsiſſe, ex quo rotunditati eius tanquam minùs Mathematicus, Antipodas denegando, detraxerit. Quod tamen illi condonandum eſt. Quare cùm rotundam eſſe non perceperit, motum etiam rotunditati competentem longè minùs admittere potuit.* I thank Prof. Harry Hine (University of St Andrews, UK) for this translation.

47 Translated in Kepler (1939, vol. 1, 8).

48 I thank Prof. Harry Hine (University of St Andrews, UK) for this translation.

49 On the polemical usage of this rediscovered episode, see Mayaud (2005, 91–121). Here see the next paragraphs and pp. 178–179.

50 The story of Virgilius, taken from Aventinus, was also recounted in other texts in the context of geographical discussion on the antipodes, such as Mazzoni, *Della difesa della Comedia di Dante* (1587, vol. 1, see above p. 131) or in ecclesiastical histories, such as *Octava Centuria ecclesiasticae historiae* by Flacius Illyricus (1564, cols. 534, 785 and 800; on this book, see p. 183).

51 I thank Prof. Harry Hine (University of St Andrews, UK) for this translation.

52 *Nec a Theologis huius artis contemtus sperandus. Quandoquidem ea immensam DEI Opt. Max. sapientiam ac potentiam ineffabilemque maiestatem et in administratione Mundi providentiam conspectui evidentiori subiiciat. Iamdudum enim explosa est crassa illa et asinina prorsus Bonefacii inscitia, Virgilium Episcopum Salaeburgensem damnantis, quod is pro concione dixerat, antipodas esse.* (I thank Prof. Harry Hine (University of St Andrews, UK) for this translation).

53 Latin original published by Granada (1996, 827), and in Rothmann (2003, 203). I thank Prof. Harry Hine (University of St Andrews, UK) for this translation.

54 For the different portrayals of the fate of Virgilius, see endnote 84.

55 On dating, see Blackwell (1991, 75).

56 Latin text in Foscarini, *Lettera*, 1615, 6–8; English translation in Salusbury (1661, vol. 1, 474–475).

57 This censure was published by Berti (1881, 72–73); translated by Blackwell (1991, 253–254).
58 The Latin autograph copy at the Archivo Generale dell'Ordine Carmelitano (Rome) was published by Boaga (1990, 204–214).
59 The text of this letter is translated by Finocchiaro (1989, 67–69) and Blackwell (1991, 265–267).
60 Two versions of the *Defensio* have been preserved, apparently from an unknown original, which were published by Berti, 1881 (short), and Franco, 1911 (long); see the discussion in Caroti, 1987, and Boaga, 1990. In the *Defensio* some of the most incisive portions were removed or softened, including the reference to the errors of early Christian authors discussed above. In discussing his translation, Blackwell seems to confuse the *Defensio* with the letter to Bellarmine published by Boaga (1990); see Blackwell (1991, x).
61 There is some debate over whether it was written before or after the condemnation of Copernicanism on 5 March 1616. The current evidence dates it shortly before that date, but arriving at the Vatican a few days after the condemnation (Blackwell, 1991, 9, 19–24; Lerner, 2005, 23). For these and other dating of the works of Campanella, I have followed Headley (1997b).
62 Blackwell, in the introduction to Campanella (1994, 5, 9), Lerner (1987).
63 Campanella, *Apologia* 4 and 5, 1622, 37, 38 (twice), 50, 57; translated in Campanella (1994, 91, 92, 110, 122).
64 On the implications of those contradictions, see pp. 185–186.
65 However, Blackwell mentions that this is not supported by the surviving fragments of Xenophanes (Blackwell in notes of his translation of Campanella, 1994, 144, note 173).
66 Campanella was correct in stating that the idea that day and night are produced by the rotation of the Sun around a high mountain in the north of Europe was defended by several ancient Christians (in fact, authors from the school of Antioch, see p. 35) who rejected the sphericity of the Earth, and it was later graphically depicted by one of them, Cosmas Indicopleustes in his 6th-century *Christian Topography* (Kominko, 2013, 67–70, 252–255); see https://brunelleschi.imss.fi.it/galileopalazzostrozzi/oggetto/CosmaIndicopleusteTopographiaChristiana.html. However, Campanella was wrong in linking Lactantius and Augustine with those authors (as we have seen in Chapter 2).
67 *Procopius Gazæus anno Domini 500. contexuit catenam expoſitionum ſuper S. Scripturas ex omnium patrum momumentis; & probat, antipodas non exſtare.* (Translated in Campanella, 1994, 77).
68 I am grateful to Sir Richard Sorabji for pointing this out to me. The Greek text appeared there in parallel with a Latin translation. See Scholten (1996, 12).
69 I give here and below the references to the two editions of *De gentilismo non retinendo*. A similar listing is found in Campanella (1636, 60; 1637, fol. eiiij[r]), also quoting Daniel 12:4. The habitability of the Equator is again discussed in Campanella (1636, 26; 1637, fol. ciij[v]) and in Campanella (1636, 55; 1637, fol. eij[v]), mentioning the favourable opinion of Pliny, Albert the Great, Avicenna and referring to the experience of sailors, in particular the Portuguese.
70 This text was added to this second edition and it is not found in the first one of 1636.
71 *Omnes ergo, qui Philoſophiam alicuius Autoris canonizant ex Theologia, exponunt fidem irriſioni, vt docet D. Aug. & S. Thomas,* opuſc. 9.
72 *Nolo alios errores manifeſtos pandere: pudenda enim PP. tegenda ſunt.* (I thank Prof. Harry Hine (University of St Andrews, UK) for this translation).
73 Columbus and Galileo are linked in a text where Campanella discussed and replied to a series of objections to his famous *Civitas solis* (*City of the Sun*), which was written in an Italian prison in 1602, and published in Latin as an appendix to the third part – on

politics – of his *Realis philosophiae epilogisticae partes quatuor*, Frankfurt, 1623, 415–464). His later discussion on *Civitas solis* appeared as *Quaestio IV de Optima Republica* at the end of the *Quaestiones* on his politics, published as part of his *Disputationum in quatuor partes suae philosophiae realis libri quatuor*, usually abbreviated as *Philosophiae realis* (Paris, 1637, 100–113, this is nearly at the end of a book of more than 1,000 pages because pagination re-starts several times in different sections; *Civitas solis* was reprinted a few pages later, in pages 145–169).

74 *8. Naturale eſt homini inquirere opera Dei, peregrinaci per Mundum, ſcientias vndique acquirere, omniaque experiri. Viuentes autem in tali Rep. [republica] ſunt ſicut monachi, qui libris ſolis ſtudent. & cum audiunt, quod in eis non legerunt, ſcandalizantur, & conturbantur. Vnde nunc vix credunt obſervationibus Galilei, ſicut pridem, nec Columbo inuentori noui Emiſpheri, quod S. Auguſtinus negarat* (translated by Headley, 1997b, 307, with slight modifications).

75 *De gentilismo non retinendo* and the *Quaestio IV* are the only two portions of the large *Philosophiae realis* that I have investigated in detail. A glance at the whole book shows many more references to Lactantius, Augustine, the antipodes, Columbus and Galileo. There are many other works by Campanella that I have not analysed, and which could be of interest.

76 Gasser's fears in 1540, which Maestlin knew (see p. 162 and endnote 15 here, and endnote 6 in Appendix, no. 13), finally became true (I owe this connection to Lerner, 2005, 11–12, 25).

77 I thank Prof. Harry Hine (University of St Andrews, UK) for this translation.

78 See Appendix, no. 2 on Aventinus and pp. 53–54 on Virgilius alleged teachings and their relation to the old debate on the antipodes and antipodeans.

79 Although not referring to this document, the words of Grant in p. 208 fit here.

80 Montaigne affirmed that a thousand years ago, "[i]t had beene flat hereſie to avouch, that there were Antipodes" ("c'eſtoit hereſie d'auouer des Antipodes" Montaigne, Essais 2.12, 1588, fol. 241[v]; this sentence was an addition to the original edition of 1580; English translation from Montaigne, 1603, 332).

81 Appendix, no. 4. Similarly, Bacon, 1606 (published in 1734) and Bacon (1620); see Appendix, no. 3 and 5.

82 See p. 179.

83 See pp. 175–177.

84 He was supposed to have been "condemned" (Ramus, 1567; Appendix, no. 15), "condemned on a charge of impiety" (Rothmann, c. 1589; Appendix, no. 16), "condemned [...] and expelled [...] from the priesthood, temple and Church" (Origanus, 1609; Appendix, no. 14), "removed from his office" (Kepler, 1620; Appendix, no. 11), "condemned for heresy" (Maestlin, c. 1621; Appendix, no. 13), "condemned" or "deposed" as "heretic" (Campanella, 1636–1637; pp. 176–177). See also Table 5.1.

85 *eò quod multi Antiſtites patrię Hispanę manifeſtũ reũ hęreſeos eſſe planè afferebãt* (translated in Wey Gómez, 2008, 112). See pp. 107 and 121 on Geraldini. In describing the previous *Junta* in Salamanca (1486–1487), Ferdinand Columbus indicated that Augustine's denial of the antipodes was mentioned, but he did not report an accusation of heresy to his father (see p. 92 and Appendix, no. 6).

86 Other ideas influential for medieval views of history were the historical division according to the four monarchies of Daniel 2 and 7 (as in Jerome's *Commentary on Daniel*), the six ages inspired by the six days of creation in Genesis 1 (as in Augustine's *City of God* 22.30) or, with an eschatological flavour, the three ages of Joachim de Fiore, etc. Outside the biblical context, the idea of a Golden Age linked to the origins of humanity and the notion of human history as stages in the life of a person (young-adult-old) were also popular. For these ideas, see Nisbet (1980[1994], 81–86).

87 Mommsen noted that mediaeval thinkers softened the 'fall' of Rome, assuming that the empire never disappeared, but "had been transferred from the Romans to the Byzantines, the Franks and eventually the Germans" (Nisbet, 1980[1994], 235). However, Italian humanists denied this *translatio imperii* to Germany and already with Petrarch longed for a return of Rome to its ancient glory (Mommsen, 1942; Russell, 1991, 65). On what we can describe as the invention of the Dark Ages, see also Grant (2001, 283–364), Shank and Lindberg (2013).

88 Ironically, the humanists reaped the same disdain they cultivated: "[...] just as the humanists proclaimed their scorn for and independence from the scholastics while drawing on many of their methods and beliefs, so too the "novatores" of the seventeenth century owe more to Renaissance humanism than their claims to bold originality suggest" (Blair and Grafton, 1992, 536).

89 One of those navigators, Pacheco, in c. 1508, regarded Portuguese discoveries as recovering ancient voyages (see Chapter 4, p. 114 and endnote 62). On the Portuguese as bringing new worlds out of darkness, see Chapter 4, p. 116 and endnote 73.

90 Before the Reformation, Erasmus gave Protestants a whole derogative vocabulary to attack scholastics and the Roman Catholic Church. Ungureanu (2019c, 107), points in this regard to the *Praise of Folly* (1511) and the *Colloquies* (1518).

91 Maas (2010, 149–151), Ungureanu (2019c, 108). References to history also appear in Luther's *95 Thesis* (1517).

92 Williamson (2008, 45), Ungureanu (2019c, 109–110); for a study of *Vitae*, see Maas (2010, 107–136; pp. 174–175 on Luther's preface).

93 Ferguson (1948, 47–48). See also Andrea (1992). Backus (2003), 1–2, traces this view to a work by Pontien Polman in 1932 that she re-evaluates and qualifies in her book.

94 Ferguson (1948, 49), Backus (2003, 374), Ungureanu (2019c, 106).

95 The original *Catalogus* was composed by Flaccius Illyricus in 1556 and enlarged in 1562, but the 1597 edition by Goulart was, as he said on the cover page, "a practically new book."

96 Although Backus does not mention that, almost the same text as the one discussed in the previous endnote is found in a history textbook originally designed in German by Johann Carion (1499–1537), with the support of Phillip Melanchthon (Wittenberg, 1532). After Carion's death in 1537, Melanchthon expanded and published it in Latin (the first part in 1558 and a second part in 1560). The text from Goulart's *Catalogus* is nearly identical to what is found in the third part of Carion's *Chronicle* (belonging to a section entitled *De rebus ecclesiae*, book 4, fol. 85r–86r), edited by Caspar Peucer after Melanchthon's death in 1560 (Peucer, 1562). For details on the editions of Carion's *Chronicle*, see Lotito (2019, 207–280; 337–351).

97 See above. On the composition of this long work, see Grafton (2001[2009], 102–108).

98 Virgilius and the antipodes were not mentioned in the long entry for Boniface in these editions by Flaccius Illyricus (1556, 115–119, and 1562, 65–67), despite referring there to Aventinus.

99 What the Renaissance was for the humanities and the Reformation for theology, the geographical and astronomical discoveries of the 15th–17th centuries were for science. Therefore, in all these disciplines, the historical tripartite scheme was adopted.

100 On this "revolt of the medievalists," see Melve (2006). For its implications in the field of the history of science, see Shank and Lindberg (2013). For a description of the periodisation process here described, see a detailed chapter in Grant (2001, 283–355).

101 He preferred Jerome (see p. 80), and noted Augustine's shortcomings with these harsh words, echoing the usual criticisms on the topic of the antipodes (here applied to his ignorance of biblical languages): "Augustine was undeniably a saint and a man of integrity endowed with a keen mind, but immensely credulous, and, moreover, lacking

the equipment of languages [...]" (Erasmus, *Annotations on the New Testament*, 1516, note 15 to John 21; quoted from Rummel, 1986, 59). On Erasmus's view of Augustine, see Visser (2011, 33–39).

102 In 1549, two books were published using Lactantius in opposite ways. In one of them, Antoine du Moulin followed the same strategy as Copernicus of attacking Lactantius, although to defend esoterism, in the preface-letter to the printer of his French translation of Jean de Indagine's *Chiromantia* (1531), published as *Chiromance & Physiognomie*. In the other, Diogo de Sá reversed the argument, attacking in the name of a philosophical theology –behind the authority of Lactantius– the scientific ambitions of mathematics (see pp. 124–125).

103 It is interesting to mention here how, already in 1837, Whewell noted the importance given by Kepler to the story of Virgilius, indicating correctly that the story of the "deposition from his bishopric" was "altogether false" (see Chapter 1, p. 11 and endnote 2).

104 Another example, slightly later and with less anger, is John Wilkins (1638), see p. 201, who used Senensis (1566), Góis (1593) and Campanella (1620). In the 1640 edition, Wilkins also referred to Aventinus (1554) and Acosta (1588). McColley (1939) has studied the influence of Campanella on Wilkins.

105 On the anti-Copernicanism of Bacon, see Jalobeanu (2014). Behind his comparison of different ages lays a disregard for the medieval period (already in *The Advancement of Learning*, 1605), and the tripartite historical scheme we discussed in pp. 180–183, found in *Novum Organum* 1.78 (Bacon, 1620, 91). For Bacon's attack on the Middle Ages, see Grant (2001, 315–317).

106 Bradner (1951), Elliott (1970[1992], 10–12), Gil (1983), Varela (2010, 509–523, 537–562), Rubiés (2017, 58–59).

107 Pagden (1986, 4–5). The first step in this process, that I know, is Juan de Castellanos, *Elegías de varones ilustres de Indias* 1.2 (1589, 18), and Lope de Vega's play *El Nuevo mundo descubierto por Cristóbal Colón*, lines 403, 683, 940, 1614, fols 32^{r}, 34^{v} and 36^{v} (composition dated 1596–1603; Saracino, 2017). Both stressed the *intention* of Columbus to find the antipodes. Castellanos listed Christians who denied the antipodes (see p. 131).

General conclusions

From the invention of the flat error to the idea of an historical conflict between science and Christianity

Answering a challenging question

This investigation aims to understand how the geographical discoveries of c. 1500, in an intensely Christian culture, could later be transformed into a formidable weapon against Christianity itself by wrongly alleging that up to the late 15th century, Christians believed the Earth was flat. In the Introduction, we explained that this historical misrepresentation has been termed the 'flat error'. How was it created?

As we saw in Chapter 1, previous scholarship, in particular Jeffrey Burton Russell's *Inventing the Flat Earth* (1991), placed the origin of the flat error in the early 19th century. This book argues instead that the origin of the flat error is much more complex and lies deeper in history.

Lactantius and Augustine were the two key historical figures around whom debates revolved from the 15th century onwards on what past Christian authorities had said regarding the shape and population of the Earth. Therefore, this research investigates the historical reception (from the 4th to the 17th centuries) of two key passages from these authors, as discussed in the introduction.

In Chapter 2, we found that Lactantius was almost forgotten during the medieval period, while Augustine was continuously remembered. It was only in the early 15th century, with Cardinal Fillastre (1414–1418), that Lactantius's flat Earth views were recovered and harshly criticised together with Augustine's well-known denial of the antipodes, as discussed in Chapter 3. Most humanists praised Lactantius's literary style, while Antonio da Rho (1444) and Galeotus Marzio (1477) furthered the critique of Lactantius on theological and philosophical grounds, also including the topic of the antipodes. However, Augustine was usually defended first due to his status as the foremost theological authority of the Western Church throughout the Middle Ages, and, second, to the fact that his denial of the antipodes was a lesser error compared with Lactantius's flat-earthism.

In Chapter 4, we considered the implications of the revolutionary geographical discoveries, from the 15th to the 17th centuries. European travellers revealed Lactantius and Augustine to be mistaken in some of their statements, leading to a diminishing of their authority. They – in particular Augustine – lost their status as revered authorities and began to be viewed as pitifully ignorant in cosmography. More relevant for the origin of the flat error, as the late 16th century approached,

DOI: 10.4324/9781003545682-7

misinterpretations regarding the extent of the alleged belief in the flat Earth of the ancient and medieval Christians began to appear.

A key moment in this process concerns the reference to the flat-earther Lactantius by Copernicus in the preface-dedication to the Pope of his *De revolutionibus* (1543). However, the scholarly interpretation of this important reference is frequently confused, as this recent example demonstrates: "Copernicus [...] mocked Lactantius, a respected Father of the Church, for asserting that the Earth is flat on the basis of the Bible" (Omodeo, 2014, 57). There are three problems there: first, Lactantius was not 'respected' by 1543; he had rather been ridiculed for more than a century; second, he was not a 'Father of the Church', but a layman criticised by some since his own times due to his defective theology; third, he had not used the Bible in discussing the shape of the Earth, but purely secular argumentation.[1] Other authors doubt that the reference to Lactantius in Copernicus was influential at all: "Whether Copernicus's mention of Lactantius was subsequently influential is unclear. It was surely not during the sixteenth to eighteenth centuries, when Copernicus's treatise was most frequently used" (Grant, 2001, 341). As discussed in detail in Chapter 1, Russell has fluctuated between highlighting the importance of the Copernican statement (Russell, 1991, 64, 70) and minimising its importance (Russell, 2001, 19; 2002).

In contrast to that previous scholarship, we see from Chapters 3 and 4 that what Copernicus did by mentioning Lactantius was to reflect what was already a commonplace in the mid-16th century. If Copernicus was creative in any way, it was to connect Lactantius's well-known mistake with the 'science and Christianity' discussion in relation to Copernicanism, and in this regard, he was very influential, as demonstrated in Chapter 5. However, criticisms of ancient Christians in regard to the flat Earth can also be found outside the specific field of Copernicanism, which complicates the analysis.

This investigation locates the 16th-century criticisms of Lactantius and Augustine within the context of the new geographical discoveries and the new astronomical model of that time. I propose that these criticisms create the conditions in which the flat error emerges, as well as nurture an environment that gives rise to the idea that Christianity conflicts with progress in science.

A new proposal for the origins of the flat error in Early Modernity

During the course of this investigation, we have seen that many identifications of early 'culprits' of the flat error are mistaken, and we sought to clarify that.[2] However, more importantly, it became obvious that the chase for monocausal precursors (or originators) of the flat error is an incomplete and misleading line of research regarding a topic that has shown an unexpected complexity. Differently from other investigations of the origins of mistaken modern views about the medieval age, it is not possible in the case of the flat error to uncover a neat single cause.[3]

In contrast, here different fields were investigated in order, first, to arrive at a more nuanced description of how the erroneous claim of a medieval Christian belief in the flat Earth took shape in Early Modernity, and, second, to demonstrate how this claim has influenced the equally false claim that science and Christianity are – and have been for centuries – at war with each other, the so-called 'conflict thesis' in science and Christianity, that we will discuss below. The different aspects investigated reveal a complex interconnection of ideas as well as different origins and forms of flat error.

In addition, the flat error emerged neither by chance nor by a malevolent machination. The claim that medieval Christians believed in a flat Earth cannot be traced back to one idea that triggered it all; instead, the origins of the claim are the consequence of the confluence of different ideas c. 1600. These include various misunderstandings, exaggerations and errors embedded within a long history that needs to be understood and unravelled. We can see the origin of the flat error as the final step in a funnel-shaped historical development of several centuries, which finally coalesced by c. 1600, not necessarily in a single form of flat error, as several variants of flat error appeared and can still be found today, as noted by Russell (see p. 18). The main findings discussed in detail in Chapters 3–5 can be summarised in three groups of observations.

The early modern view of the past and the traditional authorities

As shown in Chapter 5, an increasingly dark perception of the Middle Ages, connected with the rise of the periodisation of history, was already visible in the 14th century in Petrarch, and became customary by the 15th century. This easily led to contrasting medieval people as ignorant and moderns as informed and knowledgeable. This sense of superiority was particularly associated with the voyages of discovery from the 15th century onwards, reinforcing historical periodisation and contributing to the modern idea of progress, as was discussed in Chapter 4. In the late 16th century, this was strengthened by new revolutionary astronomical observations (novas and comets).[4]

Moving now to our more specific topic, we have often seen in Chapters 3–5 how Lactantius was vilified for his ideas on the shape and habitability of the Earth, at least since Fillastre (1414–1418), while there were sustained efforts to 'save' Augustine. This dual strategy endured until the late 16th century, as can be seen in Acosta (1588/1590). However, although up to the mid-16th century the debate revolved only around Lactantius and Augustine, over time scholars compiled extensive lists of earlier authors for and against the sphericity of the heavens and the Earth, the existence of antipodes, the habitability of the Torrid Zone, etc. As can be seen in Table 4.1, the lists were made up with a few Western Latin testimonies but mostly enlarged with Eastern Greek authors, specifically regarding the opposition to the sphericity of the heavens and/or the Earth, as this position was not really found in Western medieval authors, who often denied the antipodes but not the sphericity of the heavens and/or the Earth. The discussion of extensive lists of past authors's opinions on cosmological topics continued among Copernicans in the 17th century.

The early modern perception of a science and Christianity conflict

Although the topic of an entrenched historical conflict between science and Christianity became popular in the late 19th century, its first occurrence can be traced to Early Modernity in relation to the cosmographical debates here investigated. In fact, the flat Earth and debates about the antipodes were used as examples of how *not* to relate science and Christianity. This apparently started with Copernicus's criticism – for defensive reasons – of Lactantius in the preface-dedication of his *De revolutionibus* (1543) to the Pope.

However, the discussion did not end with this particular example, as progressively additional forgotten controversies between science and Christianity were unearthed. The most important example, recovered by Protestant author Aventinus in 1554, was the 8th-century controversy which involved Archbishop Boniface of Mainz and Pope Zacharias I against the priest Virgilius of Salzburg, an alleged defender of the existence of the antipodes. This story became a popular topic among Protestant Copernicans by the end of the century.

The practice of compiling lists of past Christian errors in cosmography, combined with the above examples of controversies between science and Christianity, led to the creation of an entire history of science and Christianity conflicts. The expansion of the criticisms of Marzio or Copernicus against Lactantius, to include other medieval and early modern authors, led in the early 17th century to a narrative of consecutive and ongoing science and Christianity conflicts, as in Kepler and Bacon, and to some extent in Campanella.

The growing confusion regarding past debates between science and Christianity

As the 16th century unfolded, it is not difficult to find many errors regarding the historical details of past relations between science and Christianity, which lead to the creation of the feeling of an historical conflict. We will review here some of the most common errors that we saw in the preceding chapters. Looking at the dates, it is possible to see how these errors become more frequent as we approach the 17th century and beyond.

Regarding Lactantius, the most common attacked Christian author among the early moderns, we find mistaken claims that Lactantius's views on the Earth and the antipodes were based on the Bible, something already implied in Copernicus and which still appears today. More confusion and discussion is found regarding Augustine, and there are by c. 1600 mistaken claims that Augustine was a flat-earther. In some cases, this was prompted by studying texts other than *City of God* 16.9, as with Acosta (1588/1590) and Góis (1593), who cite *De genesi ad litteram* 2.9, which does not seem to have been of concern before the 16th century.[5] There is also the erroneous claim, already in Brahe (1589/1590), that Augustine's denial of the antipodes implied a flat Earth. Other authors took Augustine for a flat-earther by associating Augustine with Lactantius, as in Origanus (1609), Campanella (1616) and Maestlin (c. 1621).[6] More generally, it was possible to conflate the antipodes and flat Earth debates, as in Gómara (1552, mentioning Lactantius) or Maestlin (mixing up Lactantius and Augustine).

Apart from the errors concerning individual authors, a frequent error, which became more common as new authors entered the debate during the 16th century, was the quantification of ancient Christian flat-earthers as "many," which is related to the above list-making trend but was strengthened by the use of unfounded generalisations, as we saw in Chapters 4 and 5, even in those authors who did not display those lists.[7] During the 17th century, the effect of such generalisations was reinforced when a *terminus ad quem* for the flat error was placed as recently as in the 15th century. The key here is Abulensis, also known as Tostatus (his real name was Alonso Fernández de Madrigal), a 15th-century Spanish bishop who denied the antipodes and, as other authors, used the quotation of Psalm 18/19 in Romans 10 as an argument. He appeared – correctly – in lists of authors who rejected the antipodes in Góis (1593) and Wilkins (1638), although strangely, he was the only representative of this view mentioned after those up to the 8th century. A small change in Wilkins's list of the deniers of the antipodes, ending with "...*Abulenſis* with others" (1638, 6) to "*Abulenſis*, together with all thoſe Fathers or other Authors who denied the roundneſſe of the heavens" (Wilkins, 1640 (*Discovery*), 5–6), conflated Abulensis with those who denied the sphericity of the heavens. This new version was reprinted in 1708 and in later editions. Finally, Abulensis was identified – wrongly – as a flat-earther by the Benedictine Bernard de Montfaucon (1706), as we will see below.

In Chapter 5, we also found two further errors: first the alleged idea that belief in the antipodes was a heresy in Christian Late Antiquity and the Middle Ages. The second, more specific, was framing Columbus in a science and Christianity conflict, as in later years the opposition by his contemporaries was wrongly considered to be theologically motivated, and Columbus's discoveries were wrongly framed as anti-theological ideas. Such developments were already seen arising in Castellanos (1589) and Campanella (1622, 1636–1637) and over the next centuries became the centre of the 19th-century flat error.

From the wisdom to the ignorance of the ancients and the barbarism of the medievals

The flat error arose from an argumentative manoeuvre that transformed criticism of past Christian authors, into the misconception that most of them were flat-earthers. From fair and well-grounded criticism of the past, a shift occurred to a misrepresentation and darkening of the past, motivated by the belief that there had been a denial of the sphericity of the Earth. This alleged Christian denial was seen as a *step backwards* in history as, at that point, the majority of pagan ancients (despite their mistakes in other matters) agreed with the moderns. Apart from the specific errors discussed above, we can identify two main lines of thinking behind many of the ideas that led to the flat error, viz., the 'invention' of the Dark Ages and the impact of geographical discoveries.

Indeed, the creation of the flat error was in harmony with the Renaissance and Reformation 'invention' of the medieval Dark Ages. In the early 17th century, the new idea of progress offered not only a proud view of the present and the future (which later became characteristic of the Enlightenment), but also retained from the Renaissance the concept of the Dark Ages. As historian of science Reijer Hooykaas summarised, the last years of the 16th century witnessed "a transition from looking backward to looking forward" (Hooykaas, 1979, 66).

In keeping with my discussion at the end of Chapter 4, it is my contention that the flat error is also, to a large extent, a consequence of the new geographical discoveries which challenged hitherto accepted authorities, in particular Christian authorities. In the late 16th century, while the last echoes of the geographical revolution were fading away,[8] the Copernican revolution started its development – Copernicus then connected the geographical and the astronomical debates (Chapter 5). And it is to this period that we can trace the early roots, even if fragmentary, of the flat error.

Interestingly, this idea is already in Russell and coexists in his work as an undertone with the major theme of the origin of the flat error in the early 19th century. For example: "The idea that 'Columbus showed that the world was round' is an invention. The soil was prepared for the invention by the Humanists and Protestants; its seed was the Copernican controversy" (Russell, 1991, 70).[9] Although giving less details than we provide here, historians Violaine Giacomotto-Charra and Sylvie Nony also suggested a multicausal origin for the flat error: "probably progressively born of the more or less deliberate confusion between the real ecclesiastical opposition to heliocentrism and its cosmological consequences, the fables about the representativeness of Cosmas's opinion, and a distortion of the debate over the antipodes" (Giacomotto-Charra and Nony, 2021, 148).[10]

While we agree with this general view of the historical process, here we show that there is no need to wait until the publication of Cosmas (1706), the Enlightenment or the early 19th century to see the rise of the flat error that can already be found c. 1600. This book's aim has been reconstructing this nearly forgotten complex context, recovering old discussions that lay buried for centuries with deep roots in pre-Copernican discussions on geography. This allows a detailed narrative of the origin and early development of the flat error prior to the publication of Irving's 1828 fictional biography of Columbus and even before the 1706 recovery of Cosmas's flat Earth views or the discussions following Galileo's condemnation in 1633.

The Copernican trajectory to the flat error

One reason for this transformation over nearly two centuries (beginning at least with Cardinal Fillastre in the 1410s) of routine criticisms of Lactantius and Augustine into the aggressive flat error was the theological rejection of heliocentric ideas. This was already a problem for Copernicus himself in the early 1540s and remained a problem for his followers, both Protestants such as Kepler and Catholics such as Galileo. Copernicus's first idea was to place his heliocentrism in a science and

Christianity conflict scenario, in which "babblers" would "distort" the Bible to attack Copernicanism, a practice he rejected. Thus, he reversed the hitherto common practice of using theological ideas or biblical texts to buttress natural philosophical concepts. Copernicus's second idea was to use the example of Lactantius's error as a central warning against theological interference in a new context unrelated to the geographical debates, viz., astronomy. This transfer was easy as these disciplines were not separated at the time, but part of cosmography (see pp. 159–161). As the study of the reception of Lactantius has shown in previous chapters, this was not just a chance idea by Copernicus but an intentional tactic, as Lactantius was very well known to – and resonating with – his contemporaries. By that time, Lactantius was the paradigmatic example of a mistaken Christian who erred in his attacks against learning.[11] Therefore, he was the perfect example for Copernicus's defensive strategy to neutralise theological criticisms, even though Lactantius was hardly a representative Christian theologian.

However, one historical case was not sufficient to trigger the antipathy toward medieval Christianity seen in the flat error. In 1554, the rediscovery of another – truly medieval – example of a science and Christianity conflict became the second case in the path to the flat error: the alleged papal condemnation of Virgilius for defending the antipodes. It took some time for this historical episode to be connected with the Copernican rhetorical defence against theological argumentation. The first cases I know of are an unpublished work by Rothmann in 1589 and a published book by Origanus in 1609 (the earlier precedent in Ramus, 1567, is in a different context; see p. 172 and Table 5.1). Now two examples of theological blunders prompted by overstepping into science became available to Copernicans: one by late ancient Christian authors (Lactantius and Augustine, even if many preferred to 'save' Augustine), and another by medieval 'Catholic' authorities who condemned Virgilius (Archbishop Boniface and Pope Zacharias I).

The dots were starting to be united to form a continuous line of supposed cosmographical conflicts between science and Christianity: the rejection of the sphericity of the Earth and the rejection of the antipodes and antipodeans. These could be separated chronologically (Kepler) or mingled together (Maestlin). It was then time to introduce a third chronological example of conflict by fitting Columbus into that narrative. As discussed at the end of Chapter 5, that came in a favourable time, as by the late 16th century Columbus was starting to rise to the status of science hero. In the Copernican environment of c. 1600, the discoverer of new worlds on the Earth became paired with Copernicus and later with Galileo as pathfinders of new worlds in the sky. This parallel could be exploited in other directions: as Copernicans had to confront theological opposition, the opposition faced by Columbus against his bold navigational plan could be transformed into a theological rejection of the explorer. The first 'Copernican' author who reflects these processes is Campanella (see pp. 174–177). In his *De gentilismo non retinendo* (1636–1637) and *Quaestiones* (1637) the rejection of Columbus is described in religious terms and connected with Galileo's disapproval.[12] While most of the 16th-century authors still had a clear view of the difference between acceptance of the Earth's sphericity and the existence of the antipodes, by c. 1600 this past distinction had become blurred. Earlier in this Chapter several examples of this growing confusion were identified, and although it was not explicitly said that opponents of Columbus were flat-earthers in the texts here studied,[13] this was assumed by Irving's time (1828).

But probably none of this would have triggered the flat error without the 1616 Catholic condemnation of Copernicanism, which led to the consolidation of this line of reasoning. Now Copernicans had to face a new conflict, not from the past but from their own time. This became the climactic example confirming the historical trajectory of science and Christianity conflicts and the need to free science from the hands of ignorant religious people, as Copernicus had suggested. Forgotten was his praise of educated, science-loving theologians, including the Pope himself. The condemnation of Copernican ideas in 1616 was the trigger for Campanella's strong words and the furious language of attack in Maestlin, comparing it with medieval "barbarism." When in 1634, after the inquisitorial condemnation of Galileo, Descartes compared it to the condemnation of Virgilius, he was following a well-trodden path. In this way, we can see how the origin of the flat error became entangled with the science and Christianity debates since the early modern era, suggesting a clue not only to the real reason for its rise but also for its entrenchment in contemporary Western culture.

Other trajectories

There was not only one single form of flat error. What has been described above is one historical trajectory arriving at this false claim which is strongly shaped by Copernicanism. However, an alternative route to this false claim was the misrepresentation of ancient Christians, as mostly flat-earthers, which appeared in the Jesuit Herbert Rosweide (1615) to explain the context of an early Christian text. A more blatant distortion of the role that the flat Earth played among early Christians would appear in the Benedictine Montfaucon (1706), when he provided the context to the first printing of a rediscovered 6th-century Christian text, namely *Christian Topography* by the true flat-earther Cosmas Indicopleustes, a sensation in the Enlightenment.[14]

After reviewing several early Christian authors who were accused of having "abhorred the term sphere",[15] Montfaucon made the startling claim that:

> [...]. The opinion about the flat shape of the earth has been advocated up to the fifteenth century by the majority of theologians, so that Tostatus [Abulensis] himself [...] rejected the spherical argument as though it was not secure in faith and rash.
>
> (Montfaucon, Præfatio in *Cosmæ Ægyptii Topographiam Christianam* 2, 1706, vol. 2, IV)[16]

Abulensis, who had been incorporated in Góis's list of those who rejected the antipodes (1593), is finally and completely transformed into a flat-earther, in a clear example of flat error. It is unlikely that Montfaucon was inspired by Copernican authors. More likely, he may have read theologians such as Senensis or Góis, who did mention Bede and Abulensis as rejecting the antipodes (see p. 133 and Table 4.1), although the text he was introducing (Cosmas Indicopleustes's *Christian Topography*) was chronologically and culturally close to the text introduced by Rosweide (*Vita Sancti Macarii*).

The problem in both Rosweide and Montfaucon was a faulty and careless reconstruction of the cosmographical debates in the ancient and medieval Church. This was perhaps influenced by late 16th-century fellow Catholics such as Senensis, Acosta or Góis, who gathered lists of important early Christian flat-earthers, often from the Antiochian tradition, which could lead readers to assume that such a mistaken view was the norm in the first centuries of Christianity.[17] Only the ignorance – which increased during the 18th century – of the real cosmographical debates of the late 15th century could explain how Montfaucon extended the reign of the flat-earthers up to Abulensis.[18]

Other trajectories to the flat error could be put forward, such as the example of Francis Bacon, a Protestant interested in publicising the new scientific learning but not supporting Copernicus. His approach has similarities to and differences from that of the Copernicans. In his *Novum Organum* (1620, see Appendix, no. 5, available at www.routledge.com/9781032893068) Bacon combined the confusion of denying the antipodeans and the sphericity of the Earth with an historical trajectory of science and religion conflicts beginning in pagan Greece (see pp. 135, 187). Although Kepler did not mention ancient pagan authors in his linear progression in *Astronomia nova* (1609), he did mention in the first volume of his *Epitome* (1618) the religious attack against Aristarchus of Samos by Cleanthes, a famed heliocentric story, obviously not mentioned by the geocentrist Bacon (see Appendix, no. 10).

Moreover, Bacon considered "superstition" and "blind and immoderate zeal of religion" as the root of this historical conflict, while in the *Epitome*, Kepler referred to "irrational religion." An earlier Protestant author, La Popelinière, already noticed in 1582 the point of overzealous theological positions in his criticism of Lactantius and other Christians, also concluding with a critical reflection against his contemporaries (see p. 127). Similarly, Bacon's accusation of "the Greeks" and "the ancient fathers of the Christian church" jumped over the centuries to criticise contemporary "schoolmen" (scholastics). The two texts of Kepler also ended up attacking specifically contemporary critics of Copernicus. These references to ancient Greeks connected with contemporary debates pointed to a conflict beyond Christianity and into the more broad 'science and religion conflict', where the flat earth debate was just one piece among others.

However, if alternative historical trajectories could converge in the flat error, they do not exhaust all historical possibilities, not even for Copernicans. As indicated in p. 166, several Copernicans, both Catholic and Protestant, never mentioned the shape or habitability of the Earth and never mentioned Lactantius, Augustine or Virgilius in this regard. Furthermore, such topics are not present in the strong polemics of such a vocal anti-Christian Copernican as Bruno.[19] On the side of Christian orthodoxy, we saw above that Wilkins, being a Protestant and having available the same information as Maestlin, did not delve into the flat error – even after the

1633 condemnation of Galileo – when he wrote a vigorous defence of Copernicanism in 1638.

The relevance of the flat error in the rise of the idea of an historical conflict between science and Christianity (the 'conflict thesis')

What is described above was not designed in the 16th century to discredit Christianity as such. Some authors such as Acosta were quite explicit about avoiding that, "But we ought not to take it offencively, nor efteeme leffe of the Doctors of the holy Church" (Acosta, *Historia* 1.1, 1590, 15; translated in Acosta, 1604, 4). Indeed, Acosta and most of the authors here studied were clerics or firmly committed Christians.

Furthermore, as we saw in Chapters 3–5, during the 15th and 16th centuries, those discussing the mistakes of the past tried actively to avoid an open theological conflict by 'saving' Augustine. Augustine's error was justified by pointing to his devotion to higher causes, to the limited geographical knowledge in antiquity, and to the misleading influence of leading pagan geographers and scholars of his time, while Lactantius was repudiated as ignorant.

Therefore, although Copernicus placed heliocentrism in the context of a science and Christianity conflict scenario, he did not do so to discredit Christianity, but rather to protect his science from improper theological attacks, which is why he dedicated his book to the Pope. The same intention can be seen among Protestant Copernicans. This was then a debate within Christianity itself striving to assess its own history in relation to the nascent sciences.

The final 'hatching' of the flat error by c. 1600 marked the early beginning of a breakdown in the hitherto peaceful co-existence of science and Christianity, which would become a central argument in 19th-century literature regulated by the 'conflict metaphor'.[20] As we have seen here, the origin of what happened in the 19th century had roots that go back not only to the 'Galileo trial' of 1633, but to older clashes. Those stories were put together c. 1600 to give birth to the flat error in the contexts of the first Copernican debates and the rise of religious confessionalism in Europe.

Paradoxically, although it was from the 17th century onwards that conflicts between nascent scientific ideas and established Christianity began to appear in European culture, the idea of an historical conflict between both was present since the beginning of that century. Whewell was not mistaken when in 1837 he criticised Kepler "and more modern writers" for circulating the false "story of the deposition of Virgilius" (see Chapter 1, p. 11 and endnote 2). The flat error also circulated in the 18th century to be finally incorporated by Irving and Letronne in the early 19th century.

Therefore, whoever wanted to criticise Christianity in the name of science had ready at hand an historical warehouse of examples to use defensively, or even offensively, against theologians interfering with science. Such an historical collection of real and alleged Christian offences to science, with the emergence of the flat error playing the crucial role, provided the foundation for the 'conflict thesis' to describe not only current relations between science and Christianity but also their historical interactions. What we find in the late 19th century with Draper, White and many others is an enlargement of that c. 1600 suggestion, with additional historical examples, in particular the troubles of post-1633 Copernicanism, and many more modern debates, such as contemporary Evolutionism.

If the idea of an old historical conflict between science and Christianity helped to establish the flat error in the early 17th century, during the 19th century the flat error became in turn part of the conflict thesis at large. It was the perfect example of a widespread inescapable science and Christianity conflict and of the final victory of science, with Columbus as the essential epochal character. Columbus was seen as a visionary hero, advanced beyond his time, who single-handed defended the sphericity of the Earth and the existence of antipodes and antipodeans even against clerical views. In this way Columbus became a sort of early anticlerical figure, moulded retrospectively in the light of what some later authors considered to be the figure of Galileo (which is often also wrong!). The debates before 1492 would exemplify the Christian attack on science, while Columbus's arrival in America would illustrate the victory of science and mark the beginning of the victorious historical parade for modern science, as can be seen in the mythification of the figure of Columbus explained in this way by Roland Bernhard, an historian of education who has analysed the figure of Columbus in textbooks:

> [...]. Columbus' voyage is the journey of the mythic *heros* who overcomes the resistance of dark forces, that is, the refusal of malign powers to renounce the notion that the earth is flat. [...]. Columbus' voyage west is his transition into this other world in which challenges await – such as overcoming the fear of falling off the edge of the earth – and his 'discovery of America' a glorious triumph over the 'dark medieval powers'. Columbus returns with a new consciousness, [...] propagating a scientific revolution that transforms the world.
>
> (Bernhard, 2019, 100)

Naturally, such historical reconstruction implied referring to science and Christianity in broad terms and projecting these 'entities' monolithically and anachronistically backwards in time. This has been recently criticised by historian of science Peter Harrison, as we discussed in the introduction.

Worse than that, already in the first decade of the 17th century, the historical perspectives of Kepler and Bacon included elements of the flat error in an historical genealogy of science and Christianity conflicts (real, imaginary and distorted) and gave it strong credibility. This, in turn, gave birth to the idea that medieval Christians were not only wrong but that they deliberately suppressed the ancient knowledge of the spherical Earth in their rejection of all pagan ideas. This criticism claimed that there was a medieval step backwards and, worse, that it was due to barbarism (Maestlin), fanaticism (Kepler, Bacon), etc. In this way, the flat error contributed to the more general anti-medieval folklore developed since the 15th century, which became even more intense after the condemnation of Copernicanism, as historian of science Edward Grant has remarked:

> This unfortunate condemnation not only boomeranged against the Church but also reverberated into the past to produce a negative reaction against the Middle Ages. Those who condemned the Church's action in the seventeenth century as antiscientific and a blow against freedom of thought did not trouble to distinguish between the Church of the seventeenth century and the medieval Church. [...].
>
> (Grant, 2001, 307)

All that could later be used as ammunition against the Catholic side in a polarised Europe. It has to be kept in mind that at c. 1600, there was the time of the so-called 'Black Legend' (*Leyenda Negra*), which formed part of anti-Spanish and anti-Catholic propaganda in the 16th and 17th centuries, demonising the Spanish Empire as barbaric and Catholicism as fanaticism. The flat error fitted well into that context. This strategy has worked until today and was present in the anti-Catholic bias of several late 19th-century founding fathers of the conflict thesis, such as Draper and White (Russell, 1991, 35–46; Ungureanu, 2018; 2019c). Of course, this latter backfired and has been used against all branches of Christianity and even 'religion' in general, as we will discuss at the end of this chapter.

In sum, we can see two argumentative lines among the Copernicans, which led to the idea of a science and Christianity conflict. On the one hand, there are the more or less accurate recollections of old debates surrounding Lactantius and Augustine, which served as a defence strategy in Copernicus (since 1543). On the other hand, this idea was reinforced and expanded with the construction of an historical line of science and Christianity conflicts by Kepler and others (1600s). In this way, a defence strategy for Copernicanism turned into an attack of theological discourses, as we can already see in Maestlin (c. 1621). However, these were partial and distorted polemical reconstructions of the history of the complex relations between Christianity and science. These two argumentative lines can also be found to foster the cause of the new science during the same historical period in a non-Copernican author such as Bacon (1606–1607, 1620). The conflict thesis continued and developed up to the late 19th century. Although 20th-century historians of science and Christianity denounced it (e.g. Reijer Hooykaas, Colin Russell, John Hedley Brooke, David Lindberg, Ronald Numbers, Edward Grant, Peter Harrison, Bernard Lightman, etc.), they also came to regard it as "the idea that wouldn't die" (Hardin et al., 2018). Today it is still very common to hear about the flat error in popular (and even academic) criticisms of Christianity inspired by the conflict thesis.

Generally, the idea of an historical science and Christianity conflict is dated back to the late 19th century, in particular with Draper and White, although this was already clearly put forward by biologist Thomas H. Huxley in the mid-19th century:

> Extinguished theologians lie about the cradle of every science as the strangled snakes beside that of Hercules; and history records that whenever

> science and orthodoxy have been fairly opposed, the latter has been forced to retire from the lists, bleeding and crushed if not annihilated; scorched, if not slain.
>
> (Huxley, 1860, 556)

In a recent book, physicist David Hutchings and historian of science James C. Ungureanu attempt to go backwards in time to find the origin of the conflict thesis and recognise the difficulties, "an odyssey, often of appropriately epic proportions" as the arguments "tend to run backward, through time [...] for the case can always be made for going just one more step further into the past." So, they settle the issue in this way: "Still, a choice must be made, or we will get nowhere. Here, then, is our itinerary: we shall begin in France, at the end of the 1780s" (Hutchings and Ungureanu, 2022, 28). However, what I have shown in this book gives a much earlier reconstruction of events leading to the conflict thesis, from a time when what we call today 'modern science' was appearing; a time when the idea of an open conflict of science with Christianity was unthinkable. And still, the first seeds of conflict were sown in the complex way this research shows.

It is important to emphasise that I do not intend to deny the existence of science and Christianity conflicts (real or contrived), but rather to understand their origins, their development, and how they are perceived and constructed regarding the particular topic discussed in this investigation. The medieval and early modern debate over the antipodes was an actual instance of a 'conflict', although not always between two opposite factions, rather at times an internal conflict, as in the case of Petrarch (see p. 82) or Agricola (see p. 122 and Appendix, no. 17), while the flat error describes an alleged medieval conflict over the shape of the Earth that never took place except in the minds of post-medieval people.[21]

On the other hand, I do not want to overemphasize the conflicts between modern science and Christianity, particularly regarding Copernicanism. It is important to remember that, by and large, the main problems that Copernicus faced were of a scientific nature, as his heliocentric astronomy was of a theoretical nature and did not offer new observational data. This reminds us of Hooykaas's clarification of the difference between the geographical and the astronomical phases of the Scientific Revolution:

> [...]. Whereas (before Copernicus) the seafarers had convinced everybody by *observations* proving that Ptolemy's *geography* was wrong in many respects, Copernicus could not adduce similar proofs that the physical basis of Ptolemy's *astronomy* was also wrong. Copernicus' advocacy was just a great achievement in the *art* of astronomy; it did not add new data to the 'history of nature'.
>
> (Hooykaas, 1987, 465; see also Hooykaas, 1980, 347–349)

It should be noted that both phases had to face not only scientific but also religious objections, as we have seen here.

History as ideological battleground: propaganda and counter-propaganda

By the time the flat error actually appeared c. 1600, the shape of the Earth was no longer a topic of science and Christianity conflict, but purely an intra-Christian discussion on their historical relationships in a way to put forward historical precedents of such an alleged conflict. In fact, the topics that beginning in the 17th century led to clashes between science and Christianity, and later to the conflict thesis, were new issues regarding cosmology, anthropology, our place in the cosmos, a new view of nature and a deistic deity, etc. The flat error was of less actual importance, but it was always ready to be deployed to provide, in Copernican fashion, an historical precedent to warn the theologians from quickly opposing new revolutionary ideas.

By the mid-19th century science and Christianity had become largely independent, at least much more than in the 17th century, with some in the science camp wishing to shake off what they considered the shackles of religion, particularly in the Anglo-Saxon culture. The increasing prestige of science allowed them to create a propagandistic narrative using not only science as opposed to Christian obscurantism, but also the history of science to provide an historical background for a contemporaneous cultural war. Such historical narrative included the extravagant idea that Columbus's contemporaries thought the Earth to be flat, the so-called 'flat error', as we saw in Chapter 1.

Jeffrey Burton Russell's denunciation in 1991 of the flat error has sparked a vigorous reaction against it, including criticisms of its alleged 19th-century originators: Irving and Letronne. As indicated in the Chapter 1, endnote 3, many authors of different religious and ideological convictions are content with Russell's interpretation of the origin of the flat error, pointing at Enlightenment commentators, anticlerical and Darwinist popularisers as originators of the flat error. A non-Christian scholar, such as Umberto Eco, accepted that the flat error is "the story of another false opinion, patiently constructed by modern secular thinking to defame religious thought." (Eco, 2004, 275). The connection of the flat error with Darwinism has been of particular interest for creationists and advocates of the Intelligent Design movement, as can be seen in the case of historian of science Michael Newton Keas (2019, 53–54), who is a senior fellow at Discovery Institute, the leading ID institution. Although Russell himself pointed to the possibility of earlier origins, this lead has not been followed, except for some suggestions discussed in pp. 13–16. Following this lead in detail is precisely the aim of this book.

Among those proposals that have gone beyond the 19th-century origin of the flat error, we mentioned in Chapter 1 at least one Catholic historian – James Hannam – who in 2009 ignored Russell and attributed the flat error to Protestant propaganda against the Catholic Church. In a 2010 blog he expanded the accusation: "The myth that people in the Middle Ages thought the earth is flat appears to date from the 17th century as part of the campaign by Protestants against Catholic teaching. But it gained currency in the 19th century" (Hannam, 2010). This statement has made its way into the Wikipedia entry for the flat error (https://en.wikipedia.org/wiki/Myth_of_the_flat_Earth, note 9). Hannam repeated recently the same connection:

> [...]. The origins of this thesis [referring to the flat error] can be found in the Reformation, when half of Europe rejected the authority of the Catholic Church. To justify the break with Rome, Protestant propaganda portrayed the Church in the Middle Ages as a front of ignorance and superstition. For example, Sir Francis Bacon [...].
>
> (Hannam, 2023, 299)

In this way, the denunciation of the flat error could become an element in confessional polemics today, in the same way as the accusation of flat-earthism was used in confessional polemics in the past. Time will tell if this confessional divide also consolidates in combating the flat error. As usual, things are not simple, and we can find a Protestant, such as Keas, also considering Bacon as the originator of the flat error, and criticising the "anti-Catholic polemics" of Protestants: "In this way Protestants unwittingly prepared the way for antitheists to launch a global assault on Christianity" (Keas, 2019, 56).

However, the historical situation was more complex than just Protestant anti-Catholic propaganda (which, of course, was part of the problem, as discussed above). Both Catholics and Protestants contributed to the early steps of the flat error, and Chapter 3 shows that some of the ideas they used preceded the Protestant Reformation. Although in its final steps it was becoming an increasingly powerful weapon in the hands of Protestants, Catholic clerics (such as Acosta, Rosweide, Foscarini and Campanella) were very much aware of the past conflicts and pointed to them without hesitation. They offered even more details regarding the shortcomings of the ancient and medieval Christians than were known to Protestant authors (in particular Protestants in the natural sciences).[22]

More important than the hostility of Protestants were probably the condemnations of Copernicanism and of Galileo by the authorities of their own Catholic Church. Nevertheless, there are grounds for suggesting that it could have appeared anyway, because, in parallel, there were other trajectories to the flat error, for example in discussions of late ancient Greek Christian texts by the Catholics Rosweide and Montfaucon who wrote with no intention of offending Christianity or the Catholic Church.[23]

It may seem a bit odd that Christians were fostering the idea of science in conflict with their own religious faith. In fact, as we saw, they did not believe that there was an actual conflict with their own Christian views; rather, the conflict was always seen in the 'other' for Protestants, the conflict was to do with Catholicism or medieval clerical authors; for Catholics, the conflict was due to the ignorance of ancient Christian and pagan authors without access to modern knowledge.

Therefore, the original context of the debate around the flat error was not one of anti-Christian conspiracy, but rather Christian self-criticism. The flat error stemmed from internal debates among scientific thinkers and theologians who were all Christians, and it does not appear to have been restricted by confessional or geographic boundaries. For example, Acosta, a Spanish Jesuit discussed openly the mistakes of past Christian authorities in 1588/1590. Another Spanish Catholic cleric, Juan de Castellanos, while describing the traditional Christian opposition to the antipodes, stated that explaining that was an ethical Christian duty: "To conceal such opinions / Would be the sign of a very bad Christian" (Castellanos, *Elegías* 1.2, 1589, 19).[24] However, the Italian Catholic cleric, Campanella, already in 1636, had a glimpse of the danger of airing the mistakes of his coreligionists from the past, affirming that "the shame of the Fathers must be covered over" (see p. 176). In spite of that, the 'moderns' needed to expose the mistakes of past theologians to soften the resistance of their contemporary religious authorities to Copernicanism or to other new learnings that fought for a place in European culture.

Remarkably, as mentioned above, an author strongly critical of Christianity such as Bruno does not seem to have used the flat error, which was just appearing during his lifetime, to advance his heterodox positions. This could be interpreted as reinforcing the view that the flat error was produced inside the Christian milieu, and that it took many years for the critics of Christianity to take advantage of the immense corrosive potential of such idea. Paradoxically, but not surprisingly, the flat error (and other modern misconceptions of the Christian past) ended up being reused by later critics of Christianity and openly anti-Christian authors who would only put together the pieces long crafted by Christians, just adding their own message to them (atheistic, deistic, agnostic, etc.).[25]

A last turn in this backfiring process was precisely the creation of the modern science against Christianity conflict thesis in the late 19th century, which included the flat error among other historical episodes. Ungureanu has shown how Draper and White did not have an anti-Christian agenda but rather belonged to a liberal Protestant trend, current at the time, which sought to criticise conservative religious views considered incompatible with modern science. Again, as with the authors c. 1600 here studied, Draper and White's aim was not an *external* attack on Christianity, but rather an *internal* revolt. Those 19th-century authors intended to expose the negative consequences of the dogma to stress the need to eliminate it in order to achieve both, free science and a purer Christianity.[26] However, this "was a risky strategy that ultimately backfired" (Ungureanu, 2018, 1114), because their views were "appropriated by skeptics, who reshaped them to justify and promote their own specific vision of a progressive and secular society" (Ungureanu, 2018, 1115). I believe that the key problem was not the theological position of each group (conservatism or liberalism), but the willingness to use false stories. This is still today something to be concerned about in our current polarized societies.

This investigation points again to a complex historical process, which reminds us of historian of science John Hedley Brooke's call for historians to respect the complexity of real life:

> Serious scholarship in the history of science has revealed so extraordinarily rich and complex a relationship between science and religion in the past that general theses are difficult to sustain. The real lesson turns out to be the complexity.
>
> (Brooke, 1991, 6)

The concept of 'complexity' has revolutionised the science and religion studies and has been discussed in detail in a volume edited by Bernard Lightman in 2019.[27] What we find in this book could be considered an innovative case-study in the exploration of different forms of complexity in the historical relations between science and Christianity. Instead of a conflict *between* science and Christianity, this historical analysis reveals a conflict *among* Christians regarding certain scientific and historical facts.

It is my hope that a better understanding, not only of the flaws of the flat error, but also of how it was constructed, could help us to enhance our appreciation of that complexity and also of the even more complex ways in which the science and Christianity relationship ends up being depicted and perceived by society. It should be a warning to all students of the interactions between science and Christianity to avoid confessional and factional aggressive attacks, historical mythmaking and apologetic hijacking of history, whether for or against Christianity.

Notes

1 His apologetic methodology was rather the opposite, as he criticised the use of the Bible in debating pagans who did not consider it as an authority (see p. 36).

2 In some cases, they do not really mention the existence of any flat Earth debate among ancient or medieval Christians (Dryander, see pp. 126–127, and Wilkins, see p. 201). In other cases, the references are too specific (Copernicus, see pp. 165–166), obscure (Rosweide, see p. 134) or tenuous (Bacon, see pp. 134–135) for their authors to be accused of being single-handed originators of the flat error. The role of each one in the origin of the flat error needs to be downplayed, even if they are part of such story, in particular Copernicus.

3 Examples are the search for the origins of the spurious idea that scholastics engaged in discussing how many angels could dance on the head of a pin (Harrison, 2016, and references therein) or the legend of the excommunication of Halley's Comet by Pope Callixtus III in 1456 (Stein, 1911, who refers to older research into this story; see also the exchange, not mentioned by Stein, between J.C. Dean and W.F. Rigge from 1908 to 1910 in the pages of Popular Astronomy, available at https://web.archive.org/web/20210505235612/https://dspace2.creighton.edu/xmlui/handle/10504/123061).

4 If the new geographical discoveries were not enough to destroy the ancients's worldview, the 1572 'nova' (a supernova) and the comet in 1577 proved fatal for Aristotelian science. The new star indicated that the heavens were not immutable, and the comet's path showed that it was not an atmospheric phenomenon, implying that it was a changeable heavenly phenomenon and, furthermore, that not all heavenly movements were circular. Other similar observations were the comets of 1585, 1607 and 1618, as well as another supernova in 1604. On these astronomical events, see Hellman (1944), Hooykaas (1987, 465–466), Weichenhan (2004), Granada (2012).

5 According to Acosta, Augustine had doubts about the sphericity of the Earth (see pp. 131–132). See p. 40 for references to the enduring debate on Augustine and the Earth's shape.

6 Campanella's text was published a few years later, 1622. As indicated in p. 187, although Maestlin's text was not published until 1858, it is an example of the ideas among some Copernicans in the early 17th century.

7 From the "others" in Góis (*alij*) to the "many" in Rosweide (*multorum*). Copernicans also used these generalisations, such as Origanus's "many" and "others" (*multi ... alij*). These expressions were also used for lists of those against the antipodes, such as "others" in Senensis (*cœteri*)

8 The intimate connection of the flat error c. 1600 with geographical topics shows that the impact of the explorations had, in some cases, as in this one, a rather quick effect. On the impact of the geographical discoveries in European culture, see pp. 135–140.
9 In pp. 11–12, we have a detailed analysis of the portrait of Copernicus's implication in the flat error according to Russell in his 1991 book; for his later views, see p. 13. Probably, the problem for Russell was that he could not find the authors who contributed to the flat error in the decades after Copernicus, which are here discussed.
10 "Le mythe de la Terre plate est ainsi probablement né peu à peu de la confusion plus ou moins volontaire entre l'opposition ecclésiastique réelle à l'héliocentrisme et à ses conséquences cosmologiques, les affabulations sur la représentativité de l'opinion de Cosmas et un détournement de la querelle sur les antipodes" (my translation). A similar position is stated a few pages before: "The construction of the myth of a medieval belief in a flat Earth was done in fact by small touches and by the skilful association of real facts (transatlantic voyages, heliocentric hypotheses) and equally real debates (the existence inhabitants of the antipodes), to a belief, that of the flat shape, perfectly marginal even within the Christian Church." My translation from: "La construction du mythe d'une croyance médiévale en une Terre plate s'est faite en effet par petites touches et par l'habile association de faits réels (les voyages transatlantiques, les hypothèses héliocentriques) et de débats tout aussi réels (l'existence d'habitants des antipodes), à une croyance, celle de la forme plate, parfaitement marginale même au sein de l'Église chrétienne" (Giacomotto-Charra and Nony, 2021, 143).
11 There is another contemporary author, Antoine du Moulin, who used Lactantius for the same purposes as Copernicus in 1549, showing that the idea was not exceptional (see Chapter 5, endnote 102).
12 Unrelated to astronomy, this idea is already present in Castellanos (1589; see p. 131). Although the opposition to Columbus did not come from a non-existent debate on the shape of the Earth, there was a debate on the antipodes in which theologians (Augustine and Lyra [in fact Burgos]) were mentioned according to Ferdinand Columbus (1571, Augustine only) and Geraldini (1631, both). Even though much importance to this theological opposition has been given since Irving, Columbus himself never referred to it. He often complained of the opposition he faced, but he never complained of *religious* persecution. There were important clerics both in favour and against Columbus (as later with Copernicus and Galileo). Here Campanella is referred to as 'Copernican' because he defended Copernicanism, although he did not accept it (see Chapter 5, endnote 62).
13 Although Castellanos (1589) placed Lactantius against Columbus, he only referred – correctly – to the antipodes debate (see p. 131).
14 The text Rosweide discussed was also a late ancient Greek text from Eastern Christianity, a little earlier (possibly 5th century, see p. 134) than that of Cosmas (mid-6th century, see pp. 48–49). Nevertheless, as they did not explain the context of these cosmologies correctly, in particular that of Cosmas, which was restricted to the Antiochian milieu, they were taken already in the 18th and up to the 20th centuries, as typical of medieval and Renaissance Christianity everywhere (including the Latin West).
15 *ſphæræ nomine abhorrebant*. He listed Lactantius, Augustine, Chrysostom, Severian of Gabala, Bede, "and many others" (*multique alii*), mentioning Basil as someone uninterested in the matter and curiously ignoring Virgilius. I thank Prof. Karla Pollmann (University of Tübingen, Germany) for the translations of Montfaucon.
16 *Sententia vero de plana mundi figura ad decimum-quintum uſque ſæculum a pleriſque Theologorum propugnata eſt; ita ut Toſtatus ipſe, paucis antequam alterum hæmiſphærium adiretur annis, ſphæræ opinionem quaſi in fide non tutam & temerariam rejecerit*. There is a note in the margin of this sentence, referring to Tostatus's *Commentary* on Genesis 1; for his actual ideas, see p. 201.
17 More research is needed to track the influences on Roseweide and Montfaucon regarding their views of the cosmology of early Christians as dominated by flat-earthism.
18 Vogel mentions that the scholastic eccentric model of the Earth was still known to Newton, but forgotten afterwards (Vogel, 1995, 18).

19 I am grateful to Prof. Miguel Ángel Granada (University of Barcelona, Spain) for verifying that Lactantius is not mentioned in Bruno's corpus. Strikingly, when in *La cena de le Ceneri* (1584a) Bruno summarised Copernicus's preface-dedication to the Pope, the portion on Lactantius is not even mentioned. In addition, Prof. Granada pointed out to me that when, at the end of his *De immenso et innumerabilibus* 3.9 (1591), Bruno copies very closely several lines from Copernicus's preface-dedication to the Pope, including most of the text surrounding the reference to Lactantius, he carefully excised the sentence mentioning Lactantius (Latin text in Fiorentino, 1879, vol. 1, 385).

20 This term has been used to describe the late 19th view that science and Christianity are (and have been historically) at war because they are fundamentally incompatible. See Russell (1989). Other similar terms that have been used are "warfare model", "military metaphor", "conflict thesis", "conflict model", etc.

21 For some reflections on the notion of 'conflict' in science and religion, far beyond the narrow negative view of the traditional 'conflict thesis', see Cantor (2010).

22 We should not be surprised to find agreements among authors separated by the deep confessional barrier of the years around 1600. Sometimes Catholics and Protestants converged in some ideas despite their differences. Ashworth (1986) drew attention to the similarities, rather than the differences, between some Catholics and Protestants, on the relation between science and Christianity. Although he focused on the 17th century, his conclusions could also be valid – probably with even more force – for the 16th century.

23 Their mistaken generalisations in the 17th and 18th centuries about early Christians as flat-earthers contrast with the mostly correct criticisms by authors of the 15th and 16th centuries.

24 "Pues tales opiniones encubrillas / Sería de maliſimo chriſtiano" (my translation; on Castellanos, see p. 131). The context is ambiguous, as this is put in the mouths of the critics of Columbus, listing Christian authorities against his navigation plans.

25 An interesting reflection could arise here if we realise that a similar situation was produced, to some degree, by aggressive internal self-criticism among pre-Christian philosophers, which fed Christian anti-philosophical apologetics, as discussed in pp. 34–35, 37. The problem of self-criticism and backfiring is, therefore, an old one, and probably an interesting subject of sociological study.

26 Ungureanu (2018, 2019a, 2019b, 2019c), Peterson (2021), Hutchings and Ungureanu (2022); see p. 182 on the ideal of recovering the 'original' 'purity' of Christianity.

27 In this volume Harrison and Brooke himself have emphasised that the principle of complexity is not a thesis like the 'conflict thesis', but rather an historical methodology (Harrison, 2019 and Brooke, 2019; and much earlier Brooke, 2006).

Bibliography

Authors appear listed alphabetically by surname or by their commonly used Latinised names, such as Aventinus or Vadianus.

Authors called Pseudo- are listed alphabetically under their potentially wrong name, e.g. for Pseudo-Justin, please see under (Pseudo-)Justin.

Primary sources (printed texts, translations and commentaries up to 1800)

Abulensis: see Fernández de Madrigal.

Achillini, Alessandro, *De elementis* (Bologna, 1505).

Acosta, José de, *The natura noui orbis libri duo* (Salamanca, 1588).

Acosta, José de, *Historia natural y moral de las Indias* (Seville, 1590). Spanish translation of Acosta, 1588.

Acosta, Joseph, *The Naturall and Morall Historie of the East and West Indies*, tr. Edward Grimston (London, 1604).

Aethicus Ister, *The Cosmography of Aethicus Ister*, ed. and tr. Michael W. Herren (Turnhout: Brepols, 2011).

Ailly, Pierre d', *Ymago Mundi* (Leuven, 1483).

Ailly, Pierre d', *Ymago Mundi*, ed. Edmond Buron, 3 vols (Paris: Maisonneuve Frères, 1930).

Alighieri, Dante and Giovanni Boccaccio, *Dante Alighieri Quæstio de aqua et terra y Giovanni Boccaccio Trattatello in laude di Dante Alighieri*, tr. and ed. Júlia Benavent (Valencia: Fundación Aquae, 2016).

Anghiera, Peter Martyr d', *Libretto de tutta la navigatione de Re de Spagna de le isole et terreni nuovamente trovati* (Venice, 1504).

Anghiera, Peter Martyr d', *Oceani decas* (Seville, 1511).

Anghiera, Peter Martyr d', *De orbe nouo Decades* (Alcalá de Henares, 1516).

Anghiera, Peter Martyr d', *De Nuper sub D. Carolo Repertis Insulis, simulatque incolarum moribus* (Basel, 1521).

Anghiera, Peter Martyr d', *Opus Epistolarum* (Alcalá de Henares, 1530a).

Anghiera, Peter Martyr d', *De orbe nouo* (Alcalá de Henares, 1530b).

Apian, Peter, *Cosmographicus Liber* (Landshut, 1524).

Apian, Peter, *Cosmographiae Introductio* (Ingolstadt, 1529).

Apian, Peter, *Cosmographia sive Descriptio universi orbis* (Antwerp, 1584).

Aristotle, *The Complete Works of Aristotle: The Revised Oxford Translation.* One volume, ed. Jonathan Barnes (Princeton and Chichester: Princeton University Press, 1995, sixth printing with corrections; 1st ed. 1984).

Athenagoras of Athens, *Athenagoras, Legatio and De Resurrectione*, ed. and tr. William R. Schoedel (Oxford: Oxford University Press, 1972).

Augustine of Hippo, *De civitate dei* (Subiaco, 1467, 1st ed.).

Augustine of Hippo, *De civitate dei* (Strasbourg, 1468, 2nd ed.). This edition included commentaries by Thomas Waleys and Nicholas Trivet.

Augustine of Hippo, *Opera Omnia*, ed. Johannes Amerbach, 11 vols (Basel, 1505–1506).

Augustine of Hippo, *De Civitate dei [...] emendatum [...] per [...] Joan. Lodovicum Vivem [...]*, ed. Juan Luis Vives (Basel, 1522).

Augustine of Hippo, *Opera Omnia*, ed. Erasmus of Rotterdam, 10 vols (Basel, 1528–1529).

Augustine of Hippo, *Qvintvs tomvs Opervm divi Avrelii Avgvstini Hipponensis episcopi, continens XXII Libros de Ciuitate Dei. Cvi accesservnt commentarij Io. Lodo. Viuis ab authore recogniti*, ed. Juan Luis Vives (Paris, 1531).

Augustine of Hippo, *The City of God*, tr. Marcus Dods, 2 vols (Edinburgh: T. &. T. Clark, 1871–1872).

Augustine of Hippo, *Letters*, tr. Sister Wilfrid Parsons, 5 vols (Washington, DC: The Catholic University of America Press, 1951–1956).

Augustine of Hippo, *City of God against the Pagans*, tr. Robert W. Dyson (Cambridge: Cambridge University Press, 1998).

Aventinus (Johannes Turmair), *Annalivm Boiorvm libri septem* (Ingolstadt, 1554).

Azurara, Gomes Eanes de: see Zurara.

Bacon, Francis, *Of the Proficience and Advancement of Learning, Divine and Humane* (London, 1605).

Bacon, Francis, *Instauratio magna* (London, 1620).

Bacon, Francis, *Sylva Sylvarum or A Naturall History in Ten Centuries* (London, 1627; 1st ed. 1626).

Bacon, Francis, *Francisci Baconi de Verulamio Scripta in Naturali et universali Philosophiâ*, ed. Isaac Grunter (Amsterdam, 1653).

Bacon, Francis, *Letters and Remains of the Lord Chancellor Bacon*, ed. R. Stephen (London, 1734).

Bacon, Francis, *The Works of Francis Bacon*, eds. James Spedding, Robert Leslie Ellis and Douglas Denon Heath, 7 vols (London: Longman & Co., 1857–1859).

Barbaro, Ermolao and Giovanni Pico della Mirandola, *Filosofia o eloquenza?*, ed. Francesco Bausi (Naples: Liguori, 1998).

Barnes, Robert, *Vitae romanorum pontificum* (Wittemberg, 1536).

Barros, João de, *Rópica pnefma* (Lisbon, 1532).

Barros, João de, *Asia*, 3 vols (Lisbon, 1552–1563). João Baptista Lavanha used Barros's manuscript to publish a fourth volume in Madrid, 1613.

Basurto (or Vasurto), Rodrigo, *De Natura loci et temporis* (Salamanca, 1494).

Bede, Venerable, *Bede: The Reckoning of Time*, tr. Faith Wallis (Liverpool: Liverpool University Press, 1999).

Bembo, Pietro, *Historia Veneta* (Venice, 1551; Paris, 1551).

Bernaldez, Andrés, *Historia de los Reyes Católicos D. Fernando y Dª. Isabel. Crónica inédita del siglo XV, escrita por el Bachiller Andrés Bernaldez*, ed. Miguel Lafuente y Alcántara, 2 vols (Granada, 1856).

Bodin, Jean, *Method for the Easy Comprehension of History*, tr. Beatrice Reynolds (New York: Columbia University Press, 1945).

Bodin, Jean, *Methodus, ad facilem historiarum cognitionem* (Paris, 1566).

Bonincontri, Lorenzo, *In Manilium commentum* (Rome, 1484).

Borro, Gasparino, *Commentum electum et pretiosum super tractatum sphaerae mundi* (Venice, c. 1500).

Bracciolini, Poggio and Ludovico de Varthema, *Travelers in Disguise: Narratives of Eastern Travel*, tr. John Winter Jones (Cambridge, MA: Harvard University Press, 1963).
Brahe, Tycho, *Epistolarum astronomicarum libri quorum primus hic*... (Uraniborg [Hven], 1596).
Brant, Sebastian, *Daß Narrenschyff ad Narragoniam* (Basel, 1494).
Bruno, Giordano, *La cena de le Ceneri* (London, 1584a).
Bruno, Giordano, *De la causa, principio et Uno* (London, 1584b).
Bruno, Giordano, *De l'infinito universo e Mondi* (London, 1584c).
Bruno, Giordano, *De Immenso et Innumerabilibus* (Frankfurt, 1591).
Buridan, Jean, *Quaestiones super libris quattuor De caelo et mundo*, ed. Ernest Addison Moody (New York: Kraus Reprint Co., 1970, reprinted; 1st ed. Cambridge, MA: The Medieval Academy of America, 1942).
Cajetan, Thomas, *Evangelia cvm commen. Caietani* (Venice, 1530).
Camões, Luís de, *Os Lusiadas* (Lisbon, 1572).
Campanella, Thomas, *Apologia pro Galileo* (Frankfurt, 1622).
Campanella, Thomas, *Realis philosophiae epilogisticae partes quatuor* (Frankfurt, 1623).
Campanella, Thomas, *De gentilismo non retinendo* (Paris, 1636).
Campanella, Thomas, *Disputationum in quatuor partes suae philosophiae realis libri quatuor* (Paris, 1637).
Campanella, Thomas, *A Defense of Galileo*, tr. Richard J. Blackwell (Notre Dame and London: University of Notre Dame Press, 1994).
Cartier, Jacques, *Brief recit, & succincte narration, de la navigation faicte es ysles de Canada, etc.* (Paris, 1545).
Cassius, Dio, *Roman History*, tr. Earnest Cary, Herbert Baldwin Foster, 9 vols (Cambridge, MA: Harvard University Press, 1914–1927).
Castellanos, Juan de, *Elegias de varones illvstres de Indias* (Madrid, 1589).
Castro, João de, *Roteiro de Dom Ioam de Castro da viagem que fizeram os portuguezes ao Mar Roxo no anno de 1541*, ed. Antonio Nunes de Carvalho (Paris: Baudry and Barrois, 1833).
Castro, João de, *Tratado da esfera. Da Geografia. Notação famosa. Informaçao sobre Maluco*, ed. A. Fontoura da Costa (Lisbon: Agência Geral das Colónias, 1940).
Castro, João de, *Obras completas de D. João de Castro*, eds. Armando Cortesão and Luís de Albuquerque, 4 vols (Coimbra: Academia Internacional da Cultura Portuguesa, 1968–1980). Volume 1 contains several texts:
a. *Tratado da Esfera por Preguntas e Respostas* (pp. 15–108).
a1. *Da Geographia por modo de dialogo* (pp. 109–114). This text is part of the *Tratado da Esfera*
b. *Roteiro de Lisboa a Goa* (pp. 115–279).
Celaya, Juan de, *Expositio magistri ioannis de Celaya Valentini in quattuor libros de celo & mundo Aristotelis: cum questionibus eiusdem* (Paris, 1518).
Céspedes, García de, *Regumento de navegación* (Madrid, 1606).
Cicero, Marcus Tullius, *Cicero, On the Commonwealth and On the Laws*, ed. and tr. James E. G. Zetzel (Cambridge: Cambridge University Press, 1999).
Cicero, Marcus Tullius, *Cicero: On Academic Scepticism*, tr. Charles Brittain (Indianapolis, IN and Cambridge, MA: Hackett Publishing Company, Inc., 2006).
Ciruelo, Pedro, *Uberrimum sphaere mundi commentum intersertis etiam questionibus domini Petri de Aliaco* (Paris, 1498).
Clavius, Christophorus, *In Sphaeram Ioannis de Sacrobosco commentarius* (Rome, 1570).
Clement of Rome, 'I Clement', in *The Apostolic Fathers with an English Translation*, tr. Kirsopp Lake, vol. 1 (Cambridge, MA: Harvard University Press and London: William Heinemann LTD., 1965, reprinted; 1st ed. 1912).

Cleomedes, 'De motu circulari corporum caelestium', in *Cleomedis de contemplatione orbiu excelsorum disputatio...*, ed. Carolus Valgulius (Brescia, 1497).

Cleomedes, *Cleomedes' Lectures on Astronomy: A Translation of The Heavens with an Introduction and Commentary*, tr. Alan C. Bowen and Robert B. Todd (Berkeley, Los Angeles and London: University of California Press, 2004).

Columbus, Christopher, *Select Letters of Christopher Columbus*, tr. and ed. R. M. Major (London: Hakluyt Society, 1870, 2nd revised ed.; 1st ed. 1847).

Columbus, Christopher, *Scritti di Cristoforo Colombo*, ed. Cesare de Lollis, 2 vols (Rome: Ministero della Pubblica Istruzione, 1894).

Columbus, Christopher, *Select Documents Illustrating the Four Voyages of Columbus*, ed. Lionel Cecil Jane, 2 vols (London: Hakluyt Society, 1930 and 1933).

Columbus, Christopher, *The Book of Prophecies Edited by Christopher Columbus*, ed. Roberto Rusconi, tr. Blair Sullivan (Berkeley, Los Angeles and London: University of California Press, 1997).

Columbus, Ferdinand, *Historie del S. D. Fernando Colombo; nelle quali s'ha particolare, & vera relatione della vita, & de' fatti dell'Ammiraglio D. Christoforo Colombo, suo padre* (Venice, 1571).

Columbus, Ferdinand, *The Life of the Admiral Christopher Columbus by His Son Ferdinand*, tr. Benjamin Keen (New Brunswick, NJ: Rutgers University Press, 1992, 2nd revised ed.; 1st ed. 1959).

Copernic, Nicolas, *De revolutionibus orbium coelestium / Des révolutions des orbes célestes*, eds. and trs. Michel-Pierre Lerner, Alain-Philippe Segonds and Jean-Pierre Verdet, with the collaboration of Concetta Luna, Isabelle Pantin, Denis Savoie and Michel Toulmonde, 3 vols (Paris: Les Belles Lettres, 2015).

Copernicus, Nicholas, *De revolutionibus orbium coelestium, libri VI* (Nuremberg, 1543).

Copernicus, Nicholas, *On the Revolutions*, tr. Edward Rosen, 2 vols (London and Basingstoke: Macmillan, 1978).

Copernicus, Nicholas, *Minor Works*, ed. Paweł Czartoryski, tr. Edward Rosen with the assistance of Erna Hilfstein (Warsaw, Cracow: Państwowe Wydawnictwo Naukowe, 1985). This is vol. 3 of Nicholas Copernicus Complete Works.

Copernicus, Nicholas and Georg Joachim Rheticus, *Three Copernican Treatises: The Commentariolus of Copernicus, the Letter against Werner, the Narratio prima* of Rheticus, tr. Edward Rosen (New York: Octabon Books, 1971, 3rd revised ed.; 1st ed. 1939).

Cortés, Martín, *Breve compendio de la sphera y de la arte de navegar* (Sevilla, 1551).

Digges, Thomas, *A Perfit Description of the Caelestiall Orbes* (London, 1576).

Dodoens, Rembert, *Cosmographica in astronomiam et geographiam isagoge* (Antwerp, 1548).

Dullaert, Jean, *Habes humanissime lector librorum metheororum Aristotelis...* (Paris, 1514).

Eden, Richard, *A Treatyse of the Newe India* (London, 1553).

Eden, Richard, *The Decades of the Newe Worlde or West India* (London, 1555).

Empiricus, Sextus, *Sexti Philosophi Pyrrhoniarum hypotyposeon libri III* (Geneva, 1562).

Empiricus, Sextus, *Sexti Empirici Opera quæ extant* (Geneva, 1621).

Enciso, Martín Fernández de, *Suma de Geographia* (Seville, 1519).

Enciso, Martín Fernández de, *A Briefe Description of the Portes, Creekes, Bayes, and Hauens, of the Weast India* (London, 1578).

Erasmus, Desiderius, *Adagiorum chiliades tres* (Venice, 1508).

Erasmus, Desiderius, *Omnia quam antehac emendatiora* (Basel, 1533).

Erasmus, Desiderius, *The Correspondence of Erasmus: Letters 446 to 593. 1516 to 1517*, tr. R. A. B. Mynors and D. F. S. Thomson, notes by J. K. McConica (Toronto and Buffalo: University of Toronto Press, 1977).

Faleiro, Francisco, *Tratado del esphera y del arte del marear* (Seville, 1535).

Fernandes, Valentim, in Marcondes de Souza, Thomaz Oscar, 'O ato notarial de Valentim Fernandes de 20 de Maio de 1503', *Revista de História* 16 (1958), 369–378.

Fernández de Madrigal, Alonso (Tostatus or Abulensis), *Opera omnia* (Venice, 1596).

Fernel, Jean, *Cosmotheoria* (Paris, 1528).

Ferrariis, Antonius de (Galateus), *Liber de situ elementorum* (Basel, 1558).

Filelfo, Francesco, *Francisci Philelfi Epistolarum liber primus* (Venice, 1473).

Filelfo, Francesco, *Francisci Philelfi Satyrarum Hecatostichon decades* (Milan, 1476).

Filelfo, Francesco, *Franciscus Philelphus. Epistolarum familiarium libri XXXVII* (Venice, 1502).

Filelfo, Francesco, *Francisci Philelfi satyrarum* (Venice, 1502).

Filelfo, Francesco and Francesco Sforza, Critical Edition *of Filelfo's* Sphortias, De Genuensium deditione, Oratio parentalis, *and his Polemical Exchange with Galeotto Marzio*, ed. Jeroen De Keyser (Hildesheim: Georg Olms Verlag, 2015).

Fiorentino, Francesco, ed., *Jordani Bruni Nolani opera latine conscripta publicis sumptibus edita*, 3 vols (Naples: D. Morano 1879–1891).

Flaccius Illyricus, Matthias, *Catalogus testium veritatis* (Basel, 1556).

Flaccius Illyricus, Matthias, *Catalogus testium veritatis* (Basel, 1562).

Flaccius Illyricus, Matthias, Johann Wigand and Matthaeus Richter, *Octava Centuria ecclesiasticae historiae* (Basel, 1564). This was part of the famed Ecclesiastical History known as *The Magdeburg Centuries* (Basel, 1559–1574).

Foresti da Bergamo, Giacomo Filippo, *Novissime hystoriarum omnium repercussiones* (Venice, 1503; 1st ed. Venice, 1483).

Foscarini, Paolo Antonio, *Lettera sopra l'opinione de' Pittagorici, e del Copernico* (Naples, 1615).

Frisius, Gemma, *De Principiis* (Antwerp, 1530).

Galilei, Galileo, *Sidereus nuncius* (Venice, 1610).

Galilei, Galileo, *Nov-antiqua sanctissimorum patrum, & probatorum theologorum doctrina, De Sacrae Sripturae testimoniis, in conclusionibus mere naturalibus...* (Stragsburg, 1636).

Galilei, Galileo, *Carta a Cristina de Lorena y otros textos sobre ciencia y religión* (Madrid: Alianza Editorial, 1987).

Geraldini, Alessandro, *Itinerarivm ad regiones sub æqvinoctiali plaga constitutas*, ed. Onofrio Geraldini de Catenaciis (Rome, 1631).

Geraldini, Alessandro, *Alejandro Geraldini, Periplo hasta las regiones ubicadas al sur del equinocio*, tr. Carmen González Vázquez and ed. Jesús Paniagua (León: Universidad de León, 2009).

Gilbert, William, *De magnete* (London, 1600).

Gilbert, William, *On the Magnet, Magnetick Bodies Also, and on the Great Magnet the Earth: A New Physiology, Demonstrated by Many Arguments & Experiments*, tr. Silvanus P. Thompson (London: Chiswick Press, 1900).

Giles, Peter, 'Prefatory Letter from Peter Giles to Jerome Busleyden' (1 November 1516), in More, Thomas, *Utopia* (Lovain, 1516).

Giustiniani, Augustino, *Psalterium, Hebraeum, Graecum, Arabicum & Chaldaeum, cum tribus latinis interpretationibus & glossis* (Genoa, 1516).

Glogoviensis, Johannes, *Introductorium compendiosum in Tractatum Spherae Materialis magistri Joannis de Sacrobosco* (Cracow, 1506).

Góis, Damião de, *Chronica do Principe Dom Ioam, Rei que foi destes Regnos segundo do nombre...* (Lisbon, 1567).

Góis, Manuel de, *Commentarii collegii conimbricensis societatis iesu in quattuor libros de coelo Aristotelis Stagiritae* (Lisbon, 1593).

Gómara, Francisco López de, *Primera y segunda parte de la historia general de las Indias con todo el descubrimiento y cosas notables que han acaecido dende que se ganaron ata el año de 1551. Con la conquista de México y de la nueva España* (Zaragoza, 1552).

Gomes, Diogo, 'De prima inventione Guineae', in Schmeller, J. A., 'Ueber Valentim Fernandez Alemão und seine Sammlung von Nachrichten über die Entdeckungen und Besitzungen der Portugiesen in Afrika und Asien bis zum Jahre 1508, enthalten in einer gleichseitigen portugiesischen Handschrift der köngl. Hof- und Staats-Bibliothek zu München', *Abhandlungen der Philosophisch-Philologischen Classe der Königlich-Bayerischen Akademie der Wissenschaften*, vol. 4, part. 3 (1847), 18–34.

Gorlaeus, David (David van Goorle) and Georg Joachim Rheticus (Georg Joachim von Lauchen), *Davidis Gorlaei Ultrajectini, idea physicae: cui adjuncta est epistola cuiusdam anonymi de terrae motu* (Utrecht, 1651).

Goulart, Simon, *Catalogus testium veritatis, qui ante nostram aetatem Pontifici Romano atque papismi erroribus reclamarunt* (Lyon [Geneva], 1597), 3rd ed.; for the previous two, see above under Flaccius Illyricus, Matthias.

Grynaeus, Simon and Johann Huttich, *Novus orbis regionum ac insularum veteribus incognitarum* (Basel, 1532).

Guicciardini, Francesco, *L'Historia d'Italia* (Florence, 1561). This edition only contained the first sixteen books; the final four books were first published in Venice, 1564.

Guicciardini, Francesco, *Della istoria d'Itala*, 4 vols (Fribourg [in fact, Florence], 1775–1776).

Guicciardini, Francesco, *The History of Italy by Francesco Guicciardini*, tr. Sidney Alexander (Princeton: Princeton University Press, 1984, reprinted; 1st ed. 1969).

Herodotus of Halicarnassus, *The History of Herodotus*, tr. George Campbell Macaulay, 2 vols (London and New York: Macmillan and Co., 1890).

Indagine, Jean de, *Chiromantia, physiognomia, ex aspectu membrorum hominis periasiomata, de faciebus signorum…* (Strasbourg, 1531).

Indagine, Jean de, *Chiromance & Physiognomie* (Lyon, 1549).

Isidore of Seville, *Etymologiae* (Augsburg, 1472).

Isidore of Seville, *The Etymologies of Isidore of Seville*, tr. Stephen A. Barney, W. J. Lewis, J. A. Beach and Oliver Berghof (Cambridge: Cambridge University Press, 2006).

Kepler, Johannes, *Prodromus dissertationum cosmographicarum, continens mysterium cosmographicum...* (Tübingen, 1596).

Kepler, Johannes, *Astronomia nova* (Prague, 1609).

Kepler, Johannes, *Dissertatio cum Nuncio sidereo* (Prague, 1610).

Kepler, Johannes, *Epitome astronomiae Copernicanae*, 2 vols (Linz, 1618–1620).

Kepler, Johannes, *Prodromus dissertationum cosmographicarum, continens mysterium cosmographicum...* (Frankfurt, 1621). Second edition of Kepler, 1596.

Kepler, Johannes, *Johannis Kepleri Opera omnia*, ed. Chistian Frisch, 8 vols (Frankfurt and Erlangen: Heyder & Zimmer, 1858–1871).

Kepler, Johannes, *The Birth of History and Philosophy of Science: Kepler's* A Defence of Tycho against Ursus *with Essays on Its Provenance and Significance*, ed. N. Jardine (Cambridge: Cambridge University Press, 1984).

Kepler, Johannes, *Johannes Kepler: New Astronomy*, tr. William H. Donahue (Cambridge: Cambridge University Press, 1992).

Kepler, Johannes, *Selections from Kepler's Astronomia Nova*, tr. William H. Donahue (Santa Fe, NM: Green Lion Press, 2004).

Kepler, Johannes, *Astronomia Nova: New Revised Edition*, tr. William H. Donahue (Santa Fe, NM: Green Lion Press, 2015).

La Croze, Maturin Veyssière, *Histoire du Christianisme des Indes* (La Haye, 1724).

Lactantius, Lucius Caecilius Firmianus, *Opera* (Subiaco, 1465; 1st printed edition).

Lactantius, Lucius Caecilius Firmianus, *Opera* (Rome, 1468; 2nd printed edition).

Lactantius, Lucius Caecilius Firmianus, *Divinarum institutionum Libri VII. De ira Dei, Liber I. De opificio Dei, Liber I. Epitome in libros suos liber acephalos. Phoenix. Carmen de dominica resurrectione. Carmen de passione Domini* (Basel, 1521).

Lactantius, Lucius Caecilius Firmianus, *The Works of Lactantius*, tr. William Fletcher [Ante-Nicene Christian Library, vol. 21], vol. I (Edinburgh: T. & T. Clark, 1871).

Lactantius, Lucius Caecilius Firmianus, *Divine Institutes*, tr. Anthony Bowen and Peter Garnsey (Liverpool: Liverpool University Press, 2003).

Lansberge, Philippe van, *Progymnasmatum astronomiae restitutae liber I. de motu soils* (Middelburg, 1619).

La Popelinière, Henri Lancelot Voisin de, *Les trois mondes* (Paris, 1582).

La Sale, Antoine de, *La Salade, nouvellement imprimee, Laquelle fait mension de tous les pays du monde Et du pays de la Sybille, avec la figure pour aller au mont de la belle Sibille Et aussi la figure de la mer et de la terre et plusieurs belles remonstrances* (Paris, 1521).

Las Casas, Bartolomé de, *Historia de las Indias*, eds. Marqués de la Fuensanta del Valle y José Sancho Rayon, 5 vols (Madrid: Miguel Ginesta, 1875–1876).

Las Casas, Bartolomé de, *Historia de las Indias*, ed. André Saint-Lu, 3 vols (Caracas: Biblioteca Ayacucho, 1986).

Lax, Gaspar, *Quaestiones physicales* (Zaragoza, 1527).

Leurechon, Jean, *Selectae propositiones in tota sparsim mathematica pulcherrimae quas in solemni festo sanctorum Ignatii et Xaverii et anniversaria collegii Mussipontani celebritate literaria propugnabunt auditors* (Pont-à-Mousson, 1622).

Lilio, Zaccaria, *Orbis breviarum compendium* (Florence, 1493).

Lilio, Zaccaria, 'Contra Antipodes', in *In hoc volumine continentur hi libri...*, ed. Zaccaria Lilio (Florence, 1496).

López de Velasco, Juan, *Geografía y descripción universal de las Indias*, ed. Justo Zaragoza (Madrid: Establecimiento Topográfico de Fortanet, 1894).

Lucena, Vasco Fernandes de, *Valasci Ferdinandi utriusque iurisconsulti illustrissimi Regis Portugallie oratoris ad Innocentium octavum pontificem maximum de obedientia oratio* (Rome, c. 1485).

Lucena, Vasco Fernandes de, *The Obedience of a King of Portugal*, tr. Francis M. Rogers (Minneapolis, MN: University of Minnesota Press, 1958).

Luis, António, *Panegyrica oratio elegantissima plurima rerum & historiarum copia referta Ioanni huius nominis tertio invictissimo Lusitaniarum regi nuncupata* (Lisbon, 1539).

Lyra, Nicholas, *Postilla super totam Bibliam*, 5 vols (Rome, 1471–1472).

Lyra, Nicholas, Paul of Burgos and Matthias Döring, *Postillae Nicolai de Lyra super totam bibliam cum additionibus Pauli Burgensis et replicis Matthiae Doringk* (Nuremberg, 1481, and Venice, 1481).

Macrobius, Ambrosius Theodosius, *Commentary on the Dream of Scipio*, tr. William Harris Stahl (New York: Columbia University Press, 1990; 1st ed. 1952).

Major, John, *In Secundum librum Sententiarum* (Paris, 1519).

Mandeville, John, *The Travels of Sir John Mandeville*, tr. Charles Moseley (London: Penguin, 2005, 2nd revised ed.; 1st ed. 1983).

Manegold of Lautenbach, *Liber contra Wolfelmum*, tr. Robert Ziomkowski (Paris, Leuven and Dudley: Peeters, 2002).

Mantua, Baptistus, *De penitencia* (Deventer, 1501).

Margalho, Pedro, *Phisices Compendium* (Salamanca, 1520).

Marzio da Narni, Galeotto, *De doctrina promiscua liber* (Florence, 1548).
Marzio da Narni, Galeotto, *Galeottus Martius. Invectivae in Franciscum Philelphum*, ed. László Juhász (Liepzig: B.G. Teubner, 1932).
Marzio da Narni, Galeotto, *Quel che i più non sanno (De incognitis vulgo)*, tr. Mario Frezza (Naples: R. Pironti & Figli, 1948).
Maurolico, Francesco, *Cosmographia in tres dialogos distincta* (Venice, 1543).
Maurolico, Francesco, *Theodosii sphaericorvm elementorvm libri III* (Messina, 1558).
Mazzoni, Jacopo, *Della difesa della Comedia di Dante. Distinta in sette libri*, vol. 1 (Cesena, 1587).
Medina, Pedro de, *Coloquio de Cosmographia* (1543). Manuscript at Yale University Library, Beinecke MS 559; available at: https://collections.library.yale.edu/catalog/17368482.
Medina, Pedro de, *Arte de Nauegar* (Valladolid, 1545).
Medina, Pedro de, *Regimiento de navegacion* (Sevilla, 1552).
Medina, Pedro de, *A Navigator's Universe; the Libro de cosmographia of 1538*, ed. and tr. Ursula Lamb (Chicago: The University of Chicago Press, 1972).
Medina, Pedro de, *Suma de cosmographia* (1561), Manuscript at Seville Cathedral Chapter. Colombina Library. MS 84-8-44. Published as *Svuma de cosmographía*, ed. Juan Fernández Jiménez (Valencia: Albratos Hispanofila, 1980).
Mejía, Pedro, *Silua de varia lecion* (Seville, 1540).
Mejía, Pedro, *Coloquios o dialogos* (Seville, 1547).
Mela, Pomponius, *Cosmographia Pomponii Mele*, ed. Johannes Cochlaeus (Nuremberg, 1512).
Mela, Pomponius, *Pomponii Melae Hispani, Libri de situ orbis tres, adiectis Ioachimi Vadiani Heluetii in eosdem Scholiis...*, ed. Joachimus Vadianus (Vienna, 1518).
Mela, Pomponius, *Pomponii Melae De orbis sitv libri tres, accvratisime emendati...*, ed. Joachimus Vadianus (Basel, 1522).
Melanchthon, Philipp, *Elementorum rhetorices libri duo* (Wittenberg, 1531).
Melanchthon, Philipp, *Elementorum rhetorices libri duo* (Strasbourg, 1539).
Mersenne, Marin, *Quaestiones Celeberrimae in Genesim* (Paris, 1623).
Mesmes, Jean Pierre de, *Les institutions astronomiques contenans les principaux fondements et premières causes des cours et mouvemens celestes* (Paris, 1557).
Migne, Jacques-Paul, ed., *Patrologiae cursus completus* [Patrologia latina], 221 vols (Paris: Migne, 1844–1855).
Monachus, Franciscus, *De orbis situ* (Antwerp, c. 1526–1529). Mistakenly dated as 1524.
Monachus, Franciscus, *De orbis situ* (Antwerp, c. 1565).
Montaigne, Michel de, *Essais*, 2 vols (Bourdeaux, 1580).
Montaigne, Michel de, *Essais* (Paris, 1588).
Montaigne, Michel de, *Essais* (Paris, 1595).
Montaigne, Michaell de, *The Essayes*, tr. John Florio (London, 1603).
Montfaucon, Bernardus de, ed., *Collectio nova patrum et scriptorum graecorum*, 2 vols (Paris, 1706).
Müller of Königsberg, Johann (Regiomontanus), *[Manilius's] Astronomicon* (Nuremberg, 1473).
Münster, Sebastian, *Geographia universalis vetus et nova* (Basel, 1540).
Münzer, Hieronymus, *Letter to King João II of Portugal* (14 July 1493a). Published in *Regimento do eſtrolabio 7 do quadzante…* (Lisbon, c. 1509), known as "Regimento of Munich", it was reproduced as facsimile in Bensaúde, 1914, 60–62. It was also published as *Tractado da Spera do mũdo tirada de latim em lingoagẽ poztugues Com hũa carta que huũ grãde doutor Alemam mandou a elRey de Portugall dom Ioam ho segundo. Seguefe ho*

regimento da declinaçam do sol pera per ella ſaber ho mareãte em qual parte eſta. ſ. aquem ou dalem da linea equinocial. Com ho regimento da eſtrela do norte (Lisbon, c. 1516). This text is known as "Regimento of Evora" and is reproduced with some modernised spelling by Harrisse, 1892, 393–395. Only a fragment of the beginning of the original Latin text has been found and published in Grauert, 1908, 315–319. I have not had the chance to read this paper, but I have seen Grauert's text as reproduced by Vignaud, 1911, vol. 2, 620–621.

Münzer, Hieronymus, *De inventione Africæ maritimæ et occidentalis videlicet Geneæ per infantem Heinricum Portugalliæ* (23 November 1493b). Published in Vasconcelos, 1932. Vasconcelos indicates on p. 172 that the Latin text is from its first publication by Kunstmann, 1854, 348–362.

Nebrija, Antonio de, *In Cosmographiae libros introductorium* (Salamanca, c. 1503).

Nebrija, Antonio de, *Rerum a Fernando et Elisabe Hispaniarum felicissimis regibus gestarum decades duae* (Granada, 1550).

Neckam, Alexander, *Alexandri Neckam De Naturis Rerum Libri Duo*, ed. Thomas Wright (London: Longman, Green, Longman, Roberts and Green, 1863).

Nunes, Pedro, *Tratado da Sphera* (Lisbon, 1537). Contains five texts:

a *Tratado da Sphera* (annotated translations of Sacrobosco's *Tractatus de Sphaera*).

b *Theorica do So e da Lua* (chapters on the Sun and Moon from Peuerbach's *Theoricae novae planetarum*).

c *Livro primeiro da Geographia de Claudio Ptolomeu* (first book of Ptolemy's *Geographia*).

d *Dous tratados [...] sobre a carta de marear* (two original works by Nunes):

d1 *Tratado que ho doutor Pero nunez fez sobre certas duuidas da nauegação.*

d2 *Tratado que ho doutor Pero nunez (...) em defensam da carta de marear.*

Nunes, Pedro, *Astronomici introductorii de spaera epitome per* Petrum Nonium *Salaciensem* (neither place nor date of publication, c. 1537–1542). Published in Guimarães, 1916.

Oresme, Nicolas, *Le Livre du ciel et du monde*, eds. Albert D. Menut and Alexander J. Denomy, tr. Albert D. Menut (Madison, WI: University of Wisconsin Press, 1968).

Origanus, David, *Novae motuum coelestium ephemerides Brandenburgicae* (Frankfurt on the Oder, 1609).

Origen of Alexandria, *On First Principles*, tr. John Behr, 2 vols (Oxford: Oxford University Press, 2017).

Orta, Garcia da, *Colóquios dos simples e drogas e Cousas Medicinais da Índia* (Goa, 1563).

Orta, Garcia da, *Colloquies on the Simples and Drugs of India*, tr. Clements R. Markham (London: Henry Sotheran and Co., 1913).

Oviedo, Gonzalo Fernández de, *Ouiedo de la natural hystoria de las Indias* (Toledo, 1526).

Oviedo, Gonzalo Fernández de, *La historia general delas Indias* (Seville, 1535).

Oviedo, Gonzalo Fernández de, *Historia general y natural de las Indias*, ed. José Amador de los Ríos, 4 vols (Madrid: Real Academia de la Historia, 1851-1855).

Pacheco Pereira, Duarte, *Esmeraldo de situ orbis por Duarte Pacheco Pereira*, ed. Raphael Eduardo de Azevedo Basto (Lisbon: Imprensa Nacional, 1892).

Pacheco Pereira, Duarte, *Esmeraldo de situ orbis by Duarte Pacheco Pereira*, tr. George H. T. Kimble (London: The Hakluyt Society, 1937).

Palacios Rubios, Juan López de and Matías de Paz, *De las Islas del mar Océano* and *Del dominio de los Reyes de España sobre los indios*, intro. Silvio Zavala, tr. Agustín Millares Carlo (México and Buenos Aires: Fondo de Cultura Económica, 1954).

Palacios Rubios, Juan López de, *De las Islas del mar Océano* (Libellus de insulis oceanis), tr. Paulino Castañeda Delgado, José Carlos Martín de la Hoz and Eduardo Fernández (Pamplona: Ediciones Universiadd de Navarra, S.A.-EUNSA, 2013).

Paschalici, Petri, *Oratoris ad Hemanuelem Lusitaniae regem oratio* (Venice, 1501).

Paul of Venice, *De Compositione mundi* (Venice, 1498).

Pérez de Oliva, Fernán, *Historia de la invención de las Yndias*, ed. J. J. Arrom (Bogotá: Instituto Caro y Cuervo, 1965); the digital edition available at: https://www.cervantesvirtual.com/obra/historia-de-la-invencion-de-las-yndias--0/.

Pérez de Oliva, Fernán, *Fernán Pérez de Oliva. Cosmografía Nueva*, eds. Cirilo Flórez Miguel, Pablo García Castillo, José Luis Fuertes Herreros and Leonardo Sandoval Ramón (Salamanca: Ediciones Universidad de Salamanca y Diputación Provincial de Salamanca, 1985).

Perez de Valencia, Jacob, *Commentum in Psalmos* (Valencia, 1484).

Petit, Françoise, ed., *La Chaîne sur la Genèse*, 4 vols (Lovain: Peeters, 1991).

Petrarca, *Secretum. Il mio segreto*, tr. Enrico Fenzi (Milan: Mursia Editore, 1992).

Petrarch, *Secretum*, tr. J. G. Nichols (Richmond: Alma Classics, 2015, 3rd revised ed.; 1st ed. 2002).

Peucer, Caspar, ed., *Tertia pars chronici carionis* (Wittenberg, 1562).

Phillips, William D., Jr., ed., *Testimonies from the Columbian Lawsuits* (Turnhout: Brepols, 2000).

Philoponus, John, In cap. *I Geneseos, De mundi creatione libri septem*, ed. and tr. Balthasar Corderius (Vienna, 1630).

Piccolomini, Aeneas Sylvius (Pope Pius II), *Historia rerum ubique gestarum* (Venice, 1477).

Piccolomini, Alessandro, *De la Sfera del Mondo* (Venice, 1540).

Pico della Mirandola, Giovanni Francesco, *De studio divinae et humanae philosophiae* (Bologna, 1497).

Pires, Tomé, *The Suma Oriental of Tomé Pires and The Book of Francisco Rodrigues*, tr. and ed. Armando Cortesão (London: The Hakluyt Society, 1944).

Plato, Meno *and* Phaedo, ed. David Sedley, tr. Alex Long (Cambridge: Cambridge University Press, 2011).

Pliny, *Natural History with an English Translation*, tr. H. Rackham, 10 vols (Cambridge, MA: Harvard University Press and London: William Heinemann, 1938–1962). Here I use vol. 1 (1st ed., 1938; revised 2nd ed., 1949) and vol. 2 (1st ed., 1942; revised 2nd ed., 1961).

Plutarch of Chaeronea, Lucius Mestrius, 'Concerning the Face Which Appears in the Orb of the Moon', in *Plutarch's Moralia with an English Translation*, tr. Harold Cherniss and William C. Helmbold, vol. 12 (Cambridge, MA: Harvard University Press and London: William Heinemann, 1957).

Poliziano, Angelo, *Omnia Opera Angeli Politiani* (Venice, 1498).

Polybius of Megalopolis, *The Histories*, intro. and notes Brian McGing, tr. Robin Waterfield (Oxford: Oxford University Press, 2010).

(Pseudo-)Justin Martyr, *Tou agiou Ioustinou philosophou kai martyros Zena kai Sereno...* [transliteration of Greek title], ed. Robert Estienne (Paris, 1551).

(Pseudo-)Vespucci, Amerigo, *Mundus Novus*, tr. Fra Giovanni del Giocondo (Florence, 1502–1503).

(Pseudo-)Vespucci, Amerigo, *Lettera di Amerigo Vespucci delle isole nuovamente trovate in quattro suoi viaggi* (Florence, c. 1505).

Procopius of Gaza, *Procopii Gazaei sophistae Commentarii in Octateuchum, hoc est, in priores octo Veteris Testamenti libros, ex vetustissimis orthodoxorum patrum, aliorumque scriptorum enarrationibus in unum corpus et perpetuum interpretationis contextum accuratissime concinnati*, tr. Konrad Clauser (Zurich, 1555). The same text was published under the name of Clauser with the title *Procopii Gazaei Sophistae in octateuchum sive*

priores octo Veteris Testamenti libros antiquae lectionis commentarii, by the same printer, apparently also in Zurich, 1555, but without place and date stated as in the other edition.
Ptolemy, Claudius, *Cosmographia* (Vicenza, 1475, 1st ed.).
Ptolemy, Claudius, *Cosmographia* (Bologna, 1477, 2nd ed.).
Ptolemy, Claudius, *Geographie opus nouissima traductione e Grecorum archetypis castigatissime pressum, ceteris ante lucubratorum multo prestantius* (Strasbourg, 1513).
Ptolemy, Claudius, *Almagestu[m] Cl. Ptolemei Pheludiensis Alexandrini* (Venice, 1515).
Ptolemy, Claudius, *De Geographia libri octo* (Basel, 1533).
Ptolemy, Claudius, *Ptolemy's Almagest*, tr. G. J. Toomer (London: Duckworth, 1984).
Ptolemy, Claudius, *Ptolemy's Geography: An Annotated Translation of the Theoretical Chapters*, tr. J. L. Lennart Berggren and Alexander Jones (Princeton and Oxford: Princeton University Press, 2000).
Pulci, Luigi, *Morgante Maggiore* (Florence, 1483).
Pulci, Luigi, *Morgante: The Epic Adventures of Orlando and His Giant Friend Morgante*, tr. Joseph Tusiani (Bloomington and Indianapolis, IN: Indiana University Press, 1998).
Pulci, Luigi, *Il Morgante* (Einaudi, 2000). Online edition available at: https://archive.org/details/morganteilluigipulci.
Purchas, Samuel, *Purchas His Pilgrimes in Five Books*, 4 vols (London, 1625).
Ramus, Peter, *Scipionis Somnium ex sexto M. T. Ciceronis De Repvblica Libro, Petri Rami prælectionibus explicatum* (Paris, 1550, 2nd ed.; 1st ed. 1546).
Ramus, Petrus, *Prooemium mathematicum* (Paris, 1567).
Ramus, Petrus, *Scholarum mathematicarum* (Basel, 1569).
Ramusio, Giovanni Battista, *Navigazioni e viaggi*, 3 vols (Venice, 1550–1559).
Rastell, John, *A New Iuterlude [sic] and a Mery of the Nature of the .iiii. Element[s]…*, (London, c. 1520).
Recorde, Robert, *The Castle of Knowledge* (London, 1556).
Reisch, Gregor, *Margarita Philosophica* (Freiburg, 1503).
Rheticus, Georg Joachim, *Narratio prima* (Danzig, 1540).
Rheticus, Georg Joachim, *Narratio Prima* (Basel, 1541a).
Rheticus, Georg Joachim, [1541b], in Franz Hipler, 'Die Chorographie des Joachim Rheticus', *Zeitschrift für Mathematik und Physik, Historisch-Literarische Abteilung* 21 (1876), 125–150. Also published on its own: Franz Hipler, *Die Chorographie des Joachim Rheticus* (Dresde: Druck von B.G. Teubner, 1876).
Rheticus, Georg Joachim, [1541c], *G. J. Rheticus' Treatise on Holy Scripture and the Motion of the Earth*, tr. Reijer Hooykaas (Amsterdam: North-Holland Publishing Co., 1984).
Rheticus, Georg Joachim, *Narratio Prima* (Tübingen, 1596).
Rho, Antonio da, *Three Dialogues against Lactantius. Critical Latin Edition, English Translation, Introduction, and Notes*, eds. and trs. David Rutherford and Paul Schulten (Leiden and Boston: Brill, 2023).
Ringmann, Mathias, *De ora antarctica per regem Portugalliae pridem inventa* (Stragsbourg, 1505).
Ringmann, Mathias and Martin Waldseemüller, *Cosmographiae introductio* (Saint-Dié, 1507).
Rive, Jean-Joseph, *Diverses notices calligraphiques, et typographiques* (Paris, 1785).
Rosweide, Heribert, ed., 'Vita sancti Macarii Romani servi Dei, qui inventus est iuxta Paradisum, auctoribus Theophilo, Sergio et Hygino', in *Vitae Patrum: de vita et verbis seniorum libri X: historiam eremiticam complectentes* (Antwerp, 1615).
Rothmann, Christoph, *Christoph Rothmanns Handbuch der Astronomie von 1589: kommentierte Edition der Handschrift Christoph Rothmanns „Observationum stellarum fixarum*

liber primus", Kassel 1589, eds. Miguel Ángel Granada, Jürgen Hamel and Ludolf von Mackensen (Frankfurt am Main: Harri Deutsch, 2003).

Sá, Diogo de, *De Navigatione Libri Tres* (Paris, 1549).

Sacrobosco, John of, *Ioannis de Sacro Bosco Sphericum opusculum* (Venice, 1485).

Sacrobosco, John of, *Textus Sphaerae Ioannis de Sacro Busto* (Venice, 1508).

Sacrobosco, John of, *Mundialis Sphere opusculum Joannis de sacro busto*, ed. Oronce Fine (Paris, 1524).

Salusbury, Thomas, *Mathematical Collections and Translations*, 2 vols (London, 1661).

Sanches, Francisco, *Quod Nihil Scitur* (Lyon, 1581).

Sanches, Francisco, *That Nothing Is Known*, intro., notes and bibliography by Elaine Limbrick, tr. Douglas F. S. Thomson (Cambridge: Cambridge University Press, 1988).

Santaella, Rodrigo Fernández de, *Cosmographia breve introductoria en el libro d'Marco paulo* (Seville, 1503).

Senensis, Sixtus, *Bibliotheca Sancta* (Venice, 1566).

Sevenborgen, Maximilian von (Maximilianus Transylvanus), *De Molvccis Insulis* (Cologne, January 1523; Paris, July 1523 and Rome, November 1523).

Solinus, *In C. Iulii Solini Polyistora Enarrationes*, ed. Johannes Camers (Vienna, 1520).

Staden, Hans, *Warhaftige Historia und Beschreibung eyner Landtschafft der Wilden, Nacketen, Grimmigen Menschfresser-Leuthen, in der Newenwelt America gelegen* (Marburg, 1557).

Staden, Hans, *Hans Staden's True History: An Account of Cannibal Captivity in Brazil*, ed. and tr. Neil L. Whitehead and Michael Harbsmeier (Durham, NC and London: Duke University Press, 2008).

Strabo of Amaseia, *The Geography of Strabo*, ed. and tr. Horace Leonard Jones, 8 vols (London: William Heinemann and New York: G.P. Putnam's Sons, 1917–1949).

Symcox, Geoffrey and Giovanna Rabitti, eds., and Peter D. Diehl, tr., *Italian Reports on America, 1493-1522: Letters, Dispatches, and Papal Bulls* (Turnhout: Brepols, 2001).

Teixeira, João, *Oratio habita ab insigni viro Ioanne Teixeira...* (Coimbra, 1562).

Téllez, Fernando, *Razonamiento de los embaxadores de españa en la obediencia que dieron al papa* (Rome: 1508a).

Téllez, Fernando, *Oratio obedientiae praestitae nomine invictissimae domini Ferdinandi Regis catholici & serenissimae atque catholicae Ioannae Reginae eius filiae Hispaniarum principum Ad Iulium II* (no place of publication, 1508b).

Téllez, Fernando, *In hoc libello duae continentur orationes, altera Antonii Nebrissensis hystoriographi Regis Fernandi...* (Barcelona: Juan Rosembach, 1516).

Theophilus of Antioch, *Ad Autolycum*, tr. Robert M. Grant (Oxford: Oxford University Press, 1970).

Thorndike, Lynn, ed., *The Sphere of Sacrobosco and Its Commentators* (Chicago: The University of Chicago Press, 1949).

Tolosani, Giovanni Maria, in Eugenio Garin, 'Alle origini della polemica anticopernicana', *Studia Copernicana* 6 (1973), 31–42.

Torquemada, Antonio de, *Jardín de Flores Curiosas* (Salamanca, 1570). Of this first edition there are several copies with some variants. Here I follow the copy Usoz/8309 (at the Biblioteca Nacional de España, Madrid), considered as the most likely *editio princeps*.

Torquemada, Antonio de, *The Spanish Madeuile of Miracles: Or the Garden of Curious Flowers* (London, 1600). English translation of Torquemada, 1570.

Tostatus: see Fernández de Madrigal.

Urbino, Bartholomew of, *Milleloquium veritatis* (Lyon, 1555).

Vadianus, Joachimus (Joachim von Watt), *Habes lector: hoc libello Rudolphi Agricola... ad I. Vadianum* (Vienna, 1515).

Valla, Lorenzo, 'In quartum librum elegantiarum praefatio', in *Prosatori Latini del Quattrocento*, ed. Eugenio Garin (Milan and Naples: Riccardo Ricciardini editore, 1952), 612–623.

Vega, Lope de, *Doce comedias de Lope de Vega Carpio Familiar del Santo Oficio, sacadas de sus originales. Cuarta parte* (Madrid, 1614).

Verrazzano, Giovanni da, in Greene, George W. and Joseph G. Cogswell, 'The Voyage of John de Verazzano Along the Coast of North America, from Carolina to Newfoundland, A.D. 1524', *Collections of the New-York Historical Society*, second series, vol. 1 (1841), 37–67 (English [Cogswell], pp. 41–54; Italian [Greene], pp. 55–67).

Verrazzano, Giovanni da, in Wroth, Lawrence C., *The Voyages of Giovanni da Verrazzano 1524-1528* (New Haven and London: Yale University Press, 1970). Reproduction of an original manuscript copy (pp. 96–120), transcription of the original by Frederick B. Adams, Jr. (pp. 121–132), and English translation by Susan Tarrow (pp. 133–143).

Vives, Juan Luis, *De disciplinis, libri XX* (Antwerp, 1531).

Vives, Juan Luis, *Vives: On Education. A Translation of the* De Tradendis Disciplinis *of Juan Luis Vives*, tr. Foster Watson (Cambridge: Cambridge University Press, 1913).

Volterra, Raffaello Maffei da (Volaterranus), *Commentariorum rerum urbanarum libri XXXVIII* (Rome, 1506).

Volterra, Raffaello Maffei da (Volaterranus), *De institutione christiana* (Rome, 1518).

Wilkins, John, *The Discovery of a World in the Moone* (London, 1638).

Wilkins, John, *A Discourse Concerning a New Planet* (London, 1640).

Wilkins, John, *The Mathematical and Philosophical Works of the Right Reverend John Wilkins, Late Lord Bishop of Chester* (London, 1708).

Yerva, Francisco Núñez de la, *Cosmographia pomponii cum figuris* (Salamanca, 1498).

Zúñiga, Diego de, *In Job Commentaria* (Toledo, 1584).

Zurara (or: Azurara), Gomes Eanes de, *Crónica do descobrimento e conquista de Guiné*, ed. Visconde da Carreira (Paris: J. P. Aillaud, 1841).

Zurara (or: Azurara), Gomes Eanes de, *The Chronicle of the Discovery and Conquest of Guinea, Written by Gomes Eanes de Azurara*, tr. Charles Raymond Beazley and Edgar Prestage, 2 vols (London: The Hakluyt Society, 1896 and 1899).

Secondary literature

Acerbi, Silvia, '*Iuxta paradisum*: Mundos imaginarios en la hagiografía tardoantigua de oriente y de occidente', *Studia Historica. Historia Antigua* 27 (2009), 115–128.

Akerman, James R., 'From Books with Maps to Books as Maps: The Editor in the Creation of the Atlas Idea', in *Editing Early and Historical Atlases*, ed. Joan Winearls (Toronto: University of Toronto Press, 1995), 1–4.

Alban, (Brother), 'The Conscious Rôle of Lactantius', *The Classical Weekly* 37 (1943), 79–81.

Albuquerque, Luís de, 'Science et Humanisme dans la Renaissance Portugaise', in *L'Humanisme Portugais et l'Europe: actes du XXIème Colloque international d'études humanistes, Tours, 3-13 juillet 1978* (Paris: Fondation Calouste Gulbenkian, Centre Culturel Portugais, 1984), 419–437.

Albuquerque, Luís de, 'Pedro Nunes e Diogo de Sá', in *As Navegações e sua projecção na Ciência e na Cultura*, ed. Luís de Albuquerque (Lisbon: Gradiva Publicações, 1987), 57–79.

Allegro, James J., 'The Bottom of the Universe: Flat Earth Science in the Age of Encounter', *History of Science* 55 (2017), 61–85

Almeida, Onésimo Teotónio, 'Portugal and the Dawn of Modern Science', in *Portugal, The Pathfinder: Journeys from the Medieval toward the Modern World 1300-ca. 1600*, ed. George D. Winius (Madison: The Hispanic Seminary of Medieval Studies, Ltd., 1995), 341–361.

Almeida, Onésimo Teotónio, 'Francisco Sanches - O "Elo Perdido" entre os descobrimentos e a Ciência Moderna', *Revista de História e Teoria das Ideias* 12, 2nd series (Spring 2001), 221–229. English adapted version at: https://web.archive.org/web/20241012024009/https://cvc.instituto-camoes.pt/ciencia_eng/e33.html. (the original web site is no longer available).

Almeida, Onésimo Teotónio, 'Science during the Portuguese Maritime Discoveries - A Telling Case of Interaction Between Experimenters and Theoreticians', in *Science in the Spanish and Portuguese Empires, 1500-1800*, eds. Daniela Bleichmar, Paula De Vos, Kristin Huffine and Kevin Sheehan (Stanford: Stanford University Press, 2009), 78–92.

Almeida, Onésimo Teotónio, '*Experiência a madre das cousas* - on the 'Revolution of Experience' in Sixteenth-century Portuguese Maritime Discoveries and Their Foundational Role in the Emergence of the Scientific Worldview', in *Portuguese Humanism and the Republic of Letters*, eds. Maria Berbara and Karl A. E. Enenkel (Leiden and Boston: Brill, 2012), 377–394.

Anastos, Milton V., 'Pletho and Strabo on the Habitability of the Torrid Zone', *Byzantinische Zeitschrift* 44 (1951), 7–10. Reprinted in *Studies in Byzantine Intellectual History*, ed. Milton V. Anastos (London: Variorum Reprints, 1979), chapter XVI.

Anastos, Milton V., 'Pletho, Strabo and Columbus', *Mélanges Henri Grégoire* IV (=Annuaire de l'Institut de Philologie et d'Histoire Orientales et Slaves XII), (1952), 1–18. Reprinted in *Studies in Byzantine Intellectual History*, ed. Milton V. Anastos (London: Variorum Reprints, 1979), chapter XVII.

Andrea, Alfred J., 'The Myth of the Middle Ages', *The Historian* 55 (1992), 183–188.

Anonymous, 'Columbus at Salamanca', *Catholic World* 9 (1869), 433–442.

Arnaud, Pascal, 'Mapping the Edges of the Earth: Approaches and Cartographical Problems', in *The Periphery of the Classical World in Ancient Geography and Cartography*, ed. Alexander V. Podossinov (Leuven, Paris and Walpole, MA: Peters, 2014), 31–57.

Ashworth, William B., Jr., 'Catholicism and Early Modern Science', in God *and* Nature, eds. David C. Lindberg and Ronald L. Numbers (Berkeley, Los Angeles and London: University of California Press, 1986), 136–166.

Aufgebauer, Peter, '„Die Erde ist eine Scheibe" – Das mittelalterliche Weltbild in der Wahrnehmung der Neuzeit', *Geschichte in Wissenschaft und Unterricht* 7/8 (2006), 427–441.

Axworthy, Angela, 'Oronce Fine and Sacrobosco: From the Edition of the *Tractatus de sphaera* (1516) to the *Cosmographia* (1532)', in *De sphaera of Johannes de Sacrobosco in the Early Modern Period*, ed. Matteo Valleriani (Cham: Springer, 2020), 185–264.

Backus, Irena, *Historical Method and Confessional Identity in the Era of the Reformation (1378-1615)* (Leiden and Boston: Brill, 2003).

Backus, Irena, 'Medieval history and the *Catalogus testium veritatis* of Matthias Flacius Illyricus revised by Simon Goulart'; talk delivered in absentia at the 2011 Kalamazoo Congress for Medieval Studies, Kalamazoo, Michigan, USA, 13 May 2011. Available at: https://www.academia.edu/1690312/Medieval_history_and_the_Catalogus_testium_veritatis_of_Matthias_Flacius_Illyricus_revised_by_Simon_Goulart.

Backus, Irena, 'The 'Confessionalization' of Augustine in the Reformation and Counter-Reformation', in *The Oxford Guide to the Historical Reception of Augustine*, ed. Karla Pollmann and Willemien Otten, vol. 1 (Oxford: Oxford University Press, 2013), 74–82.

Bakker, Frederik A., *Epicurean Meteorology: Sources, Method, Scope and Organization* (Leiden and Boston: Brill, 2016).

Barker, Peter, 'The Lutheran Contribution to the Astronomical Revolution: Science and Religion in the Sixteenth Century', in *Religious Values & The Rise of Science in Europe*, eds. John Hedley Brooke and Ekmeleddin İhsanoğlu (Istanbul: The Research Centre for Islamic History, Art and Culture, IRCICA, 2005), 31–62.

Barker, Peter and Bernard R. Goldstein, 'Patronage Patronage and the Production of *De revolutionibus*', *Journal for the History of Astronomy* 34 (2003), 345–368.

Barnes, Timothy D., *Constantine: Dynasty, Religion and Power in the Later Roman Empire* (Oxford: Wiley-Blackwell, 2011).

Bartlett, Robert, *The Natural and the Supernatural in the Middle Ages* (Cambridge: Cambridge University Press, 2011, 4th printing; 1st printing 2008).

Beazley, Charles Raymond, *Dawn of Modern Geography*, 3 vols (London: John Murray, 1897 and 1901 [vols 1 and 2]; Oxford: Claredon Press, 1906 [vol. 3]).

Beazley, Charles Raymond, 'Directorium ad faciendum passagium transmarinum', *American Historical Review* 12 (1907), 810–857 and 13 (1908), 66–115.

Beer, Susanna de, 'The World Upside Down: The Geographical Revolution in Humanist Commentaries on Pliny's *Natural History* and Mela's *De situ orbis* (1450-1700)', in *Neo-Latin Commentaries and the Management of Knowledge in the Late Middle Ages and the Early Modern Period (*1400-1700), eds. Karl Enenkel and Henk Nellen (Leuven: Leuven University Press, 2013), 139–197.

Benin, Stephen D., *The Footprints of God: Divine Accommodation in Jewish and Christian Thought* (Albany, NY: State University of New York Press, 1993).

Bensaúde, Joaquim, *Regimento do estrolabio e do quadrante. Tractado da spera do mundo. Reproduction fac-similé du seul exemplaire connu appartenant à la Bibliothèque Royale de Munich* (Munich: Carl Kuhn, 1914).

Berchet, Guglielmo, ed., *Fonti Italiane per la Storia della Scoperta del Nuovo Mondo*, part 2: *Narrazioni sincrone* (Rome: Ministero della pubblica istruzione, 1893).

Bernhard, Roland, *Geschichtsmythen über Hispanoamerika* (Göttingen: V&R unipress, 2013a).

Bernhard, Roland, 'Der Eingang des „Mythos der flachen Erde" in deutsche und österreichische Geschichtsschulbücher im 20. Jahrhundert', *Geschichte in Wissenschaft und Unterricht* 64 (2013b), 687–701.

Bernhard, Roland, 'Ästhetische und politische Sinnbildungsstrategien in der Geschichtskultur – Historisch Denken lernen mit dem Mythos Martin Behaim', in *Mythen in deutschsprachigen Geschichtsschulbüchern*, eds. Roland Bernhard, Susanne Grindel, Felix Hinz and Christoph Kühberger (Göttingen: V&R unipress, 2017), 91–115.

Bernhard, Roland, 'Martin Behaim of Nuremberg: The 'Real Discoverer of America'? German Textbooks of the 18th to 21st Centuries and the Myth of the Flat Earth', in *Myths in German-language Textbooks*, eds. Roland Bernhard, Susanne Grindel, Felix Hinz and Christoph Kühberger (Braunschweig: Georg Eckert Institute for International Textbook Research, 2019), 94–118. English translation of Bernhard, 2017.

Berti, D., 'Antecedenti al processo galileiano e alla condanna della dottrina Copernicana', *Atti della R. Accademia dei Lincei*, Serie terza, *Memorie della classe di scienze morali, storiche e filologiche* 10 (1881), 78–91.

Betten, Francis S., 'The Knowledge of the Sphericity of the Earth during the Earlier Middle Ages', *The Catholic Historical Review* 9 (1923), 74–90.

Biro, Jacqueline, *On Earth as in Heaven: Cosmography and the Shape of the Earth from Copernicus to Descartes* (Saarbrücken: VDM Verlag, 2009).

Bischoff, Bernhard and Michael Lapidge, *Biblical Commentaries from the Canterbury School of Theodore and Hadrian* (Cambridge: Cambridge University Press, 1994).

Bishop, Louise M., 'The Myth of the Flat Earth', in *Misconceptions about the Middle Ages*, eds. Stephen J. Harris and Bryon Lee Grigsby (New York: Routledge, 2008), 97–101.

Blackwell, Richard Joseph, *Galileo, Bellarmine, and the Bible* (Notre Dame and London: University of Notre Dame Press, 1991).

Blair, Ann, 'Tycho Brahe's Critique of Copernicus and the Copernican System', *Journal of the History of Ideas* 51 (1990), 355–377.

Blair, Ann and Anthony Grafton, 'Reassessing Humanism and Science', *Journal of the History of Ideas* 53 (1992), 535–540.

Boaga, Emanuele, 'Annotazioni e documenti sulla vita e sulle opere di Paolo Antonio Foscarini teologo 'copernicano' (1562c.-1616)', *Carmelus* 37 (1990), 173–216.

Boffito, Giuseppe, 'Intorno alla "Quaestio de aqua et terra,, attribuita a Dante', *Memorie della Real Accademia delle Scienze di Torino* 51 (1902), 73–159 and 52 (1903), 257–342.

Bouloux, Nathalie, *Culture et savoirs géographiques en Italie au XIVe siècle* (Turnhout: Brepols, 2002).

Bradner, Leicester, 'Columbus in Sixteenth-Century Poetry', in *Essays Honoring Lawrence C.* Wroth (Portland, ME: The Anthoensen Press, 1951), 15–30.

Brendecke, Arndt, *The Empirical Empire: Spanish Colonial Rule and the Politics of Knowledge* (Berlin and Boston: Walter de Gruyter, 2016). This is a revised and shortened version of *Imperium und Empirie: Funktionen des Wissens in der spanischen Kolonialherrschaft* (Cologne, Weimar and Vienna: Böhlau Verlag, 2009).

Briggs, Eric R., 'Un pionnier de la pensee libre au XVe siècle: Galeotto Marzio da Narni (1427?-1497?)', in *Aspects du libertinisme au XVIe siècle* (Paris: Librairie Philosophique J. Vrin, 1974), 75–84.

Brincken, Anna-Dorothee von den, 'Die Kugelgestalt der Erde in der Kartographie des Mittelalters', *Archiv für Kulturgeschichte* 58 (1976), 7–95.

Brincken, Anna-Dorothee von den, *Fines Terrae. Die Enden der Erde und der vierte Kontinent auf mittelalterlichen Weltkarten* (Hanover: Hahnsche Buchhandlung, 1992).

Brooke, John Hedley, *Science and Religion: Some Historical Perspectives* (Cambridge: Cambridge University Press, 1991). This book has been republished in the Canto Classics series in 2014.

Brooke, John Hedley, 'A Call for Complexity', *Science & Spirit* 1 (2006). Available at: https://web.archive.org/web/20090630152511/https://www.science-spirit.org/webexclusives.php?article_id=590 (the original web site is no longer available).

Brooke, John Hedley, 'Afterword: The Instantiation of Historical Complexity', in *Rethinking History, Science, and Religion: An Exploration of Conflict and the Complexity Principle*, ed. Bernard Lightman (Pittsburgh, PA: University of Pittsburgh Press, 2019), 235–242.

Bryce, Jackson, *The Library of Lactantius* (New York and London: Garland Publishing, 1990; 1st ed. 1973).

Burmeister, Karl Heinz, 'Georg Joachim Rheticus as a Geographer and His Contribution to the First Map of Prussia', *Imago Mundi* 23 (1969), 73–76.

Butterfield, Herbert, *The Whig Interpretation of History* (London: G. Bell and Sons, 1931).

Cantor, Geoffrey, 'What Shall We Do with the 'Conflict Thesis'?', in *Science and Religion: New Historical Perspectives*, eds. Thomas Dixon, Geoffrey Cantor and Stephen Pumfrey (Cambridge: Cambridge University Press, 2010), 283–298.

Cañizares-Esguerra, Jorge, 'Iberian Science in the Renaissance: Ignored How Much Longer', *Perspectives on Science* 12 (2004), 86–124.

Cañizares-Esguerra, Jorge, *Nature, Empire, and Nation: Explorations of the History of Science in the Iberian World* (Stanford: Stanford University Press, 2006).

Carey, John, 'Ireland and the Antipodes: The Heterodoxy of Virgil of Salzburg', *Speculum* 64 (1989), 1–10.

Caroti, Stefano, 'Un sostenitore napoletano della mobilità della terra: il padre Paolo Antonio Foscarini', in *Galileo e Napoli*, eds. Fabrizio Lomonaco and Maurizio Torrini (Naples: Guida, 1987), 81–121.

Castilla Urbano, Francisco, 'Francis Bacon y la imitación del modelo colonial hispano', *Revista de Estudios Políticos* 192 (2021), 37–65.

Caudano, Anne-Laurence, '"These Are the Only Four Seas": The World Map of Bologna, University Library, Codex 3632', *Dumbarton Oaks Papers* 70 (2016), 167–190.

Ceulemans, Reinhart, 'The Transmission, Sources and Reception of Procopius' Exegesis of Genesis: Observations in the Wake of the New Edition', *Vigiliae Christianae* 71 (2017), 205–224.

Chapman, Allan, *Slaying the Dragons: Destroying Myths in the History of Science and Faith* (Oxford: Lion Hudson, 2013).

Chiesa, Paolo, 'Marckalada: The First Mention of America in the Mediterranean Area (c. 1340)', *Terrae Incognitae* 53 (2021), 88–106.

Clemens, Raymond, 'Medieval Maps in a Renaissance Context: Gregorio Dati and the Teaching of Geography in Fifteenth-century Florence', in *Cartography in Antiquity and the Middle Ages: Fresh Perspectives, New Methods*, ed. Richard J. A. Talbert and Richard W. Unger (Leiden and Boston: Brill, 2008), 237–256.

Contente Domingues, Francisco, 'Science and Technology in Portuguese Navigation: The Idea of Experience in the Sixteenth Century', in *Portuguese Oceanic Expansion, 1400-1800*, eds. Francisco Bethencourt and Diogo Ramada Curto (Cambridge: Cambridge University Press, 2007), 460–479.

Copenhaver, Brian P., 'Hermes Theologus: The Sienese Mercury and Ficino's Hermetic Demons', in *Humanity and Divinity in the Renaissance and Reformation*, eds. John W. O'Malley, Thomas M. Izbicki and Gerald Christianson (Leiden: Brill, 1993), 149–182.

Cormack, Lesley B., 'Flat Earth or Round Sphere: Misconceptions of the Shape of the Earth and the Fifteenth-century Transformation of the World', *Ecumene* 1 (1994), 363–385.

Cormack, Lesley B., 'That Medieval Christians Taught that the Earth was Flat', in *Galileo Goes to Jail and Other Myths about Science and Religion*, ed. Ronald L. Numbers (Cambridge, MA: Harvard University Press, 2009), 28–34.

Cormack, Lesley B., 'That Before Columbus, Geographers and Other Educated People Thought the Earth Was Flat', in *Newton's Apple and Other Myths about Science*, eds. Ronald L. Numbers and Kostas Kampourakis (Cambridge, MA: Harvard University Press, 2015), 16–22.

Couprie, Dirk Leendert, 'Some Remarks on the Earth in Plato's *Phaedo*', *Hyperboreus* 11 (2005), 192–204.

Couprie, Dirk Leendert, *Heaven and Earth in Ancient Greek Cosmology: From Thales to Heraclid* (New York: Springer, 2011).

Couprie, Dirk Leendert, 'Aristotle's proofs of the sphericity of the earth', text of a paper, read at the *Aristotle 2400 Years World Conference* in Thessaloniki, Greece, 24 May 2016. Available at: https://www.academia.edu/25717942/Aristotles_Proofs_of_the_Sphericity_of_the_Earth.

Couprie, Dirk Leendert, *When the Earth Was Flat: Studies in Ancient Greek and Chinese Cosmology* (Cham: Springer, 2018).

Couprie, Dirk Leendert, 'The Reception of Presocratic Flat Earth Cosmology in Aristotle, the Doxography, and Beyond', in *Brill's Companion to the Reception of Presocratic Natural Philosophy in Later Classical Thought*, eds. Chelsea C. Harry and Justin Habash (Leiden and Boston: Brill, 2021), 289–322.

Couprie, Dirk Leendert, 'Who Was Socrates' Informant on the Sphericity of the Earth?', in *The Messages on the Shield of Achilles and Other Essays on Ancient Greek Cosmology and Related Topics*, ed. Dirk L. Couprie (Amsterdam: self-publication, forthcoming, 2025), 287–309.

Delalle, Louis-Auguste, Lettres a M. *Letronne, sur la cosmogonie des pères de l'église et de la Genèse* (Paris: Debécourt, 1835).

Delaney, Carol, 'Columbus's Ultimate Goal: Jerusalem', *Comparative Studies in Society and History* 48 (2006), 260–294.

Delaney, Carol, *Columbus and the Quest for Jerusalem* (New York: Free Press, 2011).

DeMar, Gary, *America's Christian History: The Untold Story* (Powder Springs, GA: American Vision, 2005, 2nd revised ed.; 1st ed. 1993).

DeMar, Gary, *Whoever Controls the Schools Rules the World* (Powder Springs, GA: American Vision, 2007).

Digeser, Elizabeth DePalma, 'Lactantius and Constantine's Letter to Arles: Dating the *Divine Institutes*', *Journal of Early Christian Studies* 2 (1994), 33–52.

Digeser, Elizabeth DePalma, *The Making of a Christian Empire: Lactantius and Rome* (Ithaca, NY: Cornell University, 2000).

Digeser, Elizabeth DePalma, 'Lactantius', in *Great Christian Jurists and Legal Collections in the First Millennium*, ed. Peter Reynolds (Cambridge: Cambridge University Press, 2019), 239–251.

Doncel y Ordaz, Domingo, *La Universidad de Salamanca en el tribunal de la historia* (Salamanca, 1858).

Draper, John William, *History of the Conflict Between Religion and Science* (New York and London: D. Appleton and Company, 1874).

Duzer, Chet van, *Henricus* Mar*tellus's World Map at Yale (c. 1491)* (Springer, 2019).

Eco, Umberto, *Serendipities: Language and Lunacy*, tr. William Weaver (New York and Chichester: Columbia University Press, 1998).

Eco, Umberto, 'The Power of Falsehood', in *On Literature* (Orlando, FL: Harcourt, 2004). The original Italian was published in 2002 and this particular chapter is an expanded version of the inaugural lecture for the academic year 1994-95 at the University of Bologna (Italy).

Eco, Umberto, 'La solita storia della terra piatta', *l'Espresso* (26 February 2005). Available at: https://www.didaweb.net/fuoriregistro/leggi.php?a=6873 and https://www.europaoggi.it/content/view/2381/94/.

Edson, Evelyn, *Mapping Time and Space: How Medieval Mapmakers Viewed Their World* (London: The British Library, 1997).

Edson, Evelyn, 'Maps in Context: Isidore, Orosius, and the Medieval Image of the World', in *Cartography in Antiquity and the Middle Ages: Fresh Perspective, New Methods*, eds. Richard J. A. Talbert and Richard W. Unger (Leiden and Boston: Brill, 2008), 219–236.

Elert, Werner, *The Structure of Luteranism*, vol. 1 (St Louis: Concordia Publishing House, 1962). German original from 1931.

Elford, Dorothy Jane, *Developments in the natural philosophy of William of Conches: a study of his 'Dragmaticon' and a consideration of its relationship to the 'Philosophia'*, unpublished PhD thesis (Cambridge: University of Cambridge, 1983). Available at: https://www.repository.cam.ac.uk/items/cf43b1f4-2a70-4e3c-80b1-ab57385ddd61.

Elliott, John Huxtable, 'Renaissance in Europe and America: A Blunted Impact?', in *First Images of America: The Impact of the New World on the Old*, eds. Fredi Chiapelli, Michael J. B. Allen and Robert L. Benson, 2 vols, vol. 1 (Berkeley, Los Angeles and London: University of California Press, 1976), 11–23.

Elliott, John Huxtable, *The Old World and the New, 1492-1650* (Cambridge: Cambridge University Press, 1992, reprinted; 1st ed. 1970).

Elliott, John Huxtable, 'Final Reflections: The Old World and the New Revisited', in *America in European Consciousness, 1493-1750*, ed. Karen Ordahl Kupperman (Chapel Hill, NC and London: University of North Carolina Press, 1995), 391–409.

Elliott, John Huxtable, 'Europa y América en la época moderna', in *La incorporación de las Indias al mundo occidental en el siglo XVI*, ed. F. Sánchez-Montes (Granada: Universidad de Granada, Sociedad Estatal Para la Conmemoración de los Centenarios de Felipe II y Carlos V y Fundación Hispano Británica, 2000), 23–44.

Everson, Jane E., *The Italian Romance Epic in the Age of Humanism: The Matter of Italy and the World of Rome* (Oxford: Oxford University Press, 2001).

Faes de Mottoni, Barbara, 'Lattanzio e gli Accademici', *Mélanges de l'École française de Rome. Antiquité* 94 (1982), 335–377.

Falcón y Ozcoidi, Modesto, *Cristóbal Colon y la Universidad de Salamanca* (Salamanca: Imp. y Lit. de Sebastián Cerezo, 1881).

Fear, Andrew, 'Putting the Pieces Back Together: Isdore and De Natura Rerum', in *Isidore of Seville and His Reception in the Early Middle Ages: Transmitting and Transforming Knowledge*, eds. Andrew Fear and Jamie Wood (Amsterdam: Amsterdam University Press, 2016), 75–92.

Felipe, Pablo de, 'The Antipodeans and Science and Faith Relations: The Rise, Fall and Vindication of Augustine', in *Augustine beyond the Book: Intermediality, Transmediality, and Reception*, eds. Karla Pollmann and Meredith J. Gill (Leiden and Boston: Brill, 2012), 281–311.

Felipe, Pablo de, 'The modern myth of the medieval flat earth' (Evangelische Akademie im Rheinland, 2017). Available at: https://www.theologie-naturwissenschaften.de/en/dialogue-between-theology-and-science/editorials/flat-earth-myth/; German version at: https://www.theologie-naturwissenschaften.de/startseite/leitartikelarchiv/flache-erde-mythos; French version at: https://scienceetfoi.com/le-mythe-moderne-de-la-terre-plate-medievale-1-2/ and https://scienceetfoi.com/le-mythe-moderne-de-la-terre-plate-medievale-2/; Spanish version at: https://educacioncienciayfe.wordpress.com/2020/04/17/el-gran-mito-moderno-creian-los-cristianos-de-la-edad-media-que-la-tierra-era-plana-pablo-de-felipe/.

Felipe, Pablo de, 'Curiosity in the Early Christian Era – Philoponus's Defence of Ancient Astronomy against Christian Critics', *Science and Christian Belief* 30 (2018), 38–56. See discussion and erratum in *Science and Christian Belief* 31 (2019), 78–80.

Felipe, Pablo de, *Flattening the medieval earth: the early modern origins of the "flat error"*, unpublished PhD thesis (Bristol: University of Bristol, 2022). Abstract availableat:https://research-information.bris.ac.uk/en/studentTheses/flattening-the-medieval-earth.

Felipe, Pablo de and Malcolm Jeeves, 'Science and Christianity Conflicts: Real and Contrived', *Perspectives on Science and Christian Faith* 69 (2017), 181–147.

Ferguson, Wallace Klippert, *The Renaissance in Historical Thought: Five Centuries of Interpretation* (Boston: Houghton Mifflin Company, 1948).

Fernández-Armesto, Felipe, *Columbus* (Oxford and New York: Oxford University Press, 1991).

Fernández-Armesto, Felipe, *Columbus on Himself* (Indianapolis, IN: Hackett Publishing, 2010).

Fernández de Navarrete, Martín, *Colección de los viages y descubrimientos que hicieron por mar los españoles desde fines del siglo XV*, 5 vols (Madrid: Imprenta Real, 1825–1837).

Ferrari, Leo, 'Augustine's Cosmography', *Augustinian Studies* 27 (1996), 131–180.

Ferrari, Leo, 'Cosmography', in *Augustine Through the Ages: An Encyclopedia*, ed. Allan D. Fitzgerald (Grand Rapids, MI and Cambridge: Wm. B. Eerdmans, 1999), 246.

Finocchiaro, Maurice A., ed., *The Galileo Affair: A Documentary History* (Berkeley, Los Angeles and London: University of California Press, 1989).

Finocchiaro, Maurice A., *Retrying Galileo, 1633-1992* (Berkeley, Los Angeles and London: University of California Press, 2005).

Fiske, John, *The Discovery of America*, 2 vols (Boston and New York: Houghton Mifflin Company, 1892).

Flint, Valerie I. J., 'Christopher Columbus and the Friars', in *Intellectual Life in the Middle Ages*, eds. Lesley Smith and Benedicta Ward (London and Rio Grande, OH: The Hambledon Press, 1992a), 295–310.

Flint, Valerie I. J., *The Imaginative Landscape of Christopher Columbus* (Princeton: Princeton University Press, 1992b).

Flórez Miguel, Cirilio, Pablo García Castillo and Roberto Albares Albares, eds., *El Humanismo Científico* (Salamanca: Caja de Ahorros y Monte de Piedad de Salamanca, 1988).

Flórez Miguel, Cirilo, Pablo García Castillo and Roberto Albares Albares, eds., and José Sanz Ramos, tr., *La ciencia del cielo: Astrología y filosofía naural en la Universidad de Salamanca (1450-1530)* (Salamanca: Caja de Ahorros y Monte de Piedad de Salamanca, 1989).

Flórez Miguel, Cirilo, Pablo García Castillo and Roberto Albares Albares, eds., and José Sanz Ramos, tr., *La ciencia de la Tierra: Cosmografía y cosmógrafos salmantinos del Renacimiento* (Salamanca: Caja de Ahorros y Monte de Piedad de Salamanca, 1990).

Floridi, Luciano, 'The Rediscovery and Posthumous Influence of Scepticism', in *The Cambridge Companion to Ancient Scepticism*, ed. Richard Bett (Cambridge: Cambridge University Press, 2010), 267–287.

Franco, Antonino, 'P. Paulus Antonius Foscarini', *Analecta Ordinis Carmelitarum* 2 (1911–1913), 461–648, 493–504, 524–527.

Friedman, John Block, *The Monstrous Races in Medieval Art and Thought* (Syracuse, NY: Syracuse University Press, 2000).

Galynina, Irina, 'Accessus ad Lactantium?', *Revue d'Histoire des Textes* 12 (2017), 161–196.

Garnsey, Peter, 'Lactantius and Augustine', in *Representations of Empire: Rome and the Mediterranean World*, eds. Alan K. Bowman, Hannah M. Cotton, Martin Goodman and Simon Price (Oxford: Oxford University Press/The British Academy, 2004, reprinted; 1st ed. 2002), 153–179.

Garwood, Christine, 'Alfred Russel Wallace and the Flat Earth Controversy', *Endeavour* 25 (2001), 139–143.

Garwood, Christine, *Flat Earth: The History of an Infamous Idea* (London: Thomas Dunne Books, 2008; 1st ed. 2007).

Gascoigne, J. 'Crossing the Pillars of Hercules: Francis Bacon, the Scientific Revolution and the New World', in *Science in the Age of Baroque*, eds. Ofer Gal and Raz Chen-Morris (Dordrecht: Springer, 2013), 217–238.

Gautier Dalché, Patrick, 'L'œuvre géographique du cardinal Fillastre († 1428). Représentation du monde et perception de la carte à l'aube des découvertes', *Archives d'histoire doctrinale et littéraire du Moyen Age* 59 (1992), 319–383.

Gautier Dalché, Patrick, *La Géographie de Ptolémée en occident (Ive-SVIe siècle)* (Turnhout: Brepols, 2009).

Gautier Dalché, Patrick, 'Les représentations de l'espace en Occident de l'Antiquité tardive au xvie siècle', *Annuaire de l'École pratique des hautes études (EPHE), Section des sciences historiques et philologiques* 141 (2011), 97–108. As I have been only able to read the non-paginated online version, available at: https://journals.openedition.org/ashp/991, I quote by the paragraph number shown in this electronic version.

Gautier Dalché, Patrick, *La Terre. Connaissance, Représentations, Mésure au Moyen Âge* (Turnhout: Brepols, 2013).

Giacomotto-Charra, Violaine and Sylvie Nony, *La Terre Plate. Généalogie d'une idée fausse* (Paris: Les Belles Lettres, 2021). I quote by the page number shown in the electronic version.

Gil, Juan, 'La épica latina quiñentista y el descubrimiento de América', *Anuario de Estudios Americanos* 40 (1983): 203–251.

Gil, Juan, *Mitos y utopias del descubrimiento*, 3 vols (Seville: Athenaica, 2017-2019, 2nd slightly revised ed.; 1st ed. 1989), vol. 1: Colón y su tiempo.

Gingerich, Owen, 'Astronomy in the Age of Columbus', *Scientific American* 267 (1992), 66–71.

Gingerich, Owen, 'Book Review of Jeffrey Burton Russell's *Inventing the Flat Earth: Columbus and Modern Historians*', *Speculum* 68 (1993), 885.

Gingerich, Owen, 'Scientific Cosmology Meets Western Theology: A Historical Perspective', *Annals of the New York Academy of Sciences* 950 (2001), 28–38.

Gliozzi, Giuliano, *Adamo e il nuovo mondo: La nascita dell' antropologia come ideologia coloniale, dalle genealogie bibliche alle teorie razziali (1500-1700)* (Florence: La Nuova Italia Editrice, 1977).

Goddu, André, 'Reflections on the Origin of Copernicus's Cosmology', *Journal for the History of Astronomy* 37 (2006), 37–53.

Goldie, Matthew Boyd, *The Idea of the Antipodes: Place, People, and Voices* (New York and Abingdon: Routledge, 2010).

Goldstein, Thomas, 'The Renaissance Concept of the Earth in Its Influence upon Copernicus', *Terrae incognitae* 4 (1972), 19–51.

Goldstein, Thomas, *Dawn of Modern Science: From the Arabs to Leonardo da Vinci* (Boston: Houghton Mifflin Company, 1980).

Gould, Stephen Jay, 'The Persistently Flat Earth', *Natural History* 103 (1994), 12–19. Reprinted in 'The Late Birth of a Flat Earth', in *Dinosaur in a Haystack*, ed. Stephen Jay Gould (New York: Harmony Books, 1995), 38–50.

Gould, Stephen Jay, 'Columbus and the Flat Earth: An Example of the Fallacy of Warfare Between Science and Religion', in *Rocks of Ages: Science and Religion in the Fullest of Life*, ed. Stephen Jay Gould (New York: Ballantine Books, 1999), 111–124.

Grafton, Anthony, with April Shelford and Nancy Siraisi, New Words*, Ancient Texts: The Power of Tradition and the Shock of Discovery* (Cambridge, MA and London: Harvard University Press, 1995; repr. of 1st ed. 1992).

Grafton, Anthony, 'Where Was Salomon's House? Ecclesiastical History and the Intellectual Origins of Bacon's *New Atlantis*', in *Die europäische Gelehrtenrepublik im Zeitalter des Konfessionalismus*, ed. Herert Jaumann (Wiesbaden: Harrassowitz, 2001), 21–38. Reprinted in *Worlds Made by Words: Scholarship and Community in the Modern West*, ed. Anthony Grafton (Cambridge, MA and London: Harvard University Press, 2009), 98–113.

Graham, Daniel W., *Science before Socrates: Parmenides, Anaxagoras and the New Astronomy* (New York: Oxford University Press, 2013).

Graham, Daniel W., 'The Reception of Early Greek Astronomy', in *Brill's Companion to the Reception of Presocratic Natural Philosophy in Later Classical Thought*, eds. Chelsea C. Harry and Justin Habash (Leiden and Boston: Brill, 2021), 91–110.

Granada, Miguel Ángel, 'Il *problema astronomico-cosmologico dopo Copernico e* le *Sacre Scritture*: Il ricorso di *Christoph Rothmann* e la *«teoria dell'accomodazione»', Rivista di Storia della Filosofia* 51 (1996), 789–828.

Granada, Miguel Ángel, 'Tycho Brahe, Caspar Peucer, and Christoph Rothmann on Cosmology and the Bible', in *Nature and Scripture in the Abrahamic Religions: Up to 1700*, eds. Jitse van der Meer and Scott Mandelbrote (Leiden and Boston: Brill, 2008), 563–583.

Granada, Miguel Ángel, ed., *Novas y Cometas entre 1572 y 1618: Revolución cosmológica y renovación política y religiosa* (Barcelona: Ediciones de la Universitat de Barcelona, 2012).

Granada, Miguel Ángel and Dario Tessicini, 'Copernicus and Fracastoro: The Dedicatory Letters to Pope Paul III, the History of Astronomy, and the Quest for Patronage', *Studies in History and Philosophy of Science* 36 (2005), 431–476.

Grant, Edward, ed., *A Source Book in Medieval Science* (Cambridge, MA: Harvard University Press, 1974).

Grant, Edward, 'In Defence of the Earth's Centrality and Immobility: Scholastic Reaction to Copernicanism in the Seventeenth Century', *Transactions of the American Philosophical Society* 74 (1984), 1–69.

Grant, Edward, *Planets, Stars, and Orbs: The Medieval Cosmos, 1200-1687* (Cambridge: Cambridge University Press, 1994).

Grant, Edward, *God and Reason in the Middle Ages* (Cambridge: Cambridge University Press, 2001).

Grauert, Hermann, 'Die Entdeckung eines Verstorbenen zur Geschichte der großen Länderentdeckungen', *Historisches Jahrbuch* 29 (1908), 304–333.

Gregorio, Lamberto Di, 'L'Hermes di Eratostene', *Aevum* 84 (2010), 69–144.

Guimarães, Rodolpho, 'Un opúsculo raríssimo de Pedro Nunes', *Boletim Bibliográfico da Biblioteca da Universidade de Coimbra* 3 (1916), 268–289.

Hamel, Jürgen, *Die Vorstellung von der Kugelgestalt der Erde im europäischen Mittelalter bis zum Ende des 13. Jahrhunderts – dargestellt nach den Quellen* (Münster: Lit Verlag, 1996).

Hamel, Jürgen, 'Die Kugelgestalt der Erde: Vorstellungen im europäischen Mittelalter bis zur Mitte des 13. Jahrhunderts', in *Miscellanea Kepleriana*, eds. Friederike Boockmann, Daniel A. Di Liscia and Hella Kothmann (Augsburg: Rauner, 2005), 7–26.

Hankins, James, *Plato in the Italian Renaissance*, 2 vols (Leiden: Brill, 1990).

Hankins, James, 'The Popes and Humanism', in *Rome Reborn: The Vatican Library and Renaissance Culture*, ed. Anthony Grafton (New Haven and London: Yale University Press, in association with the Biblioteca Apostolica Vaticana, 1993), 47–85. Reprinted in *Humanism and Platonism in the Italian Renaissance*, ed. James Hankins, 2 vols, vol. 1 (Rome: Edizioni di Storia e Letteratura, 2003), 469–494.

Hannam, James, *God's Philosophers: How the Medieval World Laid the Foundations of Modern Science* (London: Icon Books, 2009).

Hannam, James, 'Science versus Christianity?' (2010). Available at: https://www.patheos.com/resources/additional-resources/2010/05/science-versus-christianity.

Hannam, James, 'Did Isidore of Seville Think the Earth Is Flat? Yes, Probably', *Quodliberta* (2021). Available at: https://bedejournal.blogspot.com/2021/06/did-isidore-of-seville-think-earth-is.html.

Hannam, James, *The Globe: How the Earth Became Round* (London: Reaktion Books, 2023).

Hardin, Jeff, Ronald L. Numbers and Ronald A. Binzley, eds., *The Warfare Between Science and Religion: The Idea That Wouldn't Die* (Baltimore, MD: Johns Hopkins University Press, 2018).

Harrison, Peter, *The Territories of Science and Religion* (Chicago and London: University of Chicago Press, 2015).

Harrison, Peter, 'Angels on Pinheads and Needles' Points', *Notes and Queries* 63 (2016), 45–47.

Harrison, Peter, 'Conflict, Complexity, and Secularization in the History of Science and Religion', in *Rethinking History, Science, and Religion: An Exploration of Conflict and the Complexity Principle*, ed. Bernard Lightman (Pittsburgh, PA: University of Pittsburgh Press, 2019), 221–234.

Harrison, Peter, *Some New Word: Myths of Supernatural Belief in a Secular Age* (Cambridge: Cambridge University Press, 2024).

Harrison, Peter, William R. Shea, David C. Lindberg and Charles C. Gillispie, 'Herbert Butterfield and the Scientific Revolution: A Forum', *Historically Speaking* 8 (2006), 7–17.

Harrisse, Henry, *Christopher Columbus and the Bank of Saint George* (New York: privately printed, 1888).

Harrisse, Henry, *The Discovery of North America* (London: Henry Stevens and son, and Paris: H. Welter, 1892).

Headley, John M., *Luther's View of Church History* (New Haven and London: Yale University Press, 1963).

Headley, John M., 'The Sixteenth-Century Venetian Celebration of the Earth's Total Habitability: The Issue of the Fully Habitable World for Renaissance Europe', *Journal of World History* 8 (1997a), 1–27.

Headley, John M., *Tommaso Campanella and the Transformation of the World* (Princeton: Princeton University Press, 1997b).

Heidel, William Arthur, *The Frame of Ancient Greek Maps, with a Discussion of the Discovery of the Sphericity of the Earth* (New York: Arno Press, 1976, reprinted; 1st ed. New York: American Geographical Society, 1937).

Heilbron, John L., *Galileo* (New York: Oxford University Press, 2010).

Hellman, C. Doris, *The Comet of 1577: Its Place in the History of Astronomy* (New York: Columbia University Press, 1944).

Hiatt, Alfred, 'Petrarch's Antipodes', *Parergon* 22 (2005), 1–30.

Hiatt, Alfred, *Terra Incognita* (Chicago and London: The University of Chicago Press, 2008).

Hiatt, Alfred, '*Terra Australis* and the Idea of the Antipodes', in *European Perceptions of Terra Australis*, eds. Anne M. Scott, Alfred Hiatt, Claire McIlroy and Christopher Wortham (Farnham and Burlington, VT: Ashgate Publishing, 2011), 9–44.

Higgins, Shannon, G. J. *Rheticus and the authorship of the anonymous* Epistola de Terrae Motu, unpublished MA Thesis (Halifax, Nova Scotia: Dalhousie University, 2013). Available at: https://dalspace.library.dal.ca/bitstream/handle/10222/34639/Higgins-Shannon-MA-HIST-August-2013.pdf.

Hine, Harry Morrison, '*Seneca* and *Anaxagoras* in *Pseudo-Bede's* De *mundi celestis terrestrisque constitutione', Viator: Medieval and Renaissance Studies* 19 (1988), 111–127.

Hine, Harry Morrison, 'Seneca's *Naturales Quaestiones* 1960-2005 (Part 2) - with Addenda covering 2006', *Lustrum* 52 (2010), 7–160.

Holt-Jensen, Arild, *Geography: History and Concepts* (London, etc.: Sage Publications, 2018, 5th revised ed.; 1st ed. 1981).

Holub, Robert C., *Reception Theory: A Critical Introduction* (London and New York: Methuen, 1984).

Hooykaas, Reijer, 'The Portuguese Discoveries and the Rise of Modern Science', *Boletim da Academia Internacional da Cultura Portuguesa* 2 (1966), 88–107. Reprinted in *Selected Studies in History of Science*, ed. Reijer Hooykaas (Coimbra: Universidade Coimbra, 1983), 579–598.

Hooykaas, Reijer, *Humanism and the Voyages of Discovery in the 16th Century Portuguese Science and Letters* (Amsterdam, Oxford and New York: Noord-Holland Publishing Company, 1979).

Hooykaas, Reijer, 'Science in Manueline Style', in Castro, 1980, vol. 4, 231–426.

Hooykaas, Reijer, 'The Rise of Modern Science: When and Why?', *British Journal for the History of Science* 20 (1987), 453–473.

Howell, Kenneth J., *God's Two Books* (Notre Dame: University of Notre Dame Press, 2002).

Hutchings, David and James C. Ungureanu, *Of Popes and Unicorns. Science, Christianity, and How the Conflict Thesis Fooled the World* (Oxford: Oxford University Press, 2022).

Huxley, Thomas H., 'Darwin on the Origin of Species', *Westminster Review*, 2nd ser., 17 (1860), 541–570.

Ingremeau, Christiane, 'Les Institutions Divines de Lactance: une composition architecturale', *Vita Latina* 132 (1993), 33–40.

Iommi Echeverría, Virginia, 'Dante Alighieri's *Quaestio de aqua et terra* in the Renaissance', *Imago Temporis. Medium Aevum* 8 (2014), 177–191.

Iommi Echeverría, Virginia, 'La imagen del mundo en la *Margarita philosophica* de Gregor Reisch', *História Unisinos* 21 (2017), 224–233.

Irving, Washington, *A History of the Life and Voyages of Christopher Columbus*, 3 vols (New York: G. and C. Carvill, 1828) and 4 vols (Paris: A. and W. Galignani, 1828). Another English edition, also from 1828 that I have not seen, was published by Murray in London, as well as a French edition in Paris by Galignani.

Jalobeanu, Dana, 'A Natural History of the Heavens: Francis Bacon's Anti-Copernicanism', in *The Making of Copernicus: Early Modern Transformations of the Scientist and His Science*, eds. Wolfgang Neuber, Thomas Rahn and Claus Zittel (Leiden and Boston: Brill, 2014), 64–87.

Jones, Charles W., 'The Flat Earth', *Thought* 9 (1934), 296–307.

Kardos, Tibor, *Néhány adalék a magyarországi humanizmus történetéhez* (Pécs: Dunántúl Pécsi Egyetemi Könyvk, 1933).

Kazhdan, Alexander P., *The Oxford Dictionary of Byzantium*, 3 vols (Oxford: Oxford University Press, 1991).

Keas, Michael Newton, *Unbelievable: 7 Myths about the History and Future of Science and Religion* (Wilmington, DE: Intercollegiate Studies Institute Books, 2019).

Kendeffy, Gábor, 'Lactantius as Christian Cicero, Cicero as Shadow-like Instructor', in *Brill's Companion to the Reception of Cicero*, ed. William H. F. Altman (Leiden and Boston: Brill, 2015), 56–92.

Keyser, Paul T., 'The Geographical Work of Dikaiarchos', in *Dicaearchus of Messana: Text, Translation, and Discussion*, eds. William Fortenbaugh and Eckart Schütrumpf (London and New York: Routledge, 2017, reprinted; 1st ed. New Brunswick, NJ: Transaction Publishers, 2001), 353–372.

Kish, George, *A Source Book in Geography* (Cambridge, MA and London: Harvard University Press, 1978).

Kleinert, Andreas, '„Eine handgreifliche Geschichtslüge". Wie Martin Luther zum Gegner des copernicanischen Weltsystems gemacht wurde', *Berichte zur Wissenschaftsgeschichte* 26 (2003), 101–111.

Kominko, Maja, *The World of Kosmas: Illustrated Byzantine Codices of the Christian Topography* (Cambridge: Cambridge University Press, 2013).

Kraai, Jesse, 'The Newly-found Rheticus Lectures', *Beiträge zur Astronomiegeschichte* 1 (1998), 32–40.

Kraai, Jesse, *Rheticus' Heliocentric Providence: a study concerning the astrology, astronomy of the sixteenth century*, unpublished PhD thesis (Heidelberg: Ruprecht-Karls-Universität Heidelberg, 2000). Available at: https://archiv.ub.uni-heidelberg.de/volltextserver/3254/.

Krüger, Reinhard, 'Kosmologisches Wissen, das Konzept des Universums und die Kugelgestalt der Erde bei Ramon Llull (1232-1316)', *Zeitschrift für Katalanistik* 11 (1998), 33–78.

Krüger, Reinhard, *Eine Welt ohne Amerika, vol. 2: Das Überleben des Erdkugelmodells in der Spätantike (ca. 60 v.u.Z. - ca.* 550*)* (Berlin: Weidler, 2000a).

Krüger, Reinhard, *Eine Welt ohne Amerika, vol. 3: Das lateinische Mittelalter und die Tradition des antiken Erdkugelmodells (ca.* 550 - *ca.* 1080*)* (Berlin: Weidler, 2000b).

Krüger, Reinhard, 'Ein Mythos der Moderne: Die Erdscheibentheorie im Mittelalter und die Verfälschung des 'Hexaemeron' des Basilius von Caesarea durch Bernard de Montfaucon (1706)', *Mittellateinisches Jahrbuch* 36 (2001), 3–29.

Krüger, Reinhard, 'Ein Versuch über die Archäologie der Globalisierung – Die Kugelgestalt der Erde und die globale Konzeption des Erdraumes in Mittelalter', in *Wechselwirkungen (= Jahrbuch aus Lehre und Forschung der Universität Stuttgart)* (Stuttgart: Universitätsbibliothek der Universität Stuttgart, 2007), 28–53.

Krüger, Reinhard, Moles globosa, globus terrae *und* arenosus globus *in Spätantike und Mittelalter. Eine Kritik des Mythos von der Erdscheibe* (Berlin: Weidler, 2012).

Kuhn, Thomas, *The Copernican Revolution: Planetary Astronomy in the Development of Western Thought* (Cambridge, MA: Harvard University Press, 1970, reprinted; 1st ed. 1957).

Kunstmann, Friedrich, 'Hieronymus Münzer's Bericht über die Entdeckung der Guinea', Abhandlung der Historischen Classe der Königlich-Bayerischen Akademie der Wissenschaften 7 (1854), 291–362.

Lanman, Jonathan T., 'The Religious Symbolism of the T in T-O Maps', *Cartographica* 18 (1982), 18–22.

Lapidge, Michael, *The Anglo-Saxon Library* (Oxford: Oxford University Press, 2005).

Launay, Robert, *Savages, Romans, and Despots: Thinking about Others from Montaigne to Herder* (Chicago and London: University of Chicago Press, 2018).

Lecoq, Danielle, 'La Mappemonde du *Liber Floridus* ou La Vision du Monde de Lambert de Saint-Omer', *Imago Mundi* 39 (1987), 9–49.

Lecoq, Danièle, 'Plate ou spherique? La conception de la Terre au Moyen Age', *Comité français de cartographie* 148 (1996), 19–38.

Lee, Alexander, *Petrarch and S. Augustine* (Leiden and Boston: Brill, 2012).

Leitão, Henrique and Walter Alvarez, 'The Portuguese and Spanish Voyages of Discovery and the Early History of Geology', *GSA Bulletin* 123 (2011), 1219–1233.

Lenhart, John M., 'Antonio da Rho, Ord. Min (d.c. 1450), Humanist', *Franciscan Studies* 6 (1946), 109–110.

León Azcárate, Juan Luis de, 'El "Libro de las Profecías" (1504) de Cristóbal Colón: la Biblia y el descubrimiento de América', *Helmantica: Revista de filología clásica y hebrea* 57 (2006), 77–123.

León Azcárate, Juan Luis de, 'El "Libro de las Profecías" (1504) de Cristobal Colón: la Biblia y el descubrimiento de América', *Religión y cultura* 241–242 (2007), 360–406.

Lerner, Michel-Pierre, 'Le «livre vivant» de Dieu: La cosmologie évolutive de Tommaso Campanella', *Baroque* 12 (1987), 111–129.

Lerner, Michel-Pierre, 'Aux origines de la polémique anticopernicienne (I). L'*Opusculum quartum* de Giovanni Maria Tolosani [1547-1548]', *Revue des sciences philosophiques et théologiques* 86 (2002), 681–721.

Lerner, Michel-Pierre. 'The Heliocentric "Heresy": From Suspicion to Condemnation', in *The Church and Galileo,* ed. Ernan McMullin (Notre Dame, IN: University of Notre Dame Press, 2005), 11–37.

Lerner, Michel-Pierre, 'Aux origines de la polémique anticopernicienne (II). Martin Luther, Andreas Osiander et Philipp Melanchthon', *Revue des sciences philosophiques et théologiques* 90 (2006), 409–452.

Lerner, Michel-Pierre, '«Der Narr will die gantze kunst *Astronomiae* umkehren»: sur un célèbre Propos de table de Luther', in *Nouveau ciel, nouvelle terre. La révolution copernicienne dans l'Allemagne de la Réforme (1530-1630)*, eds. Miguel Ángel Granada and Édouard Mehl (Paris: Les Belles Lettres, 2009), 41–65.

Lerner, Michel-Pierre and Alain-Philippe Segonds, 'Sur un avertissement célèbre: l'*Ad lectorem* du *De revolutionibus* de Nicolas Copernic', *Galilaeana* 5 (2008), 113–148.

Letronne, Jean-Antoine, 'Des Opinions Cosmographiques des Pères de L'Eglise', *Revue des deux mondes* 1 (1834), 601–633.

Lévy, Carlos, 'The Sceptical Academy: Decline and Afterlife', in *The Cambridge Companion to Ancient Scepticism*, ed. Richard Bett (Cambridge: Cambridge University Press, 2010), 81–104.

Lightman, Bernard, ed., *Rethinking History, Science, and Religion: An Exploration of Conflict and the Complexity Principle* (Pittsburgh, PA: University of Pittsburgh Press, 2019).

Lindberg, David C., 'Science as Handmaiden: Roger Bacon and the Patristic Tradition', *Isis* 78 (1987), 518–536.

Lindberg, David C., 'The Medieval Church Encounters the Classical Tradition: Saint Augustine, Roger Bacon, and the Handmaiden Metaphor', in When Science *and* Christianity Meet, eds. David C. Lindberg and Ronald L. Numbers (Chicago and London: University of Chicago Press, 2003), 7–32.

Lindberg, David C., 'The Fate of Science in Patristic and Medieval Christendom', in *The Cambridge Companion to Science and Religion*, ed. Peter Harrison (Cambridge: Cambridge University Press, 2010), 21–38.

Lindberg, David C., 'Science and the Medieval Church', in *The Cambridge History of Science: Medieval Science*, eds. David C. Lindberg and Michael H. Shank, 8 vols, vol. 2 (Cambridge: Cambridge University Press, 2013), 268–285.

Lindberg, David C. and Numbers, Ronald L., 'Beyond War and Peace: A Reappraisal of the Encounter Between Christianity and Science', *Church History* 55 (1986), 338–354.

Lindgren, Uta, 'Warum wurde die Erde für eine Kugel gehalten? Ein Forschungsbericht', *Geschichte in Wissenschaft und Unterricht* 41 (1990), 562–574.

Livingstone, David N., 'The Preadamite Theory and the Marriage of Science and Religion', *Transactions of the American Philosophical Society* 82 [Part 3] (1992).

Lotito, Mark A., *The Reformation of Historical Thought* (Leiden: Brill, 2019).

Louth, Andrew and Marco Conti, eds., *Genesis* 1-11 (Downers Grove, IL: InterVarsity Press, 2001).

Lozovsky, Natalia, *The Earth Is Our World: Geographical Knowledge in the Latin West ca.* 400-1000 (Ann Arbor, MI: University of Michigan Press, 2000).

Lyon, Gregory B., 'Baudouin, Flacius, and the Plan for the Magdeburg Centuries', *Journal of the History of Ideas* 64 (2003), 253–272.

Maas, Korey D., *The Reformation and Robert Barnes: History, Theology and Polemic in Early Modern England* (Woodbridge and Rochester, NY: The Boydell Press, 2010).

Macías Villalobos, Cristobal, *Ciencia de los astros y creencias astrológicas en el pensamiento de san Agustín* (Madrid: Ediciones Clásicas and Málaga: Charta Antiqua, 2004).

Manzano Manzano, Juan, *Cristóbal Colón: Siete años decisivos de su vida* 1485-1492 (Madrid: Ediciones Cultura Hispánica, 1989, 2nd ed.; 1st ed. 1964).

Maravall, José Antonio, *Antiguos y Modernos. Visión de la historia e dia de progreso hasta el Renacimiento* (Madrid: Alianza Editorial, 1986, 2nd ed.; 1st ed. 1966; I actually use a reprint of the 2nd ed. from 1998).

Marenbon, John, *Pagans and Philosophers: The Problem of Paganism from Augustine to Leibniz* (Princeton: Princeton University Press, 2015).

Marinheiro, Cristóvão S., 'The Conimbricenses: The Last Scholastics, the First Moderns or Something in Between: The Impact of Geographical Discoveries on Late 16th Century Jesuit Aristotelianism', in *Portuguese Humanism and the Republic of Letters*, eds. Maria Berbara and Karl A. E. Enenkel (Leiden and Boston: Brill, 2012), 395–424.

Marone, Paola, 'La luna nella cultura scientifico-naturalistica di Agostino', in *La cultura scientifico-naturalistica nei padri della Chiesa (I-V sec.). XXXV Incontro di studiosi dell'antichità cristiana (Roma* 4-6 *maggio 2006)* (Rome: Institutum Patristicum Augustinianum, 2007), 701–710.

Matsen, Herbert Stanley, *Alessandro Achillini (*1463-1512*) and His Doctrine of 'Universals' and 'Transcendentals': A Study in Renaissance Ockhamism* (Lewisburg, PA: Bucknell University Press and London: Associated University Presses, 1974).

Mayaud, Piere-Noël, 'Le premier conflit et des traces en Occident: Lactance et l'évêque Virgile', in *Le conflit entre l'Astronomie Nouvelle et l'Écriture Sainte aux XVI*e *et XVII*e *siècles. Un moment de l'histoire des idées*, 6 vols, vol. 1 (Paris: Classiques Garnier, 2005), 83–121.

Mazzucco, Clementina, 'Il libro XVI del *De civitate Dei*', in *Lettura del* De civitate Dei *libri XI-XXII. Lectio Augustini XXI-XXII. Settimana Agostiniana Pavese (2005-2006)* (Rome: Institutum Patristicum Augustinianum, 2009), 153–197.

McColley, Grant, 'The Debt of Bishop John Wilkins to the *Apologia pro Galileo* of Tommaso Campanella', *Annals of Science* 4 (1939), 150–168.

McCready, William D., 'Isidore, the Antipodeans, and the Shape of the Earth', *Isis* 87 (1996), 108–127.

McLaughlin, Martin, 'Petrarch and Cicero', in *Brill's Companion to the Reception of Cicero*, ed. William H. F. Altman (Leiden and Boston: Brill, 2015), 19–38.

McMullin, Ernan, ed., *The Church and Galileo* (Notre Dame, IN: University of Notre Dame Press, 2005).

Melve, Leidulf, "The Revolt of the Medievalists': Directions in Recent Research on the Twelfth-century Renaissance', *Journal of Medieval History* 32 (2006), 231–252.

Miggiano, Gabriella, 'Galeotto Marzio da Narni. Profilo biobibliografico', *Il Bibliotecario* 33–34 (1992), 65–154.

Miggiano, Gabriella, '„Etiam tacente Christo": l'eresia laica di Galeotto Marzio', in *La civiltà ungherese e il cristianesimo. Atti del IV° Congresso Internazionale di studi ungheresi. Roma-Napoli 9-14 settembre 1996,* eds. István Monok and Péter Sárközy, vol. 1. (Budapest and Szeged: Nemzetközi Magyar Filológiai Társaság—Scriptum Rt., 1998), 208–226.

Miggiano, Gabriella, 'La fortuna di Galeotto Marzio in Europa tra cinquecento e seicento', in *L'eredità classica in Italia e Ungheria dal Rinascimento al Neoclassicismo. Atti del X Convegno di studi promosso e organizzato dall'Academica Ungherese dell Scienze in collaborazione con la Fondazione Giogio Cini di Venezia (Budapest, 18-20 ottobre 2001)*, eds. Péter Sárközy and Vanessa Martore (Budapest: Editore Universitas, 2004), 179–215.

Miggiano, Gabriella, 'Galeotto Marzio e Pierre Cuppé ai due poli di una sfida alla tradizione', *Rivista di studi ungheresi* IV (2005), 51–66.

Miggiano, Gabriella, 'Marzio, Galeotto (Galeottus Narniensis)', in *Dizionario Biografico degli Italiani*, vol. 71 (Rome: Istituto della Enciclopedia Italiana, 2008), 478–484.

Miggiano, Gabriella, 'Tibor Klaniczay e l'umanesimo italiano nel contesto europeo', *Rivista di studi ungheresi* 12 (2013), 29–59.

Mintz, Samuel I., 'The Motion of Thought: Intellectual and Philosophical Backgrounds', in *The Age of Milton: Backgrounds to Seventeenth-century Literature,* eds. Constantinos A. Patrides and Raymond B. Waddington (Manchester: Manchester University Press, 1980), 138–169.

Miralles Maldonado, José C., 'Discurso de obediencia de Antonio Agustín y Siscar al Papa Julio II en nombre del Rey Fernando el Católico (1507)', *Humanistica Lovaniensia* 65 (2016), 131–163.

Mittman, Asa Simon, 'Are the 'Monstrous Races' Races?', *Postmedieval* 6 (2015), 36–51.

Mittman, Asa Simon and Peter J. Dendle, eds., *The Ashgate Research Companion to Monsters and the Monstrous* (Farnham and Burlington, VT: Ashgate Publishing, 2012).

Mommsen, Theodor E., 'Petrarch's Conception of the 'Dark Ages'', *Speculum* 17 (1942), 226–242.

Montagnini, Leone, 'La questione della forma della terra. Dalle origini alla tarda antichità', *Studi Sull Oriente Cristiano* 12 (2009), 31–68.

Moody, Ernest A., 'John Buridan on the Habitability of the Earth', *Speculum* 16 (1941), 415–425.

Moretti, Gabriella, 'The Other World and the 'Antipodes': The Myth of Unknown Countries Between Antiquity and the Renaissance', in *The Classical Tradition and the Americas: European Images of the Americas and the Classical Tradition*, eds. Wolfgang Haase and Meyer Reinhold, 6 vols, vol. 1 (Berlin: De Gruyter, 1994), 241–284.

Morison, Samuel Eliot, *Admiral of the Ocean Sea: A Life of Christopher Columbus* (London: Oxford University Press, 1942).

Morison, Samuel Eliot, *The European Discovery of America: The Northern Voyages A.D.* 400-1600 (New York: Oxford University Press, 1971).

Morison, Samuel Eliot, *The European Discovery of America: The Southern Voyages A.D.* 1492-1616 (New York: Oxford University Press, 1974).

Moritz, Joshua M., 'The War that Never Was: Exploding the Myth of the Historical Conflict Between Christianity and Science', *Theology and Science* 10 (2012), 113–123.

Moro Abadía, Oscar and Francisco Pelayo, 'Reflections on the Concept of 'Precursor': Juan de Vilanova and the Discovery of Altamira', *History of the Human Sciences* 23 (2010), 1–20.

Nardi, Bruno, *Studi su Pietro Pomponazzi* (Florence: Felice Le Monnier, 1965).

Navarro Brotóns, Víctor, 'Humanismo y ciencia en el siglo XVI', in Antonio de Nebrija: Edad Media y Renacimiento, eds. Carmen Codoñer and Juan Antonio González Iglesias (Salamanca: Ediciones Universidad de Salamanca, 1994), 359–370.

Navarro Brotóns, Víctor and William Eamon, 'Spain and the Scientific Revolution: Historiographical Questions and Conjectures', in *Más allá de la leyenda Negra: España y la revolución científica/Beyond the Back Legend: Spain and the Scentific Revolution*, eds. Víctor Navarro Brotóns and William Eamon (Valencia: Instituto de Historia de la Ciencia y Documentación López Piñero, Universidad de Valencia, CSIC, 2007), 27–40.

Neugebauer, Otto, *A History of Ancient Mathematical Astronomy* (New York, Heidelberg, Berlin: Springer-Verlag, 1975).

Nicholson, Oliver P., 'The Source of the Dates in Lactantius' *Divine Institutes*', *Journal of Theological Studies* 36 (1985), 291–310.

Nicolaidis, Efthymios, *Sience and Eastern Orthodoxy: From the Greek Fathers to the Age of Globalization* (Baltimore, MD: Johns Hopkins University Press, 2011).

Nisbet, Robert, *History of the Idea of Progress* (New Brunswick, NJ: Transaction Publishers, 1994, 2nd ed.; 1st ed. 1980; I actually use a reprint of the 2nd ed. [London and New York: Routledge, 2017]).

Nolhac, Pierre de, *Pétrarque et l'Humanisme*, 2 vols (Paris: Librairie Honore Champion, 2nd revised ed. 1907; 1st ed. 1892).

Norelli, Enrico, 'The Authority Attributed to the Early Church in the *Centuries of Magdeburg* and the *Ecclesiastical Annals* of Caesar Baronius', *The Reception of the Church Fathers in the West: From the Carolingians to the Maurists*, ed. Irena Backus, vol. 2 (Leiden: Brill, 1997), 745–774.

Nothaft, Carl Philipp Emanuel, 'Augustine and the Shape of the Earth: A Critique of Leo Ferrari', *Augustinian Studies* 42 (2011), 33–48.

Nothaft, Carl Philipp Emanuel, 'Zaccaria Lilio and the Shape of the Earth: A Brief Response to Allegro's "Flat Earth Science"', *History of Science* 55 (2017), 490–498.

Oberman, Heiko Augustinus, *Masters of the Reformation: The Emergence of a New Intellectual Climate in Europe* (Cambridge: Cambridge University Press, 1981).

O'Daly, Gerard, *Augustine's City of God: A Reader's Guide* (Oxford: Oxford University Press, 1999).

O'Daly, Gerard, 'The Response to Skepticism and the Mechanisms of Cognition', in *The Cambridge Companion to Augustine*, eds. Eleonore Stump and Norman Kretzmann (Cambridge: Cambridge University Press, 2005; 1st printing 2001), 159–170.

Ogilvie, Brian W., 'Science and Medicine', in *The Oxford Handbook of Neo-Latin*, eds. Sarah Knight and Stefan Tilg (Oxford: Oxford University Press, 2015), 263–277.

Oliva, A., 'Alessandro Geraldini e la tradizione manoscritta dell'*Itinerarium ad regiones sub aequinoctiali plaga constitutas*', in *Alessandro Geraldini e il suo tempo. Atti del Convegno storico internazionale (Amelia,* 19-21 novembre 1992), ed. E. Menestò (Spoleto: Centro Italiano di Studi sull'Alto Medioevo, 1993), 175–209.

Omodeo, Pietro Daniel, *Copernicus in the Cultural Debates of the Renaissance: Reception, Legacy, Transformation* (Leiden and Boston: Brill, 2014).

Ortroy, Fernand van, *Bio-Bibliographie de Gemma Frisius, fondateur de l'École belge de Géographie, de son fils Corneille et de ses neveux les Arsenius* (Brussels: Académie Royale de Belgique, 1920).

Osler, Margaret J., 'The Canonical Imperative: Rethinking the Scientific Revolution', in *Rethinking the Scientific Revolution*, ed. Margaret J. Osler (Cambridge: Cambridge University Press, 2000), 3–22.

Otten, Willemien, 'The Reception of Augustine in the Early Middle Ages (c.700-c.1200). Presence, Absence, Reverence, and Other Modes of Appropriation', in *The Oxford Guide to the Historical Reception of Augustine*, ed. Karla Pollmann and Willemien Otten, vol. 1 (Oxford: Oxford University Press, 2013), 23–39.

Pagden, Anthony, 'The Impact of the New World on the Old: The History of an Idea', *Renaissance and Modern Studies* 30 (1986), 1–11.

Pagden, Anthony, *European Encounters with the New Word, from Renaissance to Romanticism* (New Haven, CT and London: Yale University Press, 1993).

Panizza, Letizia A., 'Lorenzo Valla's *De vero falsoque bono*, Lactantius and Oratorical Scepticism', *Journal of the Warbug and Courtauld Institutes* 41 (1978), 76–107.

Payne, Edward John, *History of the New World Called America* (Oxford: Claredon Press, 1892).

Peterson, Derrick, 'Scribbling in God's Two Books: Some Historical and Normative Reflections on Scripture, Theology, Natural Philosophy, and Science', in *Philosophy and the Christian: The Quest for Wisdom in the Light of Christ* (Leesburg, VA: The Davenant Press, 2018), 401–460.

Peterson, Derrick, *Flat Earths and Fake Footnotes: The Strange Tale of How the Conflict of Science and Christianity Was Written into History* (Eugene, OR: Cascade Books, 2021).

Pimentel, Juan, 'La Monarquía Hispánica y la ciencia donde no se ponía el sol', in Madrid, Ciencia y Corte, eds. Antonio Lafuente and Javier Moscoso (Madrid: CSIC, 1999).

Pimentel, Juan, 'The Iberian Vision: Science and Empire in the Framework of a Universal Monarchy, 1500-1800', *Osiris* 15, 2nd series (2000), 17–30.

Pimentel, Juan and José Pardo-Tomás, 'And Yet, We Were Modern: The Paradoxes of Iberian Science after the Grand Narratives', *History of Sciene* 55 (2017), 133–147.

Pino Díaz, Fermín del, 'Contribución del padre Acosta a la constitución de la Etnología. Su evolucionismo', *Revista de Indias* 38 (1978), 507–546.

Pollmann, Karla, 'The Proteanism of Authority: The Reception of Augustine in Cultural History form His Death to the Present', in *The Oxford Guide to the Historical Reception of Augustine*, ed. Karla Pollmann and Willemien Otten, vol. 1 (Oxford: Oxford University Press, 2013), 3–14.

Polónia, Amélia, Fabiano Bracht and Gisele C. Conceição, eds., *Connecting Worlds: Production and Circulation of Knowledge in the First Global Age* (Newcastle upon Tyne: Cambridge Scholars Publishing, 2018).

Priestley, Jessica, *Herodotus and Hellenistic Culture: Literary Studies in the Reception of the* Histories (Oxford: Oxford University Press, 2014).

Randles, William Graham Lister, 'Le Nouveau Monde, l'autre monde et la pluralité des mondes', in *Actas do Congresso Internacional de História dos Descobrimentos*, vol. 4 (Lisbon, 1961), 347–382. Reprinted in *Geography, Cartography and Nautical Science in the Renaissance: The Impact of the Great Discoveries*, ed. William Graham Lister Randles (Aldershot and Burlington, VT: Ashgate Publishing, 2000), chapter XV.

Randles, William Graham Lister, *De la Terre plate au globe terrestre. Une mutation épistomologique rapide (1480-1520)* (Paris: Armand Colin, 1980).

Randles, William Graham Lister, 'Modèles et obstacles épistémologiques: Aristote, Lactance et Ptolémée à l'époque des découvertes', in *L'Humanisme portugais et l'Europe: actes du XXIe colloque internationale d'études humanistes* (Paris: Fondation Calouste Gulbenkian and Centre Culturel Portugais, 1984), 437–443.

Randles, William Graham Lister, 'O redescobrimento da Geografia de Ptolomeu na Itália do Renascimento e o seu impacte em Espanha e Portugal durante os Descobrimentos', in *A abertura do mundo: Estudos de História dos Descobrimentos Europeus*, eds. Francisco Contente Domingues and Luís Filipe Barreto, 2 vols, vol. 1 (Lisbon: Editorial Presença, 1986–1987), 145–151. Reprinted in English as 'The Recovery of Ptolemy's Geography in Renaissance Italy and Its Impact in Spain and Portugal in the Period of the Discoveries' in *Geography, Cartography and Nautical Science in the Renaissance: The Impact of the Great Discoveries*, ed. William Graham Lister Randles (Aldershot and Burlington, VT: Ashgate Publishing, 2000), chapter V.

Randles, William Graham Lister, 'La cartographie de l'Atlantique a la veille du voyage de Christophe Colomb', in *Actas do II Colóquio Internacional de História da Madeira*. Funchal, Madeira, 1989 (Coimbra: Centro de estudos de história do Atlãntico, 1990a), 925–935. Reprinted in *Geography, Cartography and Nautical Science in the Renaissance: The Impact of the Great Discoveries*, ed. William Graham Lister Randles (Aldershot and Burlington, VT: Ashgate Publishing, 2000), chapter X.

Randles, William Graham Lister, 'The Evaluation of Columbus' 'India' Project by Portuguese and Spanish Cosmographers in the Light of the Geographical Science of the Period', *Imago Mundi* 42 (1990b), 50–64. Reprinted in *Geography, Cartography and Nautical Science in the Renaissance: The Impact of the Great Discoveries*, ed. William Graham Lister Randles (Aldershot and Burlington, VT: Ashgate Publishing, 2000), chapter IX.

Randles, William Graham Lister, *De la tierra plana al globo terrestre. Una rápida mutación epistemológica, 1480-1520* (Mexico, DF: Fondo de Cultura Económica, 1990c). Spanish translation of Randles, 1980.

Randles, William Graham Lister, 'L'Atlantico nella cartografia e nella cultura europea dal medioevo dal rinascimento', in *Cristoforo Colombo e l'Apertura degli Spazi*, ed. Guglielmo Cavallo, 2 vols, vol. 1 (Rome: Istituto Poligrafico e Zecca dello Stato, 1992), 427–448. Reprinted in English as 'The Atlantic in European Cartography and Culture from the Middle Ages to the Renaissance' in *Geography, Cartography and Nautical Science in the Renaissance: The Impact of the Great Discoveries*, ed. William Graham Lister Randles (Aldershot and Burlington, VT: Ashgate Publishing, 2000), chapter II.

Randles, William Graham Lister, 'Colomb découvreur: perceptions contemporaines de son projet et de sa réalisation', in *Atti conclusivi del XXVI Congresso Geografico Italiano (Genova, 4-9 Maggio 1992)*. (Genoa, 1993a), 83–89. Reprinted in *Geography, Cartography and Nautical Science in the Renaissance: The Impact of the Great Discoveries*, ed. William Graham Lister Randles (Aldershot and Burlington, VT: Ashgate Publishing, 2000), chapter XI.

Randles, William Graham Lister, 'La science universitaire en Europe et les découvertes portugaises: Aristotélisme doctrinaire et expérience des navigateurs', *Mare Liberum* 6 (1993b), 19–24. Reprinted in *Geography, Cartography and Nautical Science in the Renaissance: The Impact of the Great Discoveries*, ed. William Graham Lister Randles (Aldershot and Burlington, VT: Ashgate Publishing, 2000), chapter XII.

Randles, William Graham Lister, 'Classical Models of World Geography and Their Transformation Following the Discovery of America', in *The Classical Tradition and the Americas: European Images of the Americas and the Classical Tradition*, eds. Wolfgang Haase and Meyer Reinhold, 6 vols, vol. 1 (Berlin: De Gruyter, 1994), 5–76. Reprinted in *Geography, Cartography and Nautical Science in the Renaissance: The Impact of the Great Discoveries*, ed. William Graham Lister Randles (Aldershot and Burlington, VT: Ashgate Publishing, 2000), chapter I.

Randles, William Graham Lister, *The Unmaking of the Medieval Christian Cosmos, 1500-1760: From Solid Heavens to Boundless Æther* (Aldershot and Brookfield, VT: Ashgate, 1999).

Reinhardt, Thomas, 'Die Erfindung der flachen Erde: Der Mythos Kolumbus und die Konstruktion der Epochenschwelle zwischen Mittelalter und Neuzeit', *Paideuma* 53 (2007), 161–180.

Relaño, Francesc, *The Shaping of Africa: Cosmographic Discourse and Cartographic Science in Late Medieval and Early Modern Europe* (London and New York: Routledge, 2019; 1st printing Aldershot and Burlington, VT: Ashgate Publishing, 2002).

Robin, Diana, *Filelfo in Milan: Writings* 1451-1477 (Princeton: Princeton University Press, 2014).

Romm, James S., *The Edges of the Earth in Ancient Thought* (Princeton: Princeton University Press, 1992).

Rosen, Edward, 'Copernicus and the Discovery of America', *The Hispanic American Historical Review* 23 (1943), 367–371.

Rosen, Edward, 'Kepler's Defense of Tycho against Ursus', *Popular Astronomy* 54 (1946), 405–412.

Rosen, Edward, 'Kepler and the Lutheran Attitude towards Copernicanism', *Vistas in Astronomy* 18 (1975), 317–338.

Rosen, Edward, 'When Did Copernicus Write the 'Revolutionibus'?', *Sudhoffs Archiv für Geschichte der Medizin und der Naturwissenschaften* 61 (1977), 144–155.

Rosen, Edward, *Copernicus and the Scientific Revolution* (Malabar, FL: Robert E. Krieger, 1984).

Rubiés, Joan-Pau, 'Travel Writing and Humanistic Culture: A Blunted Impact?', *Journal of Early Modern History* 10 (2006), 131–168.

Rubiés, Joan-Pau, 'The Discovery of New Worlds and Sixteenth-Century Philosophy', in *The Routledge Companion to Sixteenth-Century Philosophy*, eds. Henrik Lagerlund and Benjamin Hill (New York and London: Routledge, 2017), 54–82.

Rummel, Erika, *Erasmus' Annotations on the New Testament: From Philologist to Theologian* (Toronto: Toronto University Press, 1986).

Russell, Colin A., 'The Conflict Metaphor and Its Social Origins', *Science and Christian Belief* 1 (1989), 3–26.

Russell, Jeffrey Burton, *Inventing the Flat Earth: Columbus and Modern Historians* (New York, Westport, CT, and London: Praeger, 1991).

Russell, Jeffrey Burton, 'The Flat Error: The Modern Distortion of Medieval Geography', *Mediaevalia* 13 (1993), 337–353.

Russell, Jeffrey Burton, 'The myth of the flat earth', text of a paper read at the *American Scientific Affiliation Annual Meeting* in Westmont College, 4 August 1997, summarising his book *Inventing the Flat Earth: Columbus and Modern Historians* (1997). Available at: https://www.asa3.org/ASA/topics/history/1997Russell.html.

Russell, Jeffrey Burton, 'Flat as the Earth', *Mercator's World* 6 (2001), 16–21.

Russell, Jeffrey Burton, 'Flattening the Earth', *Mercury: The Journal of the Astronomical Society of the Pacific* 31 (2002), 34–38. Available at: https://web.archive.org/web/20040108065614/https://lawrencehallofscience.org/pass/passv10/flat-earth.html (the original web site is no longer available, here it is preserved although without the original pagination).

Russell, Jeffrey Burton, *Exposing Myths about Christianity: A Guide to Answering 145 Viral Lies and Legends* (Downers Grove, IL: InterVarsity Press, 2012).

Rutherford, David, 'Antonio da Rho on Patristic authority. The status of Lactantius', in *Auctoritas Patrum II: New Contributions on the Reception of the Church Fathers in the 15th and 16th Centuries*, eds. Leif Grane, Alfred Schindler and Markus Wriedt (Mainz: Verlag Philipp von Zabern, 1998).

Rutherford, David, 'Lactantius Philosophus? Reading, Misreading, and Exploiting Lactantius from Antiquity to the Early Renaissance', in *Essays in Renaissance Thought and Letters: In Honor of John Monfasani*, eds. Alison Frazier and Patrick Nold (Leiden and Boston: Brill, 2015), 446–491.

Rutherford, David, 'The Manuscripts of Lactantius and His Early Renaissance Readers', *Studia Patristica* 80 (2017), 155–171.

Rutherford, David and Paul Schulten, 'Historical Themes in Antonio da Rho's *Dialogi tres in Lactantium*', *Rivista di storia e letteratura religiosa* 53 (2017), 601–621.

Saak, Eric L., 'The Augustinian Renaissance: Textual Scholarship and Religious Identity in the Later Middle Ages', in *The Oxford Guide to the Historical Reception of Augustine*, ed. Karla Pollmann and Willemien Otten, vol. 1 (Oxford: Oxford University Press, 2013), 58–67.

Sáenz-López Pérez, Sandra, 'El mundo como una manzana en la palma de la mano: El pomo y su relación con la cartografía medieval', *Anales de historia del arte* 23, extra nº 2 (2013), 537–549.

Sambursky, Samuel, *Physics of the Stoics* (London: Routledge and Kegan Paul, 1959).

Santos Alpalhão, Margarida and Isabel Barros Dias, 'Diálogos Portugueses: contributo para um catálogo (Idade Média-século XVIII)', *eHumanista* 33 (2016), 457–518.

Saracino, María Agostina, 'Geografías reales e imaginadas en la construcción de Colón como profeta en *El nuevo mundo descubierto por Cristóbal Colón* de Lope de Vega', *Bibliographica Americana* 13 (2017), 104–116.

Scafi, Alessandro, *Mapping Paradise: A History of Heaven on Earth* (Chicago: University of Chicago Press, 2006).

Schmitt, Charles B., *Cicero scepticus: A Study of the Influence of the* Academica *in the Renaissance* (Dordrecht: Springer Science+Business Media, B.V., 1972).

Schneid, Mathias, 'Die Lehre von der Erdrundung und Erdbewegung im Mittelalter', *Historisch-politische Blätter für das katholische Deutschland* 80 (1877), 433–451.

Scholten, Clemens, *Antike Naturphilosophie und christliche Kosmologie in der Schrift "De Opificio Mundi" des Johannes Philoponos* (Berlin and New York: Walter de Gruyter, 1996).

Schott, Jeremy M., *Christianity, Empire, and the Making of Religion in Late Antiquity* (Philadelphia, PA: University of Pennsylvania Press, 2008).

Segundo Guzmán, Miguel Ángel, 'El descubrimiento de América en la última hora del mundo: la hermenéutica franciscana', *Nuevo Mundo Mundos Nuevos* [online published 12 July 2012]. Available at: https://journals.openedition.org/nuevomundo/63661.

Segonds, Alain-Philippe and Concetta Luna, 'The Greek Text of Ptolemy, *Almagest* XIII, as Source of Book VI of Copernicus's *De revolutionibus*', *Journal for the History of Astronomy* 44 (2013), 413–427.

Shank, Michael H. and David C. Lindberg, 'Introduction', in *The Cambridge History of Science: Medieval Science*, eds. David C. Lindberg and Michael H. Shank, 8 vols, vol. 2 (Cambridge: Cambridge University Press, 2013), 1–26.

Shapin, Steven, *The Scientific Revolution* (Chicago and London: University of Chicago Press, 1996).

Simek, Rudolf, 'Die Kugelform der Erde im mittelhochdeutschen Schrifttum', *Archiv für Kulturgeschichte* 70 (1988), 91–103.

Simek, Rudolf, *Erde und Kosmos im Mittelalter: Das Weltbild vor Kolumbus* (Munich: Beck, 1992).

Simek, Rudolf, *Heaven and Earth in the Middle Ages* (Woodbridge: The Boydell Press, 1996). English translation of Simek, 1992.

Simek, Rudolf, 'Die Form der Erde im Mittelalter und die Erfinder der Scheibengestalt', in *Mittelalter und Moderne. Entdeckung und Rekonstruktion der mittelalterlichen Welt. Kongreßakten des 6. Symposiums des Mediävistenverbandes in Bayreuth 1995*, ed. Peter Segl (Sigmaringen: Thorbecke, 1997), 139–147.

Simek, Rudolf, 'Kugel oder Scheibe? Das Bild von der Erde im Mittelalter', *Spektrum der Wissenschaft Spezial* 2 (2002), 20–24.

Simek, Rudolf, 'Die Scheibengestalt der Erde im Mittelalter', in *Burgen, Länder, Orte*, eds. Ulrich Müller and Werner Wunderlich (Constance: UVK Verlagsgesellschaft, 2008), 789–802.

Simpson, James Y., *Landmarks in the Struggle Between Science and Religion* (London: Hodder and Stoughton, 1925).

Singham, Mano, 'Columbus and the Flat Earth Myth', *Phi Delta Kappan* 88 (2007), 590–592.

Sloan, Michael C., '*De civitate Dei*', in *The Oxford Guide to the Historical Reception of Augustine*, ed. Karla Pollmann and Willemien Otten, vol. 1 (Oxford: Oxford University Press, 2013), 255–261.

Smith, Sharon Off Dunlap, *Illustrations of Raoul de Praelles' translation of St. Augustine's* City of God *between 1375 and 1420*, unpublished PhD thesis, 2 vols (New York: New York University, 1974). Available at: https://vdoc.pub/documents/illustrations-of-raoul-de-praelles-translation-of-st-augustines-city-of-god-between-1375-and-1420-rkesttgbupc0.

Smith, Pamela H., 'Science on the Move: Recent Trends in the History of Early Modern Science', *Renaissance Quarterly* 62 (200), 345–375.

Stark, Rodney William, *Bearing False Witness: Debunking Centuries of Anti-Catholic History* (West Conshohocken, PA: Templeton Press, 2016).

Stein, John, 'Bartolomeo Platina', in *The Catholic Encyclopedia*, vol. 12 (New York: Robert Appleton Company, 1911), 158–159.

Stevens, Wesley M., 'The Figure of the Earth in Isidore's "De Natura Rerum"', *Isis* 71 (1980), 268–277.

Stevens, Wesley M., 'Descriptive Models of Earth and Sky', in *La conoscenza scientifica nell'alto medioevo* (Spoleto: Fondazione Centro Italiano di Studi Sull'Alto Medioevo, 2020).

Stevens, Wesley M. and William D. McCready, 'Letters to the Editor', *Isis* 87 (1996), 678–679.

Stinger, Charles L., *Humanism and the Church Fathers: Ambrogi (*1386-1439*) and Christian Antiquity in the Italian Renaissance* (Albany, NY: State University of New York Press, 1977).

Tattersall, Jill, 'The Earth: Sphere or Disc?' *The Modern Language Review* 76 (1981), 31–46.

Taylor, Eva Germaine Rimington, 'Imago Mundi', *Scottish Geographical Magazine* 47 (1931), 78–80.

Taylor, Eva Germaine Rimington, 'Some Notes on Early Ideas of the Form and Size of the Earth', *The Geographical Journal* 85 (1935), 65–68.

Taylor, Eva Germaine Rimington, 'Ideas of the Shape and Habitability of the Earth Prior to the Great Age of Discovery', *History* 22 (1937), 54–58.

Taylor, Eva Germaine Rimington, *Ideas on the Shape, Size and Movements of the Earth* (London: The Historical Association, 1943).

Taylor, Eva Germaine Rimington, 'Since When Have We Known That the Earth Is a Globe?', *Journal of Navigation* 10 (1957), 133–137.

Tedford, Margaret, 'The Anglo-Saxon Cotton Map in Context', *Imago Mundi* 72 (2020), 88–89.

Thacher, John Boyd, *Christopher Columbus: His Life, His Work, His Remains*, 3 vols (New York and London: G. P. Putnam's Sons, The Knickerbocker Press, 1903–1904).

Thomson, J. Oliver, *History of Ancient Geography* (Cambridge: Cambridge University Press, 1948).

Thorndike, Lynn, *A History of Magic and Experimental Science during the First Thirteen Centuries of Our Era*, 8 vols (New York and London: Columbia University Press and The MacMillan Company, 1923–1958).

Toth, Peter, 'New Questions on Old Answers: Towards a Critical Edition of the Answers to the Orthodox of Pseudo-Justin', *The Journal of Theological Studies* 65 (2014), 550–599.

Ungureanu, James C., 'Myths about Science and Religion: That Medieval Christians Taught that the Earth Was Flat' (2013). Available at: https://jamescungureanu.com/2013/03/14/myths-about-science-and-religion-that-medieval-christians-taught-that-the-earth-was-flat/.

Ungureanu, James C., 'Relocating the Conflict Between Science and Religion at the Foundations of the History of Science', *Zygon* 53 (2018), 1106–1130.

Ungureanu, James C., 'Science, Religion and the 'New Reformation' of the Nineteenth Century', *Science and Christian Belief* 31 (2019a), 41–61.

Ungureanu, James C., 'Science and Religion in the Anglo-American Periodical Press, 1860-1900: A Failed Reconciliation', *Church History* 88 (2019b), 120–149.

Ungureanu, James C., *Science, Religion, and the Protestant Tradition: Retracing the Origins of Conflict* (Pittsburgh, PA: University of Pittsburgh Press, 2019c).

Vannini de Gerulewicz, Marisa, *El Mar de los descubridores: documentos y relatos inéditos o poco conocidos sobre el descubrimiento y la exploración de los mares, islas y tierras del Nuevo Mundo (siglos xv-xvi)* (Caracas: Comisión Organizadora de la III Conferencia de las Naciones Unidas sobre Derecho del March, 1974).

Varela, Consuelo, *Cristobal Colón y la construcción de un mundo nuevo. Estudios, 1983-2008* (Santo Domingo: Archivo General de la Nación, 2010).

Vargas Díaz-Toledo, Aurelio, 'Uma primeira aproximação do *corpus* dos Diálogos Portugueses dos séculos xvi-xvii', *Criticón* 117 (2013), 65–130.

Vasconcelos, Basílio de, 'De inventione Africæ maritimæ et occidentalis videlicet Geneæ per infantem Heinricum Portugalliæ', *O Instituto* 83 (1932), 140–190.

Vasoli, Cesare, 'Note su Galeotto Marzio', *Acta Litteraria Academiae Scientiarium Hungaricae* 19 (1977), 51–69. Reprinted in *La cultura delle corti*, ed. Cesare Vasoli (Bologna: Capelli, 1980), 38–63.

Vassiliev, A., *Anecdota Graeco-Byzantina* (Moscow: Universitatis Caesareae, 1893).

Vignaud, Henry, *Histoire critique de la grande entreprise de Christophe Colomb*, 2 vols (Paris: H. Welter, 1911).

Visser, Arnoud S. Q., *Reading Augustine in the Reformation: The Flexibility of Intellectual Authority in Europe, 1500-1620* (Oxford: Oxford University Press, 2011).

Vogel, Klaus Anselm, 'Amerigo Vespucci und die Humanisten in Wien. Die Rezeption der geographischen Entdeckungen und der Streit zwischen Joachim Vadian und Johannes Camers über die Irrtümer der Klassiker', *Pirckheimer-Jahrbuch* 7 (1992), 53–104.

Vogel, Klaus Anselm, *Sphaera terrae: Das mittelalterliche Bild der Erde und die kosmographische Revolution*, unpublished PhD thesis (Göttingen: Georg-August-Universität, 1995). Available at: https://ediss.uni-goettingen.de/handle/11858/00-1735-0000-0022-5D5F-5.

Vogel, Klaus Anselm, 'Cosmography', tr. Alisha Rankin, in *The Cambridge History of Science: Early Modern Science*, eds. Katharine Park and Lorraine Daston, 8 vols, vol. 3 (Cambridge: Cambridge University Press, 2006a), 469–496.

Vogel, Klaus Anselm, 'European Expansion and Self-definition', tr. Alisha Rankin, in *The Cambridge History of Science: Early Modern Science*, eds. Katharine Park and Lorraine Daston, 8 vols, vol. 3 (Cambridge: Cambridge University Press, 2006b), 818–839.

Voicu, Sever J., 'Due antiocheni periferici: Le *Quaestiones et responsiones ad Orthodoxos* (CPG 6285) e Severiano di Gabala', *Augustinianum* 55 (2015), 543–558.

Weichenhan, Michael, *"Ergo perit coelum…" Die Supernova des Jahres 1572 und die Überwindung der aristotelischen Kosmologie* (Stuttgart: Franz Steiner, 2004).

Westman, Robert S., 'Proof, Poetics, and Patronage: Copernicus's Preface to *De revolutionibus*', in *Reappraisals of the Scientific Revolution*, eds. David C. Lindberg and Robert S. Westman (Cambridge: Cambridge University Press, 1990), 167–205.

Wey Gómez, Nicolás, *The Tropics of Empire: Why Columbus Sailed South to the Indies* (Cambridge, MA and London: The MIT Press, 2008).

Whewell, William, *History of the Inductive Sciences*, 3 vols (London: John W. Parker, and Cambridge: J. and J. J. Deighton, 1837).

White, Andrew Dickson, *A History of the Warfare of Science with Theology in Christendom*, 2 vols (London: Macmillan and Co., 1896).

Wilkins, Ernest H., 'On the Earliest Editions of the *Morgante* of Luigi Pulci', *The Papers of the Bibliographical Society of America* 45 (1951), 1–22.

Williams, Jack, *Robert Recorde. Tudor Polymath, Expositor and Practitioner of Computation* (London, Dordrecht, Heidelberg and New York: Springer, 2011).

Williams, John, 'Isidore, Orosius and the Beatus Map', *Imago Mundi* 49 (1997), 7–32.

Williamson, Arthur H., *Apocalypse Then: Prophecy and the Making of the Modern World* (Westport, CT and London: Praeger, 2008).

Wills, Bernard, 'Ancient Scepticism and the *Contra Academicos* of St. Augustine', *Animus* 4 (1999), 108–123.

Wolf, Jürgen, *Die Moderne erfindet sich ihr Mittelalter – oder wie aus der ‚mittelalterlichen Erdkugel' eine ‚neuzeitliche Erdscheibe' wurde* (Mainz: Akademie der Wissenschaften und der Literatur, and Stuttgart: Franz Steiner Verlag, 2004).

Wolff, Philip, 'Flach war sie nie. Im Mittelalter glaubte niemand, die Erde sei eine Scheibe. Diesen Mythos hat die Moderne geschaffen – zum eigenen Ruhm', *Süddeutsche Zeitung (SZ) – Wissen* (June 2005), 33–36.

Woods, Thomas E., Jr., 'The flat earth myth: the real myth is the idea that anyone ever believed in a flat earth' (2005). Available at: https://www.lewrockwell.com/2005/07/thomas-woods/the-flat-earth-myth/.

Woodward, David, 'Reality, Symbolism, Time, and Space in Medieval World Maps', *Annals of the Association of American Geographers* 75 (1985), 510–521.

Woodward, David, 'Geography', in *The Cambridge History of Science: Medieval Science*, eds. David C. Lindberg and Michael H. Shank, 8 vols, vol. 2 (Cambridge: Cambridge University Press, 2013), 548–568.

Wootton, David, *The Invention of Science: A New History of the Scientific Revolution* (New York: Harper Collins, 2015).

Yeo, Richard, 'Natural Philosophy (Science)', in *An Oxford Companion to the Romantic Age: British Culture* 1776-1832, eds. Iain McCalman, Jon Mee, Gillian Russell and Clara Tuite (Oxford: Oxford University Press, 1999), 320–328.

Zaccaria, Vittorio, 'Sulle opere di Pier Candido Decembrio', *Rinascimento* 7 (1956), 13–74.

Zamora, Margarita, *Reading Columbus* (Berkeley, Los Angeles and London: University of California Press, 1993).

Index of Authors and Subjects

Note: **Bold** page numbers refer to tables, *italic* page numbers refer to figures and page numbers followed by "n" refer to endnotes.

Index of Biblical Texts

Note: **Bold** page numbers refer to tables, *italic* page numbers refer to figures and page numbers followed by "n" refer to endnotes.